U0178985

MECHANICS

MECHANICS

MECHANICS

LECTURES ON THEORETICAL PHYSICS, VOL. I

BY ARNOLD SOMMERFELD

UNIVERSITY OF MUNICH

TRANSLATED FROM THE FOURTH GERMAN EDITION
BY MARTIN O. STERN
UNIVERSITY OF CALIFORNIA

ACADEMIC PRESS   New York San Francisco London

A Subsidiary of Harcourt Brace Jovanovich, Publishers

导　言

　　论及近代物理学的构建与物理学教育，这里的物理学教育指的是物理学家的培养，有一个格外突出的人物是非提不可的，那就是德国数学家、物理学家索末菲。

　　索末菲（Arnold Sommerfeld, 1868–1951）出生于东普鲁士的科尼斯堡（今俄罗斯的加里宁格勒），父亲是位热爱自然科学的医生。1875 年，索末菲进入科尼斯堡的老城学校上学，当时学校的高年级学生中有闵可夫斯基（Hermann Minkowski, 1864–1909）和维恩（Wilhelm Wien, 1864–1928）。1886 年，索末菲进入科尼斯堡大学学习数学和物理，他在该校遇到的老师有希尔伯特（David Hilbert, 1862–1943）、林德曼（Ferdinand von Lindemann, 1852–1939）和胡尔维茨（Adolf Hurwitz, 1859–1919）这些数学巨擘。1891 年，索末菲在林德曼指导下以"数学物理中的任意函数"为题获得博士学位。服完兵役以后，索末菲于 1893 年去往哥廷恩，1894 年做了数学大家克莱因（Felix Klein, 1849–1925）的助手，1895 年以"衍射的数学理论"获得讲师资格。其后，索末菲在亚琛等地任教。1906 年，索末菲到慕尼黑大学担任理论物理教授，建立了理论物理研究所，在那里一直工作到 1939 年退休。

　　索末菲是一个理论物理学家，但他首先是个数学家、实验物理学家或者说技术物理学家。索末菲对摩擦学、无线电技术、电力传输等技术领域都有可圈可点的贡献。当然，他最值得称道的是对近代物理的贡献。在关于原子中电子的四个量子数（nlm;

m_s）中的（$l\ m$）这两个量子数都是索末菲 1915 年引入的，而精细结构常数 $\alpha = \dfrac{1}{4\pi\varepsilon_0}\dfrac{e^2}{\hbar c} \sim \dfrac{1}{137}$ 则是索末菲 1916 年引入的。索末菲是无可置疑的量子力学奠基人之一，也是最早接受相对论的学者。

索末菲在哥廷恩给克莱因短时间做过助手，深受克莱因的影响。克莱因的那套研究、教学、主持研讨班、办杂志、编书等作为一代学术宗师的行为模式，索末菲全面继承了下来。1895/1896 冬季学期，克莱因就陀螺理论做了一个系列讲座，后来委托索末菲将讲座内容整理成书。1896 年秋，索末菲开始整理陀螺理论。这期间，索末菲不仅关注相关学问的理论进展，甚至还多次走访海军基地，了解陀螺仪在鱼雷制导上的实际应用。让克莱因和索末菲都没想到的是，因为索末菲作为一个技术物理教授和数学物理教授自己也是诸事缠身，这项工作整整持续了 13 年，直到 1910 年才最终完成。不过，这套四卷本的《陀螺理论》（*Über die Theorie des Kreisels*），洋洋千余页，作为处理刚体转动这个数学物理之难题的经典，任何人凭此一项成就即足以傲视物理学界。尤其值得一提的，关于克莱因发起的"数学科学百科全书"，索末菲承担了第五卷的编辑任务，一干就是 28 年（1898–1926），收录了许多数学物理名篇，其中就包括泡利（Wolfgang Pauli, 1900–1958）的成名作《相对论》（*Die Relativitätstheorie*）。

索末菲自己是近代原子理论的主要贡献者。基于他自己的研究成果以及讲课内容，索末菲于 1919 年出版了《原子构造与谱线》（*Atombau und Spektrallinien*）一书，此书被誉为原子物理的圣经。1929 年，索末菲又针对这本书补充出版了《原子构造与谱线：波动力学补充》（*Atombau und Spektrallinien:*

Wellenmechanischer Ergänzungband)。1933 年，索末菲又出版了一本近 300 页的《金属电子理论》(*Elektronentheorie der Metalle*)，这是索末菲在自己拓展了德鲁德 (Paul Drude, 1863–1906) 的自由电子理论的基础上撰写的。将这三本著作放到一起，能勾勒出一个原子理论和老量子论 [①] 构建者形象，也就能够理解为什么他的学生能成为量子力学和量子化学的奠基人。

索末菲在慕尼黑的讲课资料最后集成了六卷本《理论物理教程》(*Vorlesungen über theoretische Physik*)，其中第一卷《力学》(*Mechanik*) 出版于 1943 年，第二卷《形变介质的力学》(*Mechanik der deformierbaren Medien*) 出版于 1945 年，第三卷《电动力学》(*Elektrodynamik*) 出版于 1948 年，第四卷《光学》(*Optik*) 出版于 1950 年，第五卷《热力学与统计》(*Thermodynamik und Statistik*) 出版于 1952 年，第六卷《物理中的偏微分方程》(*Partielle Differentialgleichungen der Physik*) 出版于 1947 年。第五卷算是遗作，是由 Fritz Bopp 和 Josef Meixner 代为整理的。第六卷非常有名，早在 1949 年即有了英文版。其他各卷英译本出版于 1964 年。世界上有一些著名的数学、物理学系列教程，有必要在此介绍给我国的物理学爱好者。比索末菲早的、比较有名的有克莱因的各种数学教程，以及亥尔姆霍兹 (Hermann von Helmholtz, 1821–1894) 的系列物理教程（包括光的电磁理论、声学原理、分立质点动力学、连续介质动力学、热学）等。在索末菲之后，有他的学生泡利的九卷本《泡利物理学讲义》(*Pauli Lectures on Physics*)，是对泡利讲课材料的

---

① 我不觉得 old quantum theory 是旧量子论。它不陈旧，它只是量子理论的早期形态而已

整理，算是对索末菲《理论物理教程》的风格继承。其他有名的理论物理教程有十卷本《朗道理论物理教程》，那是朗道（Лев Дави́дович Ланда́у, 1908–1968）发起的、由多人完成的著作，除第八卷署名为 Landau, Lifshitz, Pitaevskii 外，其他各卷署名为 Landau, Lifshitz。就实验物理而言，八卷本的德语《实验物理教程》（*Bergmann-Schaefer Lehrbuch der Experimentalphysik*）对我国的物理学教育可能具有特别的意义，因为它提供了太多我们可能未加关注的实验细节。这是一套众多作者编纂的、开放式的教程，内容随时修订，多次再版。此外，还有卷数不等的、更多是针对学生设计的系列物理教程，各有千秋，恕不一一介绍。索末菲的六卷本《理论物理教程》是独特的，因为它包含着作者自己创造的学问，是作者自己在课堂上实际使用过的，最重要的是从作者培养出来的学生水平来看这套教程是卓有成效的。此外，值得强调的是，除了第五卷遗作借助他人之手才整理完成，这套系列教程是索末菲一人的功绩，没有掠他人之美。近时期加入自己的成果与思考来系统地表述物理学的，有美国物理学家赫斯特内斯（David Hestenes, 1933– ），但做不到如索末菲那样融会贯通且凭一己之力。不是凭一己之力完成的系列物理学教程，无法保证风格的统一只是小事，缺乏思想的整体内在一致性（integrity）才是最遗憾的地方。

索末菲在慕尼黑的理论物理研究所汇集了一批少年英才，他们在索末菲的指导和引领下几乎全部成为了在各自研究领域里扬名立万的人物。据说在 1928 年前后的那段时间里，德语国家的理论物理教授有三分之一都出自索末菲门下。索末菲的博士生中有四人获诺贝尔奖，分别是海森堡（Werner Heisenberg,

1901–1976）、德拜（Peter Debye, 1884–1966）、泡利和贝特（Hans Bethe, 1906–2005），博士后中有三人获得诺贝尔奖，鲍林（Linus Pauling, 1901–1994）、拉比（Isidor I. Rabi, 1898–1988）和劳厄（Max von Laue, 1879–1960），其中德拜获得的是化学奖，鲍林获得的是化学奖与和平奖。索末菲的作为大师之大导师（MacTutor of Maestros）的成就，仅英国的汤姆孙（J. J. Thomson, 1856–1940）可与之比肩。

我国古代哲人孟子认为"得天下英才而教育之"为君子之第三乐，此乃为人师之言也。索末菲是充分享受了这样的乐趣的。然而，对于所有求学者来说，人生的首要问题却是从什么人而受教的问题。我想说，"从合格之老师而受教，不亦大幸运乎？"然而，令人痛心的现实是，绝大部分的人材可能一生中都不会遇到一个合格的老师。倘若没有强大的自学能力与格外的机遇，难免早晚坠入"泯然众人矣"的遗憾。索末菲之所以有那么大的成就，与他求学的地方学术宗师云集有关；而索末菲的那些学生们之所以都各有所成就，又何尝不是因为遇到了索末菲这样的名师及其门下一众出类拔萃的同学们的缘故。笔者在博士毕业进入研究领域多年以后才悟到同这些学术巨擘之间学问的天渊之别究竟差在哪里。倘若一个人上过的学校中没有几个值得提起的老师或者同学，那你自己可要好好努力了。

值得庆幸的是，索末菲的《理论物理教程》英文版如今在我国出版了。愿这影响了无数物理学家成长的经典之作在中华大地上能收获更多的辉煌。

曹则贤

2022 年 7 月 27 日于中国科学院物理研究所

序 言

此一理论物理教程的作者，阿诺德·索末菲，是促成1910–1930年间物理学所经历之变迁的关键人物之一。没有他的满怀雄心的、不懈的努力，关于原子的量子理论，不管是其狂飙式的进展还是广泛的传播，都不可能是我们所看到的那个样子。索末菲在慕尼黑的理论物理所 [①] 里形成了一个学派，那里的原子论研习者们，德国人和外国人，初出茅庐的和小有成就的，源源不断地创作出研究论文。他的名著《原子构造与谱线》(*Atombau und Spektrallinien*) [②]，以及随后出版的《波动力学》(*Wellenmechanik*) [③]，在很长时间里是独有的关于此一基础领域之全面的、权威的论述；其后续版本令人印象深刻，乃是对自玻尔第一批文章之后原子理论的快速进展的概述。

索末菲所受的学术训练以及早期研究都根植于经典物理的数学方法，也因此他熟练掌握特别是在1926年薛定谔波动力学之后出现的那些新生的量子物理方法。自然地，也更因为其本人从经典理论的审美中所体会的愉悦，索末菲会给他的学生们以经典方法的系统训练。数学表述，还有它的物理诠释，与实验具体化之间的和谐在索末菲的教程中如浮雕一样被呈现出来，深刻地影

[①]　这里所说的 Institute，是大学里针对具体的某个教授设立的研究机构，规模很小。不是今天中文语境里的研究所

[②]　1919 年出版

[③]　1929 年出版的 *Atombau und Spektrallinien* 第二卷，全名为《波动力学增补卷》(*Wellenmechanischer Ergänzungsband*)

响了他的学生们。

当索末菲整理他的教程准备将其付梓时，他已年过七十，结束了他长达四十年的学术培养生涯。他做这件事有两重意义：保全那些支撑物理取得伟大胜利的成果以度过危机，向年轻一代的物理学家传承那些在解决经典问题过程中形成的有价值的分析工具。自1895年撰写论物理中的任意函数的博士论文时起，索末菲在完善这些工具方面扮演了一个非常活跃的角色。在他早期的技艺高超的工作中，有一项是波在边缘上衍射问题的严格解的构造，为此他扩展了黎曼在函数理论中的方法，结果是用多值空间中的镜像法得到了衍射问题的一个解。读者会在第5卷（光学）中看到相关讨论。索末菲同哥廷恩的数学大家克莱因合著的论转动刚体理论的范本，即四卷本《陀螺理论》（*Theorie des Kreisels*），跨越了其早年在哥廷恩的时期一直到其在慕尼黑的量子时代之初 [④]。该书的目的是通过将大量不同的数学内容，如函数理论、椭圆函数、四元数、Klein-Caley参数等，应用于刚体动力学问题来展现纯数学与应用数学之间的密切联系。当其在亚琛工业专科学校做工业力学岗位教授时，即1899–1905年间，索末菲对工程问题产生了浓厚的兴趣。他关于润滑流体力学、工作于同一传输线上的发电机间的相互影响、火车刹车等问题的论文，无一不是采用了一般性处理方法而因此具有持久的价值。当无线电报出现时，索末菲及其弟子的系列论文开始研究无线电的发射与传播模式。这些都是在相应领域中索末菲堪为大师的那些数学方法的卓越案例。特别地，无线电绕地球的衍射问题被转化成了

---

④ 出版时间跨度为1897–1910

复积分的讨论，有了严格解（参见第 6 卷，第 6 章）[5]。

此处罗列一个详细的、索末菲因之丰富了物理理论的成就清单有点儿不合适。读者可以参阅文后列举的一些文章。但是，不妨就索末菲作为老师，以及如今要翻译出版的这个教程，多说几句。

慕尼黑的理论物理课程分为通识课和专业课两类。前者每周四小时，确切地说是分成 45–50 分钟一节的课，冬季学期持续 13 周，夏季学期持续 11 周。（索末菲的）那六门课就构成了当前六卷本的主题。每门课都是给选修了实验物理演示课程（任课教师先为伦琴，后来是维恩）的学生的导论。在实验物理课上，学生要求获得对物理现象的实地考察，以及基于本质上非数学的处理对考察结果的定量评价。而在理论物理课上，基础内容会被重修梳理一遍，但是目标瞄向了未来可用于高等问题的形成数学处理（能力）和构建全面理论（知识）。后者从一个系列的课程到另一个系列的课程是变动的，后期会纳入一些专题因而也会引起此前学过相关内容的高年级学生的兴致。除了讲座，还有每周两个小时的讨论课。

专业课是每周两小时的讲座，主题是那些在通识课上只能略加论述的主题，或者时令的话题。索末菲的讲座一般会和他当时的研究有关系，经常有部分内容很快会出现在他的原创论文中，这方面的例子有将洛伦兹变换诠释为四维空间里的转动（第 3 卷，第 26 节），从波动光学到几何光学的过渡（第 5 卷，第 35 节）和色散介质中信号速度的讨论（第 5 卷，第 22 节）。这些主题又

---

[5] Complex integral，字面意思是复积分。但第 6 卷第 6 章那里似乎只是一些复杂的积分。

会被纳入通识课中，此前通识课中不那么有趣的部分会被拿掉。

除了讲座类课程，还有关于高级课题的讨论班、报告会。为此学生要总结分到手的课题、作报告，这常常意味着持续数星期的高强度学习。

在学生眼中，索末菲讲座最吸引人的地方是（条理）清晰：从物理一侧的处理方式，数学问题的提取，对所采用方法之简单但具有一般性的解释，重新用物理实验语汇对结果的充分讨论。他在黑板上有力的、布局得当的板书，清爽的图示，极大地帮助了学生在每一阶段之后钻研所涉及过的主题。此外，课程的标准足够高，能耗尽优秀学生的气力，还要求细心的合作。这在一个不点到、全靠自觉的大学系统里是重要的。在就一个问题从头讨论的练习中，初学者也会引起索末菲或者助教的注意，其会因自己的付出被赞赏而受到鼓舞。

无论学生岁数大小，索末菲对于真正的努力都具有特别的鉴赏力。这是德拜、泡利、海森堡（只提如今已获得诺奖的几位）这个级别的科学家早年求学生涯中投奔他的原因。但是，那些资质一般的学生也得到了很好的照料，会赋予较小的问题和较少的期待以锻炼其能力。懒散的学生不久就会自行离开。故此，索末菲的学生们形成了一个精英小团体，但人数也维持足够多，能够产生一股协助新手迅速独自扬帆起航的和风。祝愿索末菲教程的（英文）翻译广为散播这股和风，协助其他的群体去航行在发现的海洋上。

论及（教程）作者工作的其他文章

Anon., Current Biographies, 1950, pp. 537–538.

P. Kirkpatrick, *Am. J. Physics* (1949). **17**, 5, 312–316.

M. Born, *Proc. Roy. Soc., London. A.* (1952).

P. P. Ewald, *Nature* (1951). **168**, 364–366.

W. Heisenberg, *Naturwissenschaften*（自然科学）(1951). **38**, 337.

M. v. Laue, *Naturwissenschaften*（自然科学）(1951). **38**, 513–518.

埃瓦尔特 [6]

布鲁克林工业研究所，纽约

---

[6] Paul Peter Ewald (1888–1985)，有汉译埃瓦、埃瓦尔德等，比如 Ewald's diffraction sphere 就被译为埃瓦球、埃瓦尔德球。Ewald 是德国晶体学家，X–射线衍射研究晶体的先驱。1906–1907 年，Ewald 在哥廷恩大学学的数学。1907 年，Ewald 转到慕尼黑大学跟索末菲继续学习数学，1912 年以单晶中 X–射线传播定律的论文获得博士学位。博士毕业后，Ewald 接着在慕尼黑给索末菲做了一段时间的助手。

FOREWORD TO SOMMERFELD'S COURSE

P. P. EWALD

POLYTECHNIC INSTITUTE OF BROOKLYN, NEW YORK

The author of this Course on Theoretical Physics, Arnold Sommerfeld, was one of the central figures in achieving the transformation through which physics passed in the two decades from 1910 to 1930. Without his inspired and untiring efforts both the tumultuous advance and the wide-spread dissemination of the quantum theory of the atom would not have been what they were. Sommerfeld's Institute for Theoretical Physics in Munich became a school from which issued a steady stream of research papers by German and foreign, young and mature students of atomic theory. His famous book "Atombau und Spektrallinien," followed later by the companion volume "Wellenmechanik," was for a long time the only full and authoritative account of this fundamental subject; its successive editions unroll an impressive survey of the rapid development of atomic theory following Niels Bohr's first papers.

Both by his training and previous research Sommerfeld was firmly rooted in the mathematical methods of classical physics, and to this fact he owes much of his mastery of the newly created methods of quantum physics, especially after the advent of Schrödinger's wave mechanics in 1926. It was therefore natural for Sommerfeld to give his students a thorough training in classical methods, all the more so since he himself took great delight in the aesthetic beauty of classical theory. The harmony between mathematical formalism, its physical interpretation, and experimental materialization was cast in relief in Sommerfeld's lectures and deeply impressed his students.

Sommerfeld was over seventy, and retired after forty years of academic teaching, when he committed his lectures to paper. He did it with a sense of double obligation: to preserve through a crisis the achievements that had carried physics to great triumphs and to bequeath to the young generation of physicists the valuable analytical tools that had been shaped on the classical problems. Sommerfeld had taken an active part in perfecting these tools from 1895 onwards, when he wrote his doctoral thesis on Arbitrary Functions in Physics. Among his earliest brilliant work was the construction of a strict solution for the diffraction of a wave by an edge ; he extended the methods that Riemann used in the theory of functions, with the result that a solution of the diffraction problem by an image method in multi-valued space was obtained. The reader will find this discussed in Volume V on *Optics*. Extending from Sommerfeld's early period in Göttingen to the beginning of the quantum period in Munich, and in co-authorship with the great mathematician Felix Klein in Göttingen, the preparation of the four volume standard work on the theory of rotating rigid bodies, *Theorie des Kreisels*, proceeded. This work was intended to demonstrate the intimate connection between "pure" and "applied" mathematics by bringing a great variety of mathematical topics, such as the theory of functions, elliptic functions, quaternions, Klein-Caley parameters, etc., to bear on this problem of dynamics of a rigid body. While holding the chair of Technical Mechanics at the Technische Hochschule in Aachen, 1899–1905, Sommerfeld became deeply interested in engineering problems. His papers on the hydrodynamics of lubrication, on the interaction between electrical generators working on the same power line, on the braking of trains, and on other topics all adopt a general approach that gives them lasting value. With the advent of wireless telegraphy a series of papers by Sommerfeld and his pupils began on the mode of emission and of propagation of radio waves. They offer excellent examples of the mathematical methods in which Sommerfeld was a master. In particular, the diffraction problem of these waves round the earth was brought down to the discussion of complex integrals, which form a strict solution (see Volume 6, Chapter 6).

It would be out of place to elaborate a full list of the achievements with which Sommerfeld enriched physical theory. The reader may be referred to some of the articles listed below. But a few words may be added on Sommerfeld as a teacher, and on the significance of the course of lectures now being published in translation.

The courses of theoretical physics held in Munich were of two kinds, general ones and specialized ones. The former were given four hours, or more precisely periods of 45–50 minutes, a week through a 13-week winter semester and an 11-week summer semester. These six courses form the subject of the present six volumes. Each served as an introduction for students who had taken the demonstration courses on experimental physics (given, in Munich, by Röntgen, later by W. Wien). In experimental physics the student acquired a factual survey of the phenomena and of their quantitative evaluation, based on a fundamentally non-mathematical treatment. In the courses on theoretical physics the elementary ground was gone over again, but with a view to developing the mathematical handling and to constructing an integrating theory which could then be extended to advanced problems. The latter might change from one series of courses to the next, and the inclusion of topical subjects in the second halves of these courses made them most interesting even to advanced students who had already gone over the subject in their previous work. In addition to the lectures, two hours a week were devoted to discussing problems.

The specialized courses were two-hour-a-week lecture courses on subjects which could be dealt with only briefly in the general course, or had topical interest. Those given by Sommerfeld were usually connected with his own current research and often contained parts which appeared a little later as original papers. The interpretation of the Lorentz transformation as a rotation in four dimensional space (Volume 3, §26), the transition from wave optics to geometrical optics (Volume 5, §35), the discussion of the signal velocity in a dispersive medium (Volume 5, §22) are some examples. In later general courses such subjects were often included, to the exclusion of other less interesting parts of previous courses.

Besides the lecture courses, a seminar and colloquium offered instruction in advanced topics ; here the students had to review the assigned subject and to deliver the talk, which meant several weeks' intensive study.

From the student's point of view the great attraction of Sommerfeld's lectures lay in their clarity : the approach from the physical side, the formulation of the mathematical problem, the simple and yet general explanation of the mathematical methods used, and the thorough discussion of the result again in terms of physical experiment. His firm, well-distributed writing on the blackboard, and the evidence of his diagrams, helped the student to survey at the end of each period all the subjects that had been covered. Besides, the standard of the course was high enough to tax the powers of the better students and to demand vigilant cooperation. This was all the more important in a university system where there was no check on attendance and only a voluntary one on performance. In giving an original discussion of a problem in the exercises even the beginner would attract the attention of Sommerfeld or of the assistant in charge, and he would be stimulated by the appreciative understanding his effort received.

Sommerfeld had an extraordinary flair for genuine endeavor and performance, irrespective of the age of his students. That is why scientists of the rank of Debye, Pauli, Heisenberg (to name only those who now are Nobel prize men) became attached to him in their early years of study. But also the good average student was well looked after and given smaller problems or small responsibilities to exercise his forces. The indolent student soon turned away on his own account. Thus Sommerfeld's students formed in a way a select group, but their number remained large enough to create a breeze that helped the inexperienced newcomers rapidly to unfurl their own sails. May the translation of Sommerfeld's lectures carry some of this breeze afield and assist other groups in preparing to sail the ocean of discovery.

SOME ARTICLES ON THE AUTHOR'S WORK

Anon., Current Biographies, 1950, pp. 537–538 (with portrait).
P. Kirkpatrick, *Am. J. Physics* (1949). **17**, 5, 312–316. (Presentation of the Oerstedt Medal to Sommerfeld by the American Association of Physics Teachers.)
M. Born, *Proc. Roy. Soc., London. A.* (1952). (Obituary.)
P. P. Ewald, *Nature* (1951). **168**, 364–366. (Obituary Notice.)
W. Heisenberg, *Naturwissenschaften* (1951). **38**, 337.
M. v. Laue, *Naturwissenschaften* (1951). **38**, 513–518. (A full appraisal of Sommerfeld's work.)

PREFACE TO THE FIRST EDITION, SEPTEMBER 1942

The encouragement of some of my former students and the repeated suggestion of the publishers decided me to publish my general course on theoretical physics which I gave regularly for thirty-two years at the University of Munich.

This was an introductory course, and was attended not only by the physics majors of the University and the Polytechnic Institute (Technische Hochschule), but also by candidates for teachers' degrees in mathematics and physics, by students of astronomy and some few of physical chemistry— all usually in their third and fourth years. The lectures were held four times a week and supplemented by a two-hour problem period. Special courses on modern physics, which were given concurrently with these, have not been included in this series of books; their subject matter found its way into my scientific papers, summarizing articles, and other books. While it is true that quantum mechanics always hovers in the background and reference is made to it now and then, the actual substance of these lectures is classical physics.

The order of the courses, which has been kept in their publication, was

1. Mechanics
2. Mechanics of Deformable Bodies
3. Electrodynamics
4. Optics
5. Thermodynamics and Statistical Mechanics
6. Partial Differential Equations in Physics

The courses on mechanics were given in alternate years by myself and by my colleagues in mathematics. Concurrent courses in hydrodynamics, electrodynamics, and thermodynamics were taught by younger members of the staff. Vector analysis was given in a separate course so that its systematic development could be omitted from my lectures.

In print, as in my classes, I will not detain myself with the mathematical foundations, but proceed as rapidly as possible to the physical problems themselves. My aim is to give the reader a vivid picture of the vast and varied material that comes within the scope of theory from a suitably chosen mathematical and physical vantage point. With this in mind I shall not be too concerned if I have left some gaps in the systematic justification and axiomatic structure of the work. In any case I do not wish to frighten the hearer of my lectures with drawn-out investigations of a mathematical or logical nature and distract his attention from what is physically interesting. It is my belief that this attitude proved its worth in my courses ; it has therefore been retained in the printed lectures. Whereas the lectures of Planck are irreproachable in systematic formulation, I believe that I can claim for mine a greater variety of subject matter and a more flexible handling of the mathematical apparatus. Moreover, I gladly refer the reader to the more complete and often more thorough treatment of Planck, especially in thermodynamics and statistical mechanics.

The problems collected at the end of each volume should be regarded as supplementary to the text. They were handed in by the students and then presented orally during the problem periods. Elementary numerical problems, to be found so prolifically in texts and collections of problems, have, in general, not been included. The problems are numbered by chapter. Sections are numbered through in each volume and equations in each section. References within each volume to earlier equations can thus be made merely by giving the section, and equation numbers. To make it simpler to find a given section the upper inside corners of every pair of pages bear the section and chapter number.

Looking back on my years of teaching I wish to acknowledge with gratitude my special indebtedness to two men, Röntgen and Felix Klein. Röntgen not only created the external conditions for my professional activity by calling me to a privileged sphere of action; he also stood by my side and actively furthered the increasing scope of my work over a period of many years. Even earlier Felix Klein had imparted to my mathematical thinking that turn of mind which is best adapted to applications; through his mastery in the art of lecturing he exerted a strong indirect influence on my own teaching. In particular let me mention that the last part of this course was announced for the first time while I was still instructor in Göttingen and imbued with the mathematical tradition of that university symbolized in the three names Riemann—Dirichlet—Klein. At that time my course was less comprehensive than the present Volume VI, but it found much resonance in the audience. When the course was repeated in later years my students often told me that only here had they really grasped the handling and application of mathematical results, e.g., Fourier methods, applications of the theory of functions, boundary value problems.

In conclusion, let me send out these volumes with the wish that they arouse the reader's interest in our beautiful science and give him as much pleasure as the courses gave to those attending them and to me during my many years of teaching activity.

Munich, September 1942   Arnold Sommerfeld

TABLE OF CONTENTS

PAGE

FOREWORD TO SOMMERFELD'S COURSE BY P. P. EWALD.................... v

PREFACE TO THE FIRST EDITION.. ix

INTRODUCTION 1

CHAPTER I. MECHANICS OF A PARTICLE 3
    1. Newton's Axioms 3
    2. Space, Time and Reference Systems 9
    3. Rectilinear Motion of a Mass Point............................... 16
    Examples:
        (1) Free Fall Near Earth's Surface (Falling Stone) 19
        (2) Free Fall From a Great Distance (Meteor) 20
        (3) Free Fall in Air 21
        (4) Harmonic Oscillations 22
        (5) Collision of Two Particles 24
    4. Variable Masses 28
    5. Kinematics and Statics of a Single Mass Point in a Plane and in Space.. 32
        (1) Plane Kinematics 32
        (2) The Concept of Moment in Plane Statics and Kinematics 34
        (3) Kinematics in Space 36
        (4) Statics in Space; Moment of Force About a Point and About an Axis 37
    6. Dynamics (Kinetics) of the Freely Moving Mass Point; Kepler Problem;
       Concept of Potential Energy 38
        (1) Kepler Problem with Fixed Sun 38
        (2) Kepler Problem Including Motion of the Sun.................... 44
        (3) When Does a Force Field Have a Potential ?.................... 45

CHAPTER II. MECHANICS OF SYSTEMS, PRINCIPLE OF VIRTUAL WORK, AND
    D'ALEMBERT'S PRINCIPLE 48
    7. Degrees of Freedom and Virtual Displacements of a Mechanical System;
       Holonomic and Non-holonomic Constraints 48
    8. The Principle of Virtual Work.................................... 51
    9. Illustrations of the Principle of Virtual Work 54
        (1) The Lever... 54
        (2) Inverse of the Lever: Cyclist, Bridge 55
        (3) The Block and Tackle.. 56
        (4) The Drive Mechanism of a Piston Engine 57
        (5) Moment of a Force About an Axis and Work in a Virtual Rotation .. 58
    10. D'Alembert's Principle; Introduction of Inertial Forces 59
    11. Application of d'Alembert's Principle to the Simplest Problems 62
        (1) Rotation of a Rigid Body About a Fixed Axis.................... 62
        (2) Coupling of Rotational and Translational Motion 64
        (3) Sphere Rolling on Inclined Plane 64
        (4) Mass Guided Along Prescribed Trajectory 65
    12. Lagrange's Equations of the First Kind............................. 66
    13. Equations of Momentum and of Angular Momentum 69
        (1) Equation of Momentum....................................... 70
        (2) Equation of Angular Momentum 71
        (3) Proof Using the Coordinate Method........................... 73
        (4) Examples 74
        (5) Mass Balancing of Marine Engines............................ 76
        (6) General Rule on the Number of Integrations Feasible in a Closed
        System 79

Table of Contents

PAGE

14. The Laws of Friction.. 81
    (1) Static Friction 81
    (2) Sliding Friction 83

CHAPTER III. OSCILLATION PROBLEMS 87
15. The Simple Pendulum 87
16. The Compound Pendulum.. 91
    Supplement: A Rule Concerning Moments of Inertia 93
17. The Cycloidal Pendulum 94
18. The Spherical Pendulum 96
19. Various Types of Oscillations.
    Free and Forced, Damp and Undamped Oscillations 100
20. Sympathetic Oscillations 106
21. The Double Pendulum 111

CHAPTER IV. THE RIGID BODY 118
22. Kinematics of Rigid Bodies 118
23. Statics of Rigid Bodies 125
    (1) The Conditions of Equilibrium 125
    (2) Equipollence; the Reduction of Force Systems.................. 126
    (3) Change of Reference Point..................................... 127
    (4) Comparison of Kinematics and Statics.......................... 127
    Supplement: Wrenches and Screw Displacements 129
24. Linear and Angular Momentum of a Rigid Body.
    Their Connection with Linear and Angular Velocity 130
25. Dynamics of a Rigid Body. Survey of its Forms of Motion........... 133
    (1) The Spherical Top Under No Forces 134
    (2) The Symmetrical Top Under No Forces 134
    (3) The Unsymmetrical Top Under No Forces 135
    (4) The Heavy Symmetrical Top................................... 136
    (5) The Heavy Unsymmetrical Top................................. 138
26. Euler's Equations. Quantitative Treatment of the Top Under No Forces 139
    (1) Euler's Equations of Motion 139
    (2) Regular Precession of the Symmetrical Top Under No Forces and
        Euler's Theory of Polar Fluctuations............................ 142
    (3) Motion of an Unsymmetrical Top Under No Forces. Examination
        of its Permanent Rotations as to Stability 146
27. Demonstration Experiments Illustrating the Theory of the Spinning
    Top; Practical Applications....................................... 150
    (1) The Gyrostabilizer and Related Topics......................... 153
    (2) The Gyrocompass 154
    (3) Gyroscopic Effects in Railroad Wheels and Bicycles.............. 156
    Supplement: The Mechanics of Billiards 158
        (a) High and Low Shots, 158—(b) Follow Shots and Draw Shots, 159—
        (c) Trajectories with " English " Under Horizontal Impact, 160—
        (d) Parabolic Path Due to Shot with Vertical Component, 160

CHAPTER V. RELATIVE MOTION 162
28. Derivation of the Coriolis Force in a Special Case 162
29. The General Differential Equations of Relative Motion 165
30. Free Fall on the Rotating Earth; Nature of the Gyroscopic Terms...... 167

Table of Contents

PAGE

31. Foucault's Pendulum 171
32. Lagrange's Case of the Three-Body Problem......................... 174

CHAPTER VI. INTEGRAL VARIATIONAL PRINCIPLES OF MECHANICS AND LAGRANGE'S
EQUATIONS FOR GENERALIZED COORDINATES 181
33. Hamilton's Principle 181
34. Lagrange's Equations for Generalized Coordinates 185
35. Examples Illustrating the Use of Lagrange's Equations 192
    (1) The Cycloidal Pendulum 192
    (2) The Spherical Pendulum 193
    (3) The Double Pendulum 195
    (4) The Heavy Symmetrical Top 196
36. An Alternate Derivation of Lagrange's Equations 200
37. The Principle of Least Action 204

CHAPTER VII. DIFFERENTIAL VARIATIONAL PRINCIPLES OF MECHANICS 210
38. Gauss' Principle of Least Constraint 210
39. Hertz's Principle of Least Curvature 212
40. A Digression on Geodesics 214

CHAPTER VIII. THE THEORY OF HAMILTON.............................. 217
41. Hamilton's Equations 217
    (1) Derivation of Hamilton's Equations from Lagrange's Equations 218
    (2) Derivation of Hamilton's Equations from Hamilton's Principle 218
42. Routh's Equations and Cyclic Systems 222
43. The Differential Equations for Non-Holonomic Velocity Parameters 226
44. The Hamilton-Jacobi Equation 229
    (1) Conservative Systems 230
    (2) Dissipative Systems... 231
45. Jacobi's Rule on the Integration of the Hamilton-Jacobi Equation........ 233
46. Classical and Quantum-Theoretical Treatment of the Kepler Problem 235

PROBLEMS

FOR CHAPTER I

I. 1, I. 2, I. 3. Elastic collision..................................... 240
I.   4. Inelastic collision between an electron and an atom............... 240
I.   5. Rocket to the moon 241
I.   6. Water drop falling through saturated atmosphere 241
I.   7. Falling chain 241
I.   8. Falling rope 241
I.   9. Acceleration of moon due to earth's attraction 241
I. 10. The torque as vector quantity................................. 241
I. 11. The hodograph of planetary motion 242
I. 12. Parallel beam of electrons passing through the field of an ion.
       Envelope of the trajectories.................................. 242
I. 13. Elliptical trajectory under the influence of a central force directly
       proportional to the distance................................. 242
I. 14. Nuclear disintegration of lithium 242
I. 15. Central collisions between neutrons and atomic nuclei; effect of a
       block of paraffin 243
I. 16. Kepler's equation 243

Table of Contents

PAGE

FOR CHAPTER II

II. 1. Non-holonomic conditions of a rolling wheel...................... 244
II. 2. Approximate design of a flywheel for a double-acting one-cylinder
        steam engine 245
II. 3. Centrifugal force under increased rotation of the earth 246
II. 4. Poggendorff's experiment...................................... 246
II. 5. Accelerated inclined plane 246
II. 6. Products of inertia for the uniform rotation of an unsymmetrical body
        about an axis 246
II. 7. Theory of the Yo–yo.. 246
II. 8. Particle moving on the surface of a sphere...................... 247

FOR CHAPTER III

III. 1. Spherical pendulum under infinitesimal oscillations............. 247
III. 2. Position of the resonance peak of forced damped oscillations 247
III. 3. The galvanometer 247
III. 4. Pendulum under forced motion of its point of suspension........... 248
III. 5. Practical arrangement of coupled pendulums 248
III. 6. The oscillation quencher 250

FOR CHAPTER IV

IV. 1. Moments of inertia of a plane mass distribution.................. 250
IV. 2. Rotation of the top about its principal axes 250
IV. 3. High and low shots in a billiard game.   Follow shot and draw shot.. 250
IV. 4. Parabolic motion of a billiard ball 251

FOR CHAPTER V

V. 1. Relative motion in a plane 251
V. 2. Motion of a particle on a rotating straight line 251
V. 3. The sleigh as the simplest example of a non-holonomic system...... 251

FOR CHAPTER VI

VI. 1. Example illustrating Hamilton's principle........................ 252
VI. 2. Relative motion in a plane and motion on a rotating straight line 253
VI. 3. Free fall on the rotating earth and Foucault's pendulum........... 253
VI. 4. " Wobbling " of a cylinder rolling on a plane support 254
VI. 5. Differential of an automobile 254
HINTS FOR SOLVING THE PROBLEMS 256
INDEX 283

INTRODUCTION

Mechanics is the backbone of mathematical physics. Though it is true that we no longer require physics to explain all phenomena in terms of mechanical models, as was common during the last century, we are nevertheless convinced that the principles of mechanics, such as those of momentum, energy, and least action, are of the greatest importance in all branches of physics.

We call this book " Mechanics," not " Analytical Mechanics " as the mathematicians are wont to do. The latter name has its origin in the great work of Lagrange (1788), who attempted to mold the whole system of mechanics into a consistent language of mathematical equations and was proud of the fact that " one would not find a single diagram in his work." We, on the contrary, shall draw as much as possible on illustration and comparison. The reader will find in this volume many concrete applications in astronomy, physics, and even to some degree in engineering, which should help to make the principles clearer.

The exact title of the book should be " Mechanics of Systems of a Finite Number of Degrees of Freedom "; that of the second volume would accordingly be "Mechanics of Systems with an Infinite Number of Degrees of Freedom." Since, however, the concept of degrees of freedom is not too well known and can here be explained only at the beginning of the second chapter, we shall be satisfied with the customary title "Mechanics," a title hardly subject to misunderstanding.

We commence with Newton's fundamental analysis in his " Philosophiae Naturalis Principia Mathematica " (London, 1687); not that Newton lacked important predecessors, such as Archimedes, Galileo, Kepler, and Huygens, to mention only a few. It was, nevertheless, Newton who first created a firm foundation for general mechanics. Even today, apart from some changes and reinforcements, the foundation laid down by him provides us with the most natural and didactically simplest approach to general mechanics.

We shall at first investigate the mechanics of the single mass point or particle.

CHAPTER I

MECHANICS OF A PARTICLE

§ 1. *Newton's Axioms*

The laws of motion will be introduced in axiomatic form; they summarize in precise form the whole body of experience.

First law: *Every material body remains in its state of rest or of uniform rectilinear motion unless compelled by forces acting on it to change its state.*[1]

We shall at first withhold explanation of the concept of force introduced in this law. We notice that the states of *rest* and of *uniform* (*rectilinear*) *motion* are treated on equal footing and are regarded as natural states of the body. The law postulates a tendency of the body to remain in such a natural state; this tendency is called the *inertia* of the body. One often speaks of *Galileo's law of inertia* instead of Newton's first law in referring to the above axiom. We must say in this connection that while it is perfectly true that Galileo arrived at this law long before Newton (as a limiting result of his experiments with sliding bodies on planes of vanishing inclination), we find it characteristic of Newton that the law holds top position in his system. Newton's word " body " will, for the time being, be replaced by the words " particle " or " mass point."

To formulate the first law mathematically we shall make use of definitions 1 and 2 preceding it in the " Principia."

Definition 2: *The quantity of motion is the measure of the same, arising from the velocity and the quantity of matter conjunctly.*[2]

The " quantity of motion " is hence the product of two factors, the velocity, whose meaning is geometrically evident,[3] and the " quantity of

---

[1] We mention here, and in connection with what is to follow, the book *Die Mechanik in ihrer Entwickelung* (8th ed., F. A. Brockhaus, Leipzig, 1923; translated into English under the title *The Science of Mechanics*, Open Court Publishing Co., LaSalle, Ill., 1942) by Ernst Mach. The study of this excellent critical history is recommended to all students of mechanics, especially since in our book we must restrict ourselves to the concepts of mechanics in a form ready for use and cannot delve into the origin and gradual clarification of these concepts. This should not be interpreted to mean, however, that we agree with Mach's positivistic philosophy as it is developed in Chapter IV, 4, of his book, with its attendant overemphasis of the Economy Principle, the denial of atomic theory and the preference for formal continuity theories.

[2] *Newton's Principia*, translated by Andrew Motte.—TRANSLATOR.

[3] Evident, that is, once a reference system has been chosen in which the velocity is to be measured.

3

matter," which is to be explained physically. Newton attempts the latter in his definition 1, in which he says that the quantity of matter is measured by its density and volume conjunctly. This is obviously only a mock definition, since density itself cannot be defined in any other way than by the amount of matter in unit volume. In the same definition 1, Newton also states that instead of " quantity of matter " he will use the word *mass*. We shall follow him in this, but shall postpone the physical definition of mass (as well as that of force) until later.

The quantity of motion accordingly becomes the product of mass and velocity. Like the latter it is a directed magnitude, a *vector*. We write[4]

$$(1) \qquad\qquad \mathbf{p} = m\mathbf{v}$$

and formulate the first law of motion in its final form:

$$(2) \qquad\qquad \mathbf{p} = \text{constant in the absence of forces.}$$

We shall put the law of inertia thus formulated at the head of our mechanics. It is the result of an evolution extending over many centuries, and is by no means as self-evident as it appears to us today. The philosopher Kant, for instance, says in his paper, " Thoughts on the True Estimation of Living Forces," written in 1747, long after Newton: " There exist two kinds of motions; those which have ceased after a certain time, and those which persist." The motions that in Kant's opinion cease by themselves are, according to modern ideas — and those of Newton — motions which are attenuated by frictional forces and finally destroyed.

The expression " quantity of motion " is unfortunately chosen in that it does not take into account the vector character of $m\mathbf{v}$. Thus a better term would be the word " impulse," which conveys the idea of a push of a certain magnitude in a definite direction that the given $m\mathbf{v}$ is able to impart, by collision, to some body initially at rest. Since the term " impulse " is, however, used in a somewhat different sense in mechanics, we shall have to retain the name " quantity of motion," or, in modern language, *momentum* for the vector $\mathbf{p}$. Instead of the law of inertia and Newton's first law of motion we can then speak of the *law of conservation of momentum*.

We shall now discuss Newton's **second law,** the real law of motion: *The change in motion is proportional to the force acting and takes place in the direction of the straight line along which the force acts.*

---

[4] We assume that the reader is familiar with the elements of vector algebra. Since, however, vector operations originated in close association with mechanics (including the mechanics of fluids), we shall often have occasion to explain vector concepts simultaneously with mechanical concepts.

As regards notation, vectors will be designated throughout by bold-faced letters; thus $\boldsymbol{\omega}$ for the angular velocity when regarded as a (axial) vector. In diagrams, overhead arrows will occasionally be used.

By " change in motion " Newton undoubtedly means the change with time of the previously defined momentum **p**, hence the vector $\dot{\mathbf{p}}$ (the dot is Newton's notation for the " fluxion " $\dot{\mathbf{p}} = \frac{d\mathbf{p}}{dt}$). If we designate the force by the letter **F**, our second law can then be written as

$$(3) \qquad\qquad \dot{\mathbf{p}} = \mathbf{F}.$$

Since we called **p** the momentum, this law expresses the manner in which momentum changes with time, and can, for brevity, be called simply the *law of momentum.*

Unfortunately the law is often, especially in the mathematical literature, designated as " Newton's law of acceleration." It is of course true that if we treat m as a constant, (3) combined with (1) is identical to

$$(3a) \qquad\qquad m\dot{\mathbf{v}} = \mathbf{F}: \quad \text{Mass} \cdot \text{Acceleration} = \text{Force.}$$

But mass is not always constant; it is variable in the theory of relativity for example, where Newton's formulation (3) prophetically turns out to be the correct one. We shall treat a series of examples with variable mass in § 4, where we shall have a closer look at the interrelation between formulations (3) and (3a). Incidentally, the mechanical system which is next in simplicity to that of a single mass point, namely, the rotating rigid body, leads us to an equation of motion along the lines of (3), in the form " rate of change of moment of momentum (angular momentum) = moment of force (torque) "; a description in terms of angular acceleration, similar to (3a), is not possible. An effect similar to the non-constancy of mass in relativity must be taken into account: the moment of inertia, here replacing the mass, changes with changing location of the axis of rotation in the body.

We must now seek to get a clear idea of the concept of *force*. Kirchhoff[5] wanted to degrade it to a quantity defined by the product of mass and acceleration. Hertz[6], too, tried to eliminate and replace it by coupling the system under consideration with other, generally hidden systems interacting with the former. Hertz carried out this program with admirable consistency. His method, however, hardly produced fruitful results; and it is especially unsuitable for the beginner.

We are of the opinion that we have at least a *qualitative* notion of " force " which we acquire quite directly through the feeling we experience when using our muscles. In addition the earth has provided us with the comparison standard of gravity, with which we can measure all other forces *quantitatively.* For this purpose we need merely balance the effect

---

[5] Gustav Kirchhoff, Vol. I of his *Vorlesungen über mathematische Physik,* p. 22.
[6] Heinrich Hertz, *Miscellaneous Papers,* Vol. III, *Principles of Mechanics,* Macmillan, New York, 1896.

of a given force by a suitable weight. (By means of a pulley and string we can let the vertical force of gravity act in a direction opposed to the given force.) If, in addition, we procure a number of equally heavy bodies, a "set of weights," we obtain a tentative scale with which to measure forces quantitatively.

The same is true for the concept of force as for all other physical concepts and names: word definitions have very little meaning; physically significant definitions are obtained as soon as we prescribe a way of measuring the quantity in question. Such a prescription need not contain the details of practical procedure, but merely state a way to measure the quantity in principle.

The above prescription, making use of gravity, has given a concrete content to the right member of our law of momentum (3); it has thereby become a real physical statement. It is true that the left member still contains the mass m, up to now undefined. This does not mean that the definition of mass is the only content of the law. For the law brings out that it is $\dot{\mathbf{p}}$, not $\mathbf{p}$ itself or perhaps $\ddot{\mathbf{p}}$ which is determined by the force. We shall see in § 4 how the definition of mass is obtained in case it is variable, the relativistic mass serving as example.

Third law: *Action always equals reaction*, or: *the forces two bodies exert on each other are always equal and opposite in direction.*

This is the principle of action and reaction. It says that for every pressure there is a pressure in the opposite direction. Forces always occur paired in nature. The falling stone attracts the earth just as strongly as the earth attracts the stone.

This law makes possible the transition from the mechanics of single mass points to that of compound systems ; it is therefore fundamental to the entire field of structural statics, to name but one example.

We shall call the rule of the *parallelogram of forces* our **Fourth law,** even though in Newton it appears merely as an addition or corollary to the other laws of motion. The fourth law states that two forces applied to the same mass point compound to act like the diagonal of the parallelogram formed by them: *forces add like vectors.* This seems self-evident since we equated the force **F** to the vector $\dot{\mathbf{p}}$ in the Second Law. Actually, however, the Fourth Law, as Mach emphasizes, contains the axiom that each force acting on a mass point causes it to change its motion as if this force were the only one acting there. The parallelogram of forces hence establishes axiomatically the independence of the effects of several forces acting together at the same point, or, more generally, the *principle of superposition of forces.* Of course the last statement as well as the laws of motion preceding it are nothing but an idealization and a precise formulation of our whole body of experience.

Having introduced the concept of force, we shall at this point introduce that of *work* with the definition

(4) $$dW = \mathbf{F} \cdot \mathbf{ds} = F \, ds \, \cos(\mathbf{F}, \mathbf{ds}).$$

Thus the work does not equal " force times distance " as often stated, but " component of force along path times path length" or " force times component of path length along force."

From the statement, " forces add vectorially," follows immediately the complementary statement that " work adds algebraically." Indeed

$$\mathbf{F_1} + \mathbf{F_2} + \cdots = \mathbf{F}$$

leads to

(5) $$\mathbf{F_1} \cdot \mathbf{ds} + \mathbf{F_2} \cdot \mathbf{ds} + \cdots = \mathbf{F} \cdot \mathbf{ds}$$

by scalar multiplication with the distance **ds**. Here **F** is the resultant force. The definition of the scalar product contained in (4) automatically sees to it that, for example, in the first product of (5) only ds_1, the component of the distance in the direction of the force $\mathbf{F_1}$, occurs. Hence we can also write instead of (5) that

(6) $$dW_1 + dW_2 + \cdots = dW,$$

as stated above.

Related to the concept of work is that of *power*; power is the work done in unit time.

In concluding these introductory remarks we shall have to agree on how to measure the mechanical quantities that we have introduced. Here we have a choice of two systems of units, the *physical* (or *absolute*) and the *practical* (or *gravitational*) metric systems. The difference between them is that in the absolute system the gram (or kilogram) serves as unit *mass*, whereas in the gravitational system the kilogram (or gram) serves as unit *force*. In the latter case we speak of a kilogram-weight and write

$$1 \text{ kg-weight} = g \cdot \text{kg-mass}.$$

The gravitational acceleration g is, however, a function of the location on the earth, being greater at the poles than at the equator because of a smaller distance from the center of the earth as well as because of diminished centrifugal force. Hence the kg-weight is dependent on location; a kg-weight sample cannot be transported. The gravitational system is therefore unsuitable for precise measurements. The physical system has, in contrast, been distinguished by the title, " absolute system of units." We have nevertheless become so accustomed to the gravitational system that in many cases where we should really say " mass," the word " weight " has once and for all made its way into our scientific language. Thus we talk of specific weight when we should say specific mass or density; and of atomic and molecular weights — which surely have nothing to do with the acceleration due to gravity.

Gauss, the originator of absolute measurements, decided in favor of the absolute system after some hesitation. Initially he, too, was in favor of introducing force as the basic unit, since it played a more direct role than mass in his measurements of terrestrial magnetism. On the other hand he wanted these measurements to encompass the entire surface of our globe; he therefore saw himself forced to adopt as unit a quantity whose value would not depend on location.

Below we have put the two systems next to each other and at the same time introduced the derived units dyne, erg, joule, watt, and horse power (HP):

Absolute system (CGS)	*Gravitational system (MKS)*
cm, g-mass, sec	kg-weight, m, sec
1 kg-weight $= 9.81 \cdot 10^5$ g cm sec^{-2} $= 9.81 \cdot 10^5$ dyne	1 g-mass $= \dfrac{1}{1000} \dfrac{\text{kg}}{g} \dfrac{1}{g}$ sec^2 m^{-1}
1 erg $= 1$ dyne $\cdot$ 1 cm	1 unit of work $= 1$ kg $\cdot$ 1 m
1 joule $= 10^7$ erg	
1 mkg-weight $= 1000 \cdot g \cdot 100$ erg $= 9.81 \cdot 10^7$ erg $= 9.81$ joule	
1 watt $= 1$ joule sec^{-1}	1 unit of power $= 1$ kg m sec^{-1}
1 kilowatt $= 1000$ joule sec^{-1} $= \dfrac{1 \text{ HP}}{0.736} = 1.36$ HP	1 HP $= 75$ kg m sec^{-1} $= 75 \cdot 1000 \cdot 100 \cdot 981$ erg sec^{-1} $= 75 \cdot 9.81$ watt $= 0.736$ kw

It should be noted that according to a decision of the pertinent international commissions the CGS system was to be replaced by an absolute MKS system beginning with the year 1940. In this new system the meter takes the place of the centimeter, and the kilogram that of the gram as unit of mass, while the second is retained as unit of time. This is in agreement with a proposal of G. Giorgi, which shows its advantages fully only in electro-dynamics with the addition of a fourth independent electrical unit (see Vol. III of this series). In mechanics the proposed change would have the advantage that in the definition of the joule and the watt the bothersome powers of ten are eliminated. With the new larger units M and K the units of work and power become

$$1 \ M^2KS^{-2} = 10^7 \ cm^2 \ g \ sec^{-2} = 1 \ joule,$$
$$1 \ M^2KS^{-3} = 10^7 \ cm^2 \ g \ sec^{-3} = 1 \ watt.$$

The unit of force in the new system, called the newton, is thus

$$1 \text{ newton} = 1 \text{ MKS}^{-2} = 10^5 \text{ cm g sec}^{-2} = 10^5 \text{ dyne.}$$

This, too, can be regarded as an advantage of the Giorgi system, since now the new unit of force is brought closer to the convenient gravitational unit, the kg-weight, whereas the old unit of force, the dyne, is inconveniently small for most practical uses.

§ 2. *Space, Time and Reference Systems*[7]

Newton's views about space and time seem to us moderns quite unrealistic and appear to contradict his declared intention to base his analysis on fact alone. He states:

" Absolute space, in its own nature, without regard to anything external, remains always similar and immovable.

"Absolute, true, and mathematical time, of itself, and from its own nature, flows equably without regard to anything external, and by another name is called duration."

From these two quotations one would conclude that Newton worried but little where absolute time was to be taken from, and how an immovable absolute space was to be distinguished from one moving uniformly with respect to it. This is all the more surprising since he put the states of rest and of uniform motion on the same footing in his first law. On the other hand Newton tried to clarify the distinction between absolute and relative motion by his famous pail experiment.[8] In this experiment a pail is suspended from a twisted thread and filled with water. The pail is then suddenly released and as the thread untwists acquires a rotation about its axis of symmetry. The surface of the water remains at first level, although the relative velocity between pail and water is great. Gradually the water is set in motion by friction with the walls of the pail, climbs up the wall, and its surface assumes the familiar hollow paraboloidal shape. Finally a steady state is reached in which the relative motion between pail and water is zero; the " absolute " motion of the water in space has, on the other hand, increased to a maximum, and with it the curvature of the surface.

Actually the experiment only shows that the rotating pail does not furnish a suitable reference system from which the motion of the water can be understood. Is the earth such an unsuitable reference system ?

---

[7] The beginner to whom the following somewhat abstract considerations seem un-familiar may postpone the study of this section and of § 4 until a later time.

[8] " I have performed this experiment myself," says Newton, probably with reference to the natural philosophers, perhaps his compatriot Francis Bacon, who was wont to describe the results of experiments he had not performed.

It, too, rotates and furthermore describes an orbit around the sun. In general, what are the requirements that an ideal reference system has to satisfy in mechanics? By a reference system is meant a frame in space and time which will enable us to read off the position of mass points and the passage of time; we might take a Cartesian system of coordinates x, y, z, and a time scale, t.

In practice we shall have to rely on the astronomers for this choice. The fixed stars furnish sufficiently constant directions for our coordinate axes, and the sidereal day furnishes a sufficiently constant interval of time. Theoretically, on the other hand, we are forced to recognize a disagreeable tautology: that reference frame is an ideal one in which the Galilean law of inertia holds with sufficient accuracy, for a sufficiently force-free body. Thus the first law is degraded to a formal identity or to the rank of definition. The only positive, not purely formal content that the law retains is the assertion that reference systems of the required properties do exist. All our experience indicates that one such system is approximated by astronomical determinations of position and time.

We mean essentially the same thing when we say that the laws of mechanics presuppose the existence of an *inertial frame*, i.e., an imaginary structure whose axes are trajectories of bodies moving purely under inertia.

The question now arises to what extent this ideal system of reference is determined. Is there only one such system x, y, z, t, or are there perhaps infinitely many such systems? Newton's first law gives the answer at once, for it states that any two systems x, y, z, t and x', y', z', t' are equivalent if they differ only by a uniform translational motion. In mathematical form

(1)
$$x' = x + \alpha_0 t$$
$$y' = y + \beta_0 t$$
$$z' = z + \gamma_0 t$$
$$t' = t.$$

We can generalize the transformation (1) by performing a rotation on the spatial system x, y, z about its origin, which amounts to replacing x, y, z in (1) by new space coordinates ξ, η, ζ such that

(2)
$$\xi^2 + \eta^2 + \zeta^2 = x^2 + y^2 + z^2.$$

This condition defines an arbitrary *orthogonal transformation*. With α_k, β_k, γ_k, the direction cosines, it yields

(3)

	x	y	z
ξ	α_1	α_2	α_3
η	β_1	β_2	β_3
ζ	γ_1	γ_2	γ_3

This scheme can be read equally well from left to right as from top to bottom. Because of (2) the α, β, γ satisfy the well-known relations

(4) $$\sum \alpha_k^2 = \sum \beta_k^2 = \sum \gamma_k^2 = 1, \quad \sum \alpha_k \beta_k = \cdots = 0, \text{ etc.}$$

If now we replace the x, y, z in the right member of (1) by the ξ, η, ζ of (3), we obtain the generalized transformation scheme[9]

(5)

	x	y	z	t
x'	α_1	α_2	α_3	α_0
y'	β_1	β_2	β_3	β_0
z'	γ_1	γ_2	γ_3	γ_0
t'	0	0	0	1

The fact that the primed system x', y', z', t' is just as good a reference frame for the purposes of classical mechanics as the unprimed system x, y, z, t is called the *principle of relativity of classical mechanics*. In what follows (5) will be called a *Galilean transformation*. It is a linear transformation in the four coordinates; it is orthogonal in the first three, and leaves the time coordinate invariant ($t' = t$). The last statement means that the principle of relativity of classical mechanics leaves intact the absolute character of time as postulated by Newton.

A new situation arises, however, in the field of electrodynamics, particularly in the electromagnetic theory of optical phenomena. Maxwell's equations, which form the basis of this field, require that the process of the propagation of light *in vacuo* with the velocity c be independent of the frame of reference from which this process is observed. The front of a spherical wave whose source is at the origin of coordinates is given by the equation

(6) $$x^2 + y^2 + z^2 = c^2 t^2 \quad \text{or} \quad x'^2 + y'^2 + z'^2 = c^2 t'^2$$

respectively, depending on whether we are describing the wave front in the unprimed or the primed system. It is now convenient to change the names of the coordinates in the following manner:

(7) $$x = x_1, \; y = x_2, \; z = x_3, \; ict = x_4,$$

where i is the imaginary unit; we introduce a corresponding change of notation for the primed coordinates. Equations (6) then read

(8) $$\sum_1^4 x_k^2 = 0, \quad \sum_1^4 x_k'^2 = 0,$$

---

[9] Note that this table can be read from left to right but no longer from top to bottom, since the transformation is no longer orthogonal.—TRANSLATOR.

and the fact that the propagation of light does not depend on the choice of reference frame demands that[10]

(9)
$$\sum_1^4 x_k'^2 = \sum_1^4 x_k^2.$$

Whereas Eq. (2) was an orthogonal transformation in three-dimensional space, we are dealing in (9) with an orthogonal transformation in four-dimensional space. True, the fourth coordinate is imaginary. This, however, will not affect the existence of equations analogous to (3), (4) and (5). The relation between the x_k and the x'_k arising from (5) is in general called a *Lorentz transformation*, after the great Dutch theoretical physicist Hendrik Antoon Lorentz. We write it in the form of the general scheme

(10)

	x_1	x_2	x_3	x_4
x'_1	α_{11}	α_{12}	α_{13}	α_{14}
x'_2	α_{21}	α_{22}	α_{23}	α_{24}
x'_3	α_{31}	α_{32}	α_{33}	α_{34}
x'_4	α_{41}	α_{42}	α_{43}	α_{44}

This table shows at once that the time coordinate (in the imaginary form x_4) is now involved in a change of reference system to the same extent as the space coordinates. As a necessary consequence of the invariance requirement (9) the absoluteness of time is now destroyed.

More instructive than the general Lorentz transformation is the special one which we obtain when we leave two space coordinates, say x_1 and x_2, unchanged, and transform only x_3 and x_4.

Then all the α_{ij} of the first and second rows of columns in (10) must vanish, except for

$$\alpha_{11} = \alpha_{22} = 1,$$

because $x'_1 = x_1$, $x'_2 = x_2$ (as read from left to right as well as from top to bottom). Furthermore we have the conditions analogous to (4),

(11)
$$\alpha_{33}^2 + \alpha_{34}^2 = \alpha_{33}^2 + \alpha_{43}^2 = \alpha_{43}^2 + \alpha_{44}^2 = \alpha_{34}^2 + \alpha_{44}^2 = 1,$$
and therefore

$$\alpha_{33}^2 = \alpha_{44}^2, \quad \alpha_{34}^2 = \alpha_{43}^2.$$

Letting $\delta = \pm 1$, we can write

(11a)
$$\alpha_{34} = \delta \alpha_{43}$$

---

[10] For one of the Eqs. (8) must be the consequence of the other. In view of the linearity of the relation between them, one of the expressions (8) must be proportional to the other. Since the relation is a reciprocal one, the factor of proportionality must be unity.

and we must then put

(11b)
$$\alpha_{44} = -\delta \alpha_{33}$$

because of the other orthogonality condition $\alpha_{33}\alpha_{34} + \alpha_{43}\alpha_{44} = 0$. We now make use of (11a, b) to solve for the primed coordinates in terms of the unprimed. At the same time, with the help of (7), we go back to our original coordinates z, t, z', t' to obtain

(12)
$$z' = \alpha_{33}\left(z + i\delta c\frac{\alpha_{43}}{\alpha_{33}}t\right),$$
$$t' = -\delta \alpha_{33}\left(t + i\frac{\delta}{c}\frac{\alpha_{43}}{\alpha_{33}}z\right).$$

The first of these equations shows that

(12a)
$$-i\delta c\frac{\alpha_{43}}{\alpha_{33}} = v$$

must be identified with the velocity with which the z'-axis moves parallel to the z-axis in the positive direction of the latter, as observed from the unprimed system. With the help of (12a) Eqs. (12) become

(13)
$$z' = \alpha_{33}\ (z - vt),$$
$$t' = -\delta \alpha_{33}\left(t - \frac{v}{c^2}z\right).$$

Finally we must determine α_{33}. To this end we use Eq. (9) which, in the original coordinates, now simplifies to $z'^2 - c^2 t'^2 = z^2 - c^2 t^2$. Let us introduce here the values of z' and t' from (13). The factor of $2vzt$ vanishes on the left. Comparison of the factors of z^2 and t^2 on the left and right yields

$$\alpha_{33}^2 = \frac{1}{1 - v^2/c^2}.$$

In the limit $c \to \infty$ (13) must of course reduce to the Galilean transformation (1) with $\alpha_0 = \beta_0 = 0$ and $\gamma_0 = -v$. To this end we must put $\delta = -1$ and must choose the positive sign of α_{33}. We then obtain the characteristic two-dimensional Lorentz transformation

$$z' = \frac{z - vt}{(1 - \beta^2)^{\frac{1}{2}}},$$

(14)
$$t' = \frac{t - \frac{v}{c^2}z}{(1 - \beta^2)^{\frac{1}{2}}},$$

where $\beta = \frac{v}{c}$, $(1 - \beta^2)^{\frac{1}{2}} > 0$.

The relativization of the time in (14) and the change of scale of the space coordinate z, as embodied in the denominator $\left(1 - \frac{v^2}{c^2}\right)^{\frac{1}{2}}$, are, as we have seen, a result of the fact that the velocity of light c is finite, a fact with which the principle of relativity of classical mechanics is incompatible.

If it be true that all electrodynamic effects are propagated with finite velocity c, it follows that for such effects the Galilean transformation must always be replaced by a Lorentz transformation, either in the general form (10), or the specialized form (14). We call this fact the *principle of relativity of electrodynamics*. It is evident, however, that mechanics too has to adapt itself to the fact of the finite propagation velocity of light. Now all velocities occurring in ordinary mechanics are quite small compared to c. This is the reason why for the purposes of mechanics we can, as a rule, neglect the change of scale of the space and time coordinates indicated by (14).

The wealth of physical facts embodied in the Lorentz transformation will be discussed in the third volume of this series. Here we shall only investigate the changes that we have to make in the concept of the fundamental quantity $\mathbf{p}$, the momentum, as a result of our new relativity principle.

We have called $\mathbf{p}$ a vector. This means that the three components of $\mathbf{p}$ transform just like the coordinates themselves [i.e., the components of the radius vector $\mathbf{r}=(x, y, z)$] in a change of the system of coordinates. We therefore say that $\mathbf{p}$ is covariant to $\mathbf{r}$.

This is valid only from the viewpoint of the Galilean transformation, where the time is regarded as absolute. From the viewpoint of the Lorentz transformation the radius vector is a four-component quantity, a *four-vector*

(15) $\mathbf{x} = (x_1, x_2, x_3, x_4)$.

Our relativistic momentum will similarly have to be a four-vector, i.e., must be covariant to $\mathbf{x}$, if it is to have a meaning in relativity theory. We arrive at this four-vector in the following manner:

(*a*) (15) being a four-vector, the coordinate distance between two neighboring points

(16) $\mathbf{dx} = (dx_1, dx_2, dx_3, dx_4) = (dx_1, dx_2, dx_3, ic\,dt)$

is also a four-vector.

(*b*) The magnitude of this distance is certainly invariant under a Lorentz transformation. Apart from a factor ic, it is given by

(17) $d\tau = \left[dt^2 - \frac{1}{c^2}(dx_1^2 + dx_2^2 + dx_3^2) \right]^{\frac{1}{2}}$

We follow Minkowski in calling $d\tau$ the element of *proper time*; in contrast to dt it is relativistically invariant. We shall factor out dt in (17) and introduce the ordinary velocity v of three dimensions, to obtain

(17a) $$d\tau = dt\left(1 - \frac{v^2}{c^2}\right)^{\frac{1}{2}} = dt(1 - \beta^2)^{\frac{1}{2}}.$$

(*c*) Division of the four-vector (16) by the invariant (17a) yields another four-vector; we call it the four-vector velocity

(18) $$\frac{1}{(1-\beta^2)^{\frac{1}{2}}}\left(\frac{dx_1}{dt}, \frac{dx_2}{dt}, \frac{dx_3}{dt}, ic\right).$$

(*d*) Earlier we derived the momentum vector **p** by multiplying the velocity three-vector by a mass *m* independent of the reference frame. We shall similarly deduce the momentum four-vector **p** from the four-vector (18) by multiplication by a mass factor independent of the frame of reference. We shall call this mass factor the *rest mass* m_0 and obtain

(19) $$\mathbf{p} = \frac{m_0}{(1-\beta^2)^{\frac{1}{2}}}\left(\frac{dx_1}{dt}, \frac{dx_2}{dt}, \frac{dx_3}{dt}, ic\right).$$

It is proper to call the quantity in front of the parenthesis the moving mass (since it reduces to the rest mass for $\beta = 0$), or simply the mass. We therefore assert that

(20) $$m = \frac{m_0}{(1-\beta^2)^{\frac{1}{2}}}.$$

This expression was first derived by Lorentz in 1904 under very special assumptions (deformable electron). The derivation from the principle of relativity makes such special assumptions unnecessary. Eq. (20) has been confirmed by many precision experiments with fast electrons. Together with optical experiments, notably that of Michelson and Morley, it forms the basis of the theory of relativity. Here we have proceded in inverse order and deduced Eq. (20) from the principle of relativity in what appears to be a very formal procedure. This is not only logically admissible, but especially serviceable in view of the brevity of these introductory explanations. In § 4 we shall discuss what changes in the further application of Newton's laws of motion will have to be made as a result of the velocity-dependence of mass.

At this point we should, if only sketchily, bring to a conclusion the question of the permissible frames of reference; to this end we must pass from the *special theory of relativity* treated so far to the *general theory of relativity* (Einstein, 1915). In special relativity there are allowed reference systems which are obtained from one another by Lorentz transformations, and forbidden ones, such as, for example, those that are accelerated with respect to the former. In general relativity all possible frames of reference are admitted. Transformations between them need no longer be linear and orthogonal as in (10), but can instead be given by arbitrary functions $x'_k = f_k(x_1, x_2, x_3, x_4)$. Hence we are dealing with systems which are moving and are being deformed with respect to each other in any way desired. As a result space and time lose any vestiges of the absolute character which they held in Newton's fundamental analysis. They become merely classification schemes for physical events. Euclidean geometry no longer suffices for this classification and must be replaced by the much more general metric geometry advanced by Riemann. The task then arises to give such a form to physical laws that they will remain valid in all frames of reference here considered, i.e., a form that remains invariant under arbitrary point transformations $x'_k = f_k(x_1 \cdots x_4)$ of four-dimensional space. The positive content of the general theory of relativity is precisely the possibility of this task. We cannot in this volume delve into the mathematically very involved form which the laws of mechanics take on in their invariant formulation. Suffice it to say that the general theory leads to a derivation and a more precise formulation of Newtonian gravitation.

We conclude with a remark about the name, theory of relativity. The positive achievement of the theory is not so much the complete relativization of space and time, but the proof that the laws of nature are independent of the choice of reference system, i.e., that events in nature are invariant under any change in the observer's view point. The names, "theory of the invariance of natural events," or, as occasionally proposed, "viewpoint theory," would be more appropriate than the customary name, general theory of relativity.

§ 3. Rectilinear Motion of a Mass Point

Let the motion of the particle take place along the x-axis. Only the x-components of any forces present will have any effect. Let X denote the resultant of these components.

We have $\mathbf{v} = v = \dfrac{dx}{dt}$ and $p = m \dfrac{dx}{dt}$. Then

$$(1) \qquad \dot{p} = X$$

and, with constant m,

$$(2) \qquad m \frac{d^2x}{dt^2} = X.$$

We wish to study the integration of this equation of motion for the three cases: X is given as a pure function of the time, $[X = X(t)]$, of the position, $[X = X(x)]$, or of the velocity, $[X = X(v)]$.

(a) $X = X(t)$.

Immediate integration yields

$$(3) \qquad v - v_0 = \frac{1}{m} \int_{t_0}^{t} X(t)\, dt = \frac{1}{m} Z(t).$$

Here $Z(t)$ is by definition the time integral of the force and equals the change in momentum during the time from t_0 to t.

A second integration leads to the equation of the trajectory,

$$(4) \qquad x - x_0 = v_0(t - t_0) + \frac{1}{m} \int_{t_0}^{t} Z(t)\, dt.$$

(b) $X = X(x)$.

This is the typical case of the *force field* given as a function of position. Integration is achieved by use of the principle of the conservation of energy.

We multiply (2) on both sides by $\frac{dx}{dt}$,

$$(5) \qquad m \frac{dx}{dt} \frac{d^2x}{dt^2} = X \frac{dx}{dt}.$$

The left member is now a complete differential,

$$\frac{d}{dt} \left\{ \frac{m}{2} \left(\frac{dx}{dt} \right)^2 \right\}.$$

In agreement with the general definition of (1.4) we write for the right member $dW = X\,dx$ and call dW the work done over the path dx. The equation thus arising says that *the change in kinetic energy equals the work done*.

For we define

$$(6) \qquad T = E_{\text{kin}} = \frac{m}{2} v^2$$

as the *kinetic energy* or energy of motion of the mass point; the older name, live force (Leibniz), shows the ambiguity of the word force (he distinguished live force, *vis viva*, i.e., kinetic energy, and motor force, *vis motrix*, our present-day force; even Helmholtz, as late as 1847, entitled a treatise dealing with the conservation of energy " Concerning the Conservation of Force ").

To the definition of the kinetic energy we add that of the potential energy V,

$$(7) \qquad dV = -dW = -X\,dx, \quad V = E_{\text{pot}} = -\int^{x} X\,dx.$$

In one-dimensional particle mechanics this definition suffices; in the case of two- or three-dimensional force fields the existence of V depends on the character of the fields (cf. § 6, Part 3). According to (7) V is determined only to within an additive constant.

With these definitions the integrated Eq. (5) yields the law of the conservation of energy,

(8) $T + V = \text{constant} = E.$

Here E is the energy constant or *total energy*.

The principle of the conservation of energy possesses not only an exceedingly great physical importance, but also remarkable mathematical power. For it performs, as we have seen, not only the first integration of the equation of motion (hence its alternate name, " integral of energy "), but at once makes possible — at least in the present case (b) — a second integration as well. If we write (8) in the form

$$\left(\frac{dx}{dt}\right)^2 = \frac{2}{m}[E - V(x)],$$

we can solve for dt,

$$dt = \left[\frac{m}{2(E-V)}\right]^{\frac{1}{2}} dx,$$

so that

(9) $t - t_0 = \left(\frac{m}{2}\right)^{\frac{1}{2}} \int\limits_{x_0}^{x} \frac{dx}{(E-V)^{\frac{1}{2}}}.$

Thus t is a known function of x, and therefore x can also be expressed in terms of t. (9) is then the completely integrated equation of motion.

(c) $X = X(v)$.

Now the equation of motion reads

$$m\frac{dv}{dt} = X(v)$$

which we rewrite as

$$dt = \frac{m\, dv}{X}$$

thereby at once obtaining

(10) $t - t_0 = m \int\limits_{v_0}^{v} \frac{dv}{X} = F(v).$

This also allows us to solve for v in terms of t, $v = f(t)$, so that

$$\frac{dx}{dt} = f(t),$$

from which we conclude that

$$x - x_0 = \int\limits_{t_0}^{t} f(t)\, dt.$$

Examples

1. Free Fall Near Earth's Surface (Falling Stone)

We take the positive x-direction as vertically upward. The force is constant,

$$(11) \qquad X = -mg,$$

i.e., independent of t, x, and v. Here all three methods of integration (a), (b), (c) can be applied.

We shall carry out (a) and (b), and postulate explicitly that the " gravitational mass " and the " inertial mass " be equal,

$$(12) \qquad m_{\text{inert}} = m_{\text{grav}}.$$

m_{inert} is the mass defined by the Second Law; m_{grav} is the mass occurring in the law of gravitation and hence also in our force equation (11).

Bessel recognized the necessity for testing Eq. (12) experimentally, by means of pendulum experiments.[11] A much more precise experimental proof was furnished by Eötvös with his torsion balance. Later on, Eq. (12) gave the first impulse to Einstein's theory of gravitation.

(a) $\ddot{x} = -g$. With suitable choice of the integration constants $(v = 0$ and $x = h$ for $t = 0)$ we end up with

$$\dot{x} = -gt, \qquad x = h - \frac{g}{2}t^2.$$

(b) Since $dW = -mgdx$, $V = mgx$ and $T + mgx = E$. If $v = 0$ at $x = h$, we must have $E = mgh$; therefore

$$\frac{m}{2}v^2 + mgx = mgh.$$

From this we get for the particular value $x = 0$ that $v^2 = 2gh$, or

$$(13) \qquad v = (2gh)^{\frac{1}{2}}.$$

Inverting this equation we obtain

$$(13a) \qquad h = \frac{v^2}{2g},$$

which is the height to which an arbitrary mass must be raised in order to attain, falling through this height in the gravitational field, a specified velocity v. The introduction of this height h instead of the velocity v is convenient, especially in certain engineering problems, such as the

---

[11] Incidentally we would like to direct the reader's attention to an interesting sentence occurring in Newton's *Mechanics*. At the beginning of this work, under Definition 1, Newton says: " Through very carefully performed experiments with pendula I have verified that mass and weight are proportional."

height to which water rises in a Pitot tube,[12] the pressure head in a centrifuge, etc. The height to which the water surface climbs in Newton's pail experiment is similarly given by (13a).

2. Free Fall From a Great Distance (Meteor)

Now the force of attraction is no longer constant. Instead we must use the law of gravitation

$$(14) \qquad m\frac{d^2r}{dt^2} = -\frac{mMG}{r^2},$$

where m is the mass of the meteor, M that of the earth, G the gravitational constant. Instead of the coordinate x we have introduced the distance r of the meteor from the center of the earth. Since the force is now a function of r, method of integration (b) should be used.

In particular, for the surface of the earth, with a the earth's radius, (14) yields

$$mg = \frac{mMG}{a^2},$$

so that mMG can be eliminated from (14),

$$\frac{d^2r}{dt^2} = -g\frac{a^2}{r^2}.$$

With this notation (7) yields

$$dV = -dW = mga^2\frac{dr}{r^2},$$

so that the potential energy, with zero level at infinity, becomes

$$(15) \qquad V(r) = -mg\frac{a^2}{r}.$$

Eq. (8) therefore gives

$$\frac{m}{2}\left(\frac{dr}{dt}\right)^2 - \frac{mga^2}{r} = W = -\frac{mga^2}{R},$$

where R is some hypothetical initial distance from the earth's center at which the falling mass was in a state of rest. We thus obtain

$$(16) \qquad \frac{dr}{dt} = a\left[2g\left(\frac{1}{r}-\frac{1}{R}\right)\right]^{\frac{1}{2}}$$

and, corresponding to (9),

$$(16a) \qquad t = \frac{1}{a(2g)^{\frac{1}{2}}}\int\frac{dr}{\left(\frac{1}{r}-\frac{1}{R}\right)^{\frac{1}{2}}}.$$

---

[12] A hollow tube used in fluid flow to measure the dynamic pressure. It is often used on airplanes as an airspeed indicator. Cf. Glazebrook, *Dictionary of Applied Physics* V, p. 2.—Translator.

We need not do the integration indicated in (16a) in detail, since only two special cases of (16) are of interest to us:

(a) $$R = \infty, \; r = a.$$

The meteor reaches the earth with the velocity

$$\frac{dr}{dt} = (2ga)^{\frac{1}{2}},$$

i.e., a free fall in the earth's gravitational field from infinity results in the same velocity at the surface of the earth which would be achieved by a free fall from a height $h = a$ equal to the earth's radius under a constant acceleration due to gravity g [cf. Eq. (13)].

(b) $$R = a + h. \quad h \ll a. \quad r = a.$$

Here we are concerned with a first order correction to the velocity of fall (13), taking into consideration the decreasing gravitational acceleration, but assuming that the meteor falls from not too great a height. From (16) we derive

$$\frac{dr}{dt} = \left[2ga\left(1 - \frac{1}{1 + \dfrac{h}{a}}\right)\right]^{\frac{1}{2}} = (2ga)^{\frac{1}{2}}\left(\frac{h}{a} - \frac{h^2}{a^2} + \cdots\right)^{\frac{1}{2}}$$

$$= (2ga)^{\frac{1}{2}}\left(\frac{h}{a}\right)^{\frac{1}{2}}\left(1 - \frac{1}{2}\frac{h}{a} + \cdots\right) = (2gh)^{\frac{1}{2}}\left(1 - \frac{1}{2}\frac{h}{a} + \cdots\right).$$

3. Free Fall in Air

We shall assume that the air resistance is proportional to the square of the velocity. This assumption, introduced by Newton, agrees quite well with experience if the falling body is not too small and its velocity is neither comparable to that of sound, nor vanishingly small. The resultant force is then

$$X(v) = -mg + av^2,$$

where the signs indicate that the air resistance opposes the force of gravitation. Here method (c) of p. 18 applies, and the equation of motion becomes

(17) $$\frac{dv}{dt} = -g + \frac{a}{m}v^2.$$

If we put $\dfrac{a}{mg} = b^2$, it goes over into

$$\frac{dv}{dt} = -g(1 - b^2 v^2).$$

From this we obtain the analogue of (10) with $t_0 = 0$,

$$-g\,dt = \frac{dv}{2}\left(\frac{1}{1-bv} + \frac{1}{1+bv}\right), \quad -gt = \frac{1}{2b}\cdot\ln\left(\frac{1+bv}{1-bv}\right),$$

so that

$$\frac{1+bv}{1-bv} = e^{-2bgt}$$

and

(18) $$bv = \frac{e^{-2bgt}-1}{e^{-2bgt}+1} = -\frac{\sinh bgt}{\cosh bgt} = -\tanh bgt,$$

where sinh, cosh, and tanh are the hyperbolic functions. $|bv|$ hence grows monotonically from 0 at $t=0$, and approaches the value 1 as $t \to \infty$. The limiting value of v itself is

$$|v| = \frac{1}{b} = \left(\frac{mg}{a}\right)^{\frac{1}{2}}.$$

This can also be read at once out of Eq. (17), since for the above limiting value $\frac{dv}{dt}$ becomes equal to zero.

We make use of Eq. (18) to obtain the first order correction due to air resistance which must be added to the formula derived for a free fall *in vacuo*. From the series expansion

$$\tanh \alpha = \frac{\sinh \alpha}{\cosh \alpha} = \frac{\alpha + \dfrac{\alpha^3}{6}}{1 + \dfrac{\alpha^2}{2}} = \alpha\left(1 - \frac{\alpha^2}{3}\right)$$

we obtain, according to (18), with $\alpha = bgt$,

$$v = -gt\left(1 - \frac{(bgt)^2}{3}\right).$$

4. HARMONIC OSCILLATIONS

Harmonic oscillations occur whenever a restoring force X proportional to the displacement x acts on a mass point m. We call the proportionality factor k, so that

$$X = -kx$$

and the equation of motion with constant m is

(19) $$m\frac{d^2x}{dt^2} = -kx.$$

Since the force is a given function of the coordinate [case (b) of p. 17], we make use of the rule given there and apply the integral of energy.

We must therefore first determine the potential energy of the harmonic binding force. We have

$$dW = X\,dx = -\frac{k}{2}d(x^2),$$

so that, according to (7), with a suitable choice of the zero of V,

$$V = -\int_0^x dW = \frac{k}{2}x^2.$$

The equation of energy is then

$$mv^2 + kx^2 = 2E.$$

As our initial conditions we may choose

(19a) $\qquad\qquad$ at $t=0:\quad \begin{cases} x=a \\ v=\dot{x}=0. \end{cases}$

As a result $2E$ takes the value ka^2, and

$$\left(\frac{dx}{dt}\right)^2 = \frac{k}{m}(a^2 - x^2),$$

$$\left(\frac{k}{m}\right)^{\frac{1}{2}} dt = \frac{dx}{(a^2 - x^2)^{\frac{1}{2}}}.$$

A quadrature, incorporating the initial conditions (19a), yields

(20) $\qquad\qquad \omega t = \sin^{-1}\left(\frac{x}{a} - \frac{\pi}{2}\right) \text{with } \omega = \left(\frac{k}{m}\right)^{\frac{1}{2}}.$

An inversion finally gives

(21) $\qquad\qquad x = a\,\sin\left(\omega t + \frac{\pi}{2}\right) = a\,\cos\,\omega t.$

The physical meaning of the abbreviation ω is therefore clear. It is the circular frequency, i.e., the number of vibrations in 2π units of time; T being the period of oscillation, ν the frequency[13], we have the relation

(22) $\qquad\qquad \omega = \frac{2\pi}{T} = 2\pi\nu.$

With the help of this abbreviation (19) can also be written

(23) $\qquad\qquad \ddot{x} + \omega^2 x = 0.$

The equation of energy has the advantage that it always leads to the desired end, no matter how the force X depends on x. In our case, however, where X is linear in x, another much more elegant method exists. It is

---

[13] As opposed to ω, ν is the frequency, the number of vibrations in *unit* time.

based on the immediately plausible rule that a homogeneous linear differential equation with constant coefficients of any order (x being the dependent, t the independent variable) can always be solved by putting

(24) $$x = Ce^{\lambda t},$$

provided λ is chosen to be one of the roots of an algebraic equation obtained from our differential equation. This furnishes a *particular* solution. The *general* solution is obtained by a superposition of all such particular solutions in the form

(24a) $$x = \sum C_j e^{\lambda_j t}.$$

The algebraic equation in λ is obtained by substitution of (24) in (23), and is here of second degree,

$$\lambda^2 + \omega^2 = 0 \text{ with the roots } \lambda = \pm i \omega.$$

The general solution is therefore

(24b) $$x = C_1 e^{i\omega t} + C_2 e^{-i\omega t}.$$

Constants C_1, C_2 are determined by the boundary conditions (19a):

$$\dot{x} = 0 \quad , \quad C_1 i \omega - C_2 \omega i = 0: \quad C_1 = C_2.$$

$$x = a \quad , \quad a = C_1 + C_2 = 2C_1: \quad C_1 = \frac{a}{2}.$$

The final solution of the problem is, in agreement with (21),

$$x = a \cos \omega t.$$

We shall later (Chapter III, § 19) make extensive use of this method for damped, forced, coupled, etc., oscillations, provided these can be described by linear differential equations. The title, "harmonic oscillations," which we have given to this part, calls attention to the fact that the restoring force is linear in the coordinate, so that the resulting motion can be represented by a single constant frequency ω. The method fails in case the binding force is anharmonic, i.e., non-linear; in that event one has to resort to the less elegant method of the energy integral.

5. Collision of Two Particles

Before the collision (cf. Fig. 1) let the masses m and M have velocities v_o and V_o respectively; after the collision they proceed with velocities v and V.

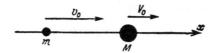

Fɪɢ. 1.   Collision of two masses M and m; velocities before collision v_0 and V_0, after collision V and v.

No matter what the nature of the collision, whether elastic or inelastic, Newton's axiom " action$=$reaction " must be valid for the forces transmitted between m and M, and also for the time integral Z of these forces. Therefore, according to Eq. (3),

(25) $$m(v - v_0) = Z = -M(V - V_0),$$

and hence also

(25a) $$mv + MV = mv_0 + MV_0.$$

This equation states that the total momentum of the system is conserved.

Let us now introduce in (25a) the coordinate of the center of mass of the system,

(25b) $$\xi = \frac{mx + MX}{m + M}.$$

We obtain

$$\dot{\xi} = \dot{\xi}_0.$$

This result says that the collision has no effect on the velocity of the center of mass.

Thus the center of mass of a shell fired *in vacuo* continues undisturbed in its parabolic path, even if at some point along the path the shell bursts into splinters each of which seems to follow an independent trajectory of its own.

So far we have two unknowns, v and V, and only one equation (25a). In order to find the complete solution of the collision problem a second relation is evidently necessary. We define an *elastic collision* as an interaction in which the kinetic energy as well as the momentum is conserved. We then require

(26) $$\frac{m}{2} v^2 + \frac{M}{2} V^2 = \frac{m}{2} v_0^2 + \frac{M}{2} V_0^2,$$

or

$$m(v^2 - v_0^2) = M(V_0^2 - V^2).$$

But from (25)

$$m(v - v_0) = M(V_0 - V).$$

Division of these two equations yields

$$v + v_0 = V_0 + V,$$

or

(26a) $$V - v = -(V_0 - v_0).$$

This equation states that the relative velocity of one mass with respect to the other after the collision is equal and opposite to that before the collision.

The combination of Eqs. (25a) and (26a),

$$mv + MV = mv_0 + MV_0$$

$$v - V = -v_0 + V_0$$

now completely determines the velocities after the collision,

(27)
$$v = \frac{m - M}{m + M} v_0 + \frac{2M}{m + M} V_0$$

$$V = \frac{M - m}{m + M} V_0 + \frac{2m}{m + M} v_0.$$

Notice that the determinant Δ of this " transformation " from initial values v_0, V_0 to final values v, V has the absolute value 1. For

$$\Delta = \begin{vmatrix} \dfrac{m - M}{m + M} & \dfrac{2M}{m + M} \\ \dfrac{2m}{m + M} & \dfrac{M - m}{m + M} \end{vmatrix} = -\left(\frac{M - m}{m + M}\right)^2 - \frac{4mM}{(m + M)^2} = -1.$$

This means that if we allow the initial velocities to have a certain range of values, the transformed surface element in v-V space has the same area as the initial surface element; the transformation is *area-preserving* (cf. Fig. 2a). This law is important in collision processes in the kinetic theory of gases and is related to Liouville's theorem (cf. Vol. V, this series).

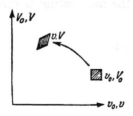

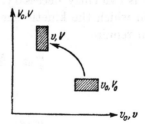

FIG. 2a. Velocity domains before and after collision. The mapping is area-preserving.

FIG. 2b. In the case of two equal masses, $m = M$, the mapping is not only area-preserving, but also angle-preserving.

Let us consider the case of two equal masses such as two billiard balls, $m = M$. Eqs. (27) become

(27a) $$v = V_0, \quad V = v_0.$$

Now the transformation is not only area-preserving, but angle-preserving as well; cf. Fig. 2b, in which the transformed rectangle is obtained from the initial one by interchange of its sides. In particular, if, in a central collision (head-on collision), in a billiard game, one ball is initially at rest, then the other one transmits all its velocity to the former and thereby comes to rest [cf. (27a) with $V_0 = 0$].

If, on the other hand, one mass is very great compared to the other, $M \gg m$, the large mass retains practically all its original velocity after the collision, while the small mass follows the large one with a velocity equal to that of the large one minus the original relative velocity. For with $m \ll M$ the Eqs. (27) can be simplified to

(27b) $$v = -v_0 + 2V_0 = V_0 - (v_0 - V_0), \quad V = V_0.$$

To complete the discussion of collisions we shall briefly go into inelastic collisions. In atomic physics one investigates inelastic collisions (" collisions of the second kind "), in which the colliding particle, say an electron, loses part of its energy in order to " excite " the atom with which it collides; such an excited atom has been raised from its ground state to a level of higher energy. Since in this type of process part of the initial energy is lost as far as the motion after collision is concerned, this motion can no longer be calculated using the formulas of elastic collisions. (Cf. problems I.1 to I.4.)

We shall here limit ourselves to the " completely inelastic collision," which is often considered in engineering problems. Such a collision is defined by the condition

$$v = V,$$

i.e., after the collision both masses m and M proceed at the same velocity, as if rigidly coupled together. The equation of momentum, as emphasized earlier, retains its validity under all circumstances; it becomes

(28) $$(m + M)v = mv_0 + MV_0$$

and alone is sufficient to determine the sole unknown v. We would like to know the energy lost in this collision, which is

$$\frac{m}{2}v_0^2 + \frac{M}{2}V_0^2 - \frac{m + M}{2}v^2,$$

or, after a simple elimination of v with the help of (28),

(28a) $$\frac{\mu}{2}(v_0 - V_0)^2.$$

The loss of energy equals the kinetic energy of a certain *reduced mass* μ which moves with the original relative velocity. μ is defined by

(28b) $$\frac{1}{\mu}=\frac{1}{m}+\frac{1}{M}, \text{ therefore } \mu=\frac{mM}{m+M}.$$

The theorem contained in Eqs. (28a, b) was first advanced by General Lazarus Carnot. (General Carnot was a mathematician and the organizer of universal military service during the French Revolution, as well as the father of Sadi Carnot, whose name has become famous in thermodynamics.)

§ 4. Variable Masses

The illustrations cited below will aid us in a critical evaluation of Newton's second law. We put this law in the form (1.3), "change of momentum equals force," rather than in the less general one (1.3a), "mass · acceleration equals force." We shall now learn how the rate of change of momentum is to be understood. We shall show that even in the case of variable mass the general form (1.3) may under certain circumstances reduce to (1.3a).

Let us consider a familiar example: a sprinkler wagon wets the asphalt on a hot summer day. The power of the motor is barely great enough to overcome the combined friction of the ground and wheels, of the air, and in the axle bearings. The vehicle therefore behaves as if under no forces. Let m be the mass of water in the tank at any instant+the constant mass of the empty vehicle. Let the amount of water squirted out per unit time be $\mu=-\dot{m}$, its exit velocity toward the rear, q as seen from the wagon, or $v-q$ as seen from the street, v being the speed of the vehicle.

If we were to use the formula (1.3) mechanically, we would obtain

(1) $$\dot{\mathbf{p}}=\dot{p}=\frac{d}{dt}(mv)=0$$

from which would follow

(1a) $$m\dot{v}=\mu v.$$

The acceleration of the wagon would then be independent of the exit velocity q. This is paradoxical, since one would expect the recoil (cf. gun) from the outgoing water jet to have some effect.

Actually we have not used the correct expression for the rate of change of momentum meant in (1.3), for it should consist not only of the member taken into account in (1), but also of a term giving the momentum contained in the water jets. This latter is $\mu(v-q)$ per unit time. Explicitly,

$$p_t=mv_t, \quad p_{t+dt}=(m+dm)(v+dv)+\mu\,dt(v-q)$$

so that the corrected rate of change of momentum becomes

(2) $$\dot{p}=\frac{d}{dt}(mv)+\mu(v-q)=0,$$

or, remembering $\mu = -\dot{m}$ and simplifying,

$$(3) \qquad\qquad m\dot{v} = \mu q.$$

From the viewpoint of (1.3a), the recoil of the water leaving the vehicle acts as an accelerating force on the latter, just as in the reaction water wheel used in rotary lawn sprinklers.

Instead of the sprinkler wagon we could have chosen as our example the interplanetary rocket, with which one might reach the moon. The rocket would be propelled by the expulsion of explosive gases. See problem I.5.

We generalize this result in two statements which are equivalent to Eqs. (2) and (3), respectively, of our illustrative example:

Either we take the viewpoint of (1.3), where we must then add to the change in momentum contained in the body in question the momentum convectively given off or added per unit time. The latter is to be calculated in the same frame of reference as the momentum of the body under investigation; the sign of $\dot{m}$ takes care of the correct sign for this term. The equation of motion then becomes

$$(4) \qquad\qquad \frac{d}{dt}(m\mathbf{v}) - \dot{m}\mathbf{v}' = \mathbf{F},$$

where $\mathbf{v}'$ is the convective velocity. In our case we had $-\dot{m} = \mu$ and $|\mathbf{v}'| = |\mathbf{v}| - q$.

Or we take the viewpoint of (1.3a), in which case we must, however, add the recoil momentum gained or lost per unit time as a kind of external force. We then obtain the equation of motion in a form analogous to (3),

$$(5) \qquad\qquad m\dot{\mathbf{v}} = \mathbf{F} + \dot{m}\mathbf{v}_{rel}.$$

$\mathbf{v}_{rel}$ is the relative velocity of the convective momentum with respect to the body under observation, measured positive in the same sense as $\mathbf{v}$. In our example we had $|\mathbf{v}_{rel}| = -q$ and again $-\dot{m} = \mu$.

Two special cases deserve our attention:

(a) $\mathbf{v}' = 0$. The elements of mass gained or lost have zero velocity and therefore do not carry any momentum. In that case the equation of motion has the Newtonian form $\dot{\mathbf{p}} = \mathbf{F}$. Examples: water drop, chain, problems I.6 and I.7.

(b) $\mathbf{v}' = \mathbf{v}$ or, equivalently, $\mathbf{v}_{rel} = 0$. The equation of motion has the form, mass · acceleration = force, in spite of the fact that the mass involved is variable. Example : rope hanging over the edge of a table, problem I.8.

In case (b) the Carnot energy loss, Eq. (3.28a), is zero; therefore the equation of energy applies in the usual form. In case (a) the form of the energy conservation law valid for a given problem is not obvious and must first be investigated.

We conclude these instructive remarks with the problem of the relativistic variation of mass. We shall talk specifically about the electron, even though equation (2.20) is of course valid not only for it, but for all masses. Here the variation of mass is a purely internal affair of the electron; there is no question of any momentum gained from or lost to the surroundings. As in case (a) the equation of motion is therefore $\dot{\mathbf{p}} = \mathbf{F}$, i.e., in view of (2.20),

(6)
$$\frac{d}{dt}\left(\frac{m_0\mathbf{v}}{[1-\beta^2]^{\frac{1}{2}}}\right) = \mathbf{F}.$$

Let us first consider the rectilinear motion of an electron; $\mathbf{F}$ acts longitudinally, that is, in the direction of $\mathbf{v}$, so that $\mathbf{F} = F_{\text{long}}$ and $\mathbf{v} = v$.

We shall change Eq. (6) to the form "mass · acceleration = force," a customary procedure in the early part of the century, though unnecessarily complicated. To this end we carry out the differentiation on the left,

(6a)
$$\frac{m_0\dot{v}}{(1-\beta^2)^{\frac{1}{2}}} + m_0 v\frac{d}{dt}(1-\beta^2)^{-\frac{1}{2}} = \frac{m_0}{(1-\beta^2)^{\frac{1}{2}}}\left(\dot{v} + \frac{v\beta\dot{\beta}}{1-\beta^2}\right).$$

Now $\beta = v/c$ so that

$$\dot{\beta} = \frac{\dot{v}}{c} \quad\text{and hence}\quad v\beta\dot{\beta} = \beta^2 v.$$

Consequently Eq. (6a) becomes

(6b)
$$\frac{m_0\dot{v}}{(1-\beta^2)^{\frac{1}{2}}}\left(1 + \frac{\beta^2}{1-\beta^2}\right) = \frac{m_0}{(1-\beta^2)^{\frac{3}{2}}}\dot{v} = F_{\text{long}}.$$

The *longitudinal mass* multiplying the acceleration $\dot{v}$ is therefore

(7)
$$m_{\text{long}} = \frac{m_0}{(1-\beta^2)^{\frac{3}{2}}}.$$

If, on the other hand, $\mathbf{F}$ acts transversely, i.e., normal to the trajectory, only the direction, not the magnitude of the velocity is altered. In that case $\dot{\beta}$ is zero; (6) simply yields

$$\frac{m_0}{(1-\beta^2)^{\frac{1}{2}}}\dot{v} = F_{\text{trans}}.$$

For this reason one introduced at the time a *transverse mass* different from the longitudinal mass and given by

(8)
$$m_{\text{trans}} = \frac{m_0}{(1-\beta^2)^{\frac{1}{2}}}.$$

In view of these complications we emphasize that the above distinction between two kinds of masses becomes unnecessary if we use only the rational form (6) of the equation of motion.

Next we wish to determine the form of the equation of energy in relativity theory. Let us therefore multiply (6) by $\frac{dx}{dt} = v = \beta c$. In the right member we obtain

(9) $$F\frac{dx}{dt} = \frac{dW}{dt} = \text{work done, or power.}$$

In the left member we have

$$m_0 c^2 \beta \frac{d}{dt}\left(\frac{\beta}{[1-\beta^2]^{\frac{1}{2}}}\right) = m_0 c^2 \beta \dot{\beta}(1-\beta^2)^{-\frac{3}{2}}.$$

We can at once convince ourselves that this is a total derivative in t, viz.,

(10) $$m_0 c^2 \frac{d}{dt}\frac{1}{(1-\beta^2)^{\frac{1}{2}}}.$$

Since (10) must be equal to (9), the rate of doing work, (10) must be the time rate of change of the kinetic energy T. We therefore have

$$T = m_0 c^2\left(\frac{1}{[1-\beta^2]^{\frac{1}{2}}} + \text{const.}\right).$$

We must put the constant equal to -1, because T, by its nature, must vanish as β vanishes. Hence the *relativistic kinetic energy* is

(11) $$T = m_0 c^2\left(\frac{1}{[1-\beta^2]^{\frac{1}{2}}} - 1\right).$$

In view of (2.20) we can also write this as

(12) $$T = c^2(m - m_0).$$

In words: *the difference in energy between a moving electron and one at rest* (which is nothing but the kinetic energy or " live force ") *equals the difference between the masses of the electron in motion and at rest, multiplied by* c^2. Thus we have verified for the simplest case the law of the equivalence of mass and energy (law of the " inertia of energy "). This important law is fundamental to the whole field of atomic weight determinations and to nuclear physics and its applications to cosmology.

For the sake of completeness we point out that for small β, (11) can be expanded in a series which yields to a first approximation the elementary expression for T,

$$T = m_0 c^2\left(\frac{1}{2}\beta^2 + \frac{3}{8}\beta^4 + \cdots\right) = \frac{m_0}{2}c^2\beta^2\left(1 + \frac{3}{4}\beta^2 + \cdots\right) \to \frac{m_0}{2}v^2,$$

as is to be expected.

§ 5. *Kinematics and Statics of a Single Mass Point in a Plane and in Space*

Kinematics deals with the geometry of motions regardless of their physical realization. Statics[14] is concerned with forces, their composition and equivalence, without regard to the motions caused by them.

(1) Plane Kinematics

We shall begin by writing down the formulas for the decomposition and composition of velocity and acceleration in Cartesian coordinates.

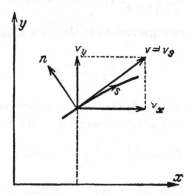

Fɪɢ. 3. Decomposition and composition of velocities in a plane; intrinsic coordinates s and n.

Velocity:

(1) $$\mathbf{v} = (v_x, v_y) = \left(\frac{dx}{dt}, \frac{dy}{dt}\right) = (\dot{x}, \dot{y});$$

(2) $$|\mathbf{v}| = (\dot{x}^2 + \dot{y}^2)^{\frac{1}{2}} = v.$$

Acceleration:

(3) $$\dot{\mathbf{v}} = (\dot{v}_x, \dot{v}_y) = \left(\frac{d^2x}{dt^2}, \frac{d^2y}{dt^2}\right) = (\ddot{x}, \ddot{y});$$

(4) $$|\dot{\mathbf{v}}| = (\ddot{x}^2 + \ddot{y}^2)^{\frac{1}{2}}.$$

Instead of decomposing velocity and acceleration in Cartesian coordinates we can also decompose them in terms of the *intrinsic coordinates* of the curve described by our mass point. Let s be the length of arc, subscript s denoting the path direction varying from point to point along the curve,

---

[14] The name statics is actually not appropriate, for it refers only to equilibrium, whereas the content of statics applies to problems of motion as well as of equilibrium. The correct name would be dynamics. This term has been in historical usage for the study of motions caused by forces, and is therefore not available for the field which its name implies, i.e., the analysis of forces.

subscript n denoting the direction normal to s at any point of the curve. We then have

(5) $$v_s = \pm v, \; v_n = 0.$$

This is trivial. The decomposition of $\dot{\mathbf{v}}$ into $\dot{\mathbf{v}}_s$ and $\dot{\mathbf{v}}_n$ is, however, significant. If we let α be the angle between the tangent to the path and the x-direction, we have

(6) $$\dot{v}_s = \dot{v}_x \cos\alpha + \dot{v}_y \sin\alpha$$

for the tangential acceleration, and

(7) $$\dot{v}_n = -\dot{v}_x \sin\alpha + \dot{v}_y \cos\alpha$$

for the normal acceleration.

   Now

$$\cos\alpha = \frac{dx}{ds} = \frac{\dot{x}}{\dot{s}} = \frac{v_x}{v}, \qquad \sin\alpha = \frac{dy}{ds} = \frac{\dot{y}}{\dot{s}} = \frac{v_y}{v},$$

so that

(8) $$\dot{v}_s = \frac{1}{v}(v_x \dot{v}_x + v_y \dot{v}_y) = \frac{1}{2v}\frac{d}{dt}(v_x^2 + v_y^2)$$

$$= \frac{1}{2v}\frac{d}{dt}v^2 = \frac{dv}{dt} = |\dot{v}|.$$

This equation states that the tangential acceleration is the change in magnitude of the velocity, no matter what its change in direction may be.
   Eq. (7), on the other hand, yields

(9) $$\dot{v}_n = \frac{1}{v}(v_x \dot{v}_y - v_y \dot{v}_x) = \frac{1}{v}(\dot{x}\ddot{y} - \dot{y}\ddot{x}) = v^2 \cdot \frac{\dot{x}\ddot{y} - \dot{y}\ddot{x}}{(\dot{x}^2 + \dot{y}^2)^{\frac{3}{2}}} = \frac{v^2}{\rho},$$

where $\dfrac{1}{\rho}$ is the curvature of the path.[15]
   The normal acceleration therefore does not depend on the change of velocity, but only on the velocity itself and on the shape of the trajectory. If, in particular, $\dfrac{dv}{dt} = 0$, the acceleration is normal to the velocity and hence to the path.
   We shall now derive the same relations in a direct, differential-geometrical fashion by means of the *hodograph*[16] introduced by Hamilton.

---

[15] Cf., for example, Franklin, *Treatise on Advanced Calculus*, p. 295.—TRANSLATOR.

[16] The name hodograph = path writer is misleading; it should really be called velocity writer, or better, polar diagram of velocity.

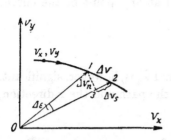

FIG. 4a. Hodograph of motion in a plane. Velocities v_1 and v_2 are laid off from the pole O in the polar diagram.

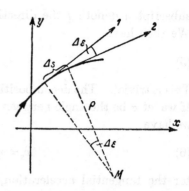

FIG. 4b. Trajectory and radius of curvature of motion in a plane.

The meaning of the hodograph becomes clear when we compare Figs. 4a and 4b. Fig. 4b shows the trajectory in the xy-plane. The velocities at two of its neighboring points, Δs apart, are indicated as tangents to the path; their included angle is $\Delta \epsilon$. The same angle $\Delta \epsilon$ also occurs at the center of curvature M. ρ being the radius of curvature,

$$(10) \qquad\qquad \Delta s = \rho \Delta \epsilon.$$

The same two velocities are plotted in Fig. 4a from a common origin O, with directions preserved. Consider the two neighboring vectors $\vec{O1}$ and $\vec{O2}$ with $\Delta \epsilon$ as their included angle. Projection of 1 on $\vec{O2}$ gives point 3. $\Delta \mathbf{v} = \vec{12}$ is decomposed into $\Delta v_s = \vec{32}$ and $\Delta v_n = \vec{13}$. We therefore obtain, in agreement with (8) and (9),

$$\dot{v}_s = \frac{\vec{32}}{\Delta t} = \frac{v_2 - v_1}{\Delta t} = \frac{\Delta v}{\Delta t} = \frac{dv}{dt},$$

$$\dot{v}_n = \frac{\vec{13}}{\Delta t} = \frac{\Delta \epsilon \cdot v}{\Delta t} = \frac{\Delta \epsilon}{\Delta s} v^2 = \frac{v^2}{\rho},$$

the latter by recalling (10). Cf. problem I.9.

(2) The Concept of Moment in Plane Statics and Kinematics

The moment of a vector quantity $\mathbf{E}$ about a given point of reference O is defined as the *vector product* of the radius vector $\mathbf{r}$ from O to the point of application P of the vector $\mathbf{E}$ by the vector $\mathbf{E}$ itself, i.e.,

$$(11) \qquad\qquad \mathbf{N} = \mathbf{r} \times \mathbf{E}.$$

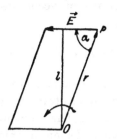

FIG. 5.   Moment of an arbitrary vector quantity $\vec{E}$ about an arbitrary point O.

N is therefore represented by the area of the parallelogram formed by **r** and **E** together with the sense of rotation of **r** into **E** as indicated by an arrow in Fig. 5.   In magnitude,

(11a) $$|\mathbf{N}| = l|\mathbf{E}| = r|\mathbf{E}|\sin\alpha,$$

where l is the " lever arm " of **E** about O.   If we take for **E** a force **F**, we obtain the *moment of the force* **F**, or *torque*

(12) $$\mathbf{L} = \mathbf{r} \times \mathbf{F}.$$

The moment of a force **F** is a fundamental concept of statics, whose discovery goes back to Archimedes himself.   Let us denote the Cartesian components of **F** by X and Y.   Elementary vector algebra readily gives

(12a) $$L_z = xY - yX.$$

The concept of moment is of importance in kinematics and kinetics as well.   Let us still restrict ourselves to problems in a plane, and form

the moment of velocity $= \mathbf{r} \times \mathbf{v}$
the moment of acceleration $= \mathbf{r} \times \dot{\mathbf{v}}$
the moment of momentum $=$ angular momentum $= \mathbf{r} \times \mathbf{p} = m(\mathbf{r} \times \mathbf{v})$

In Cartesian coordinates, with (12a) as a model, we have

(13) $$\mathbf{r} \times \mathbf{v} = x\dot{y} - y\dot{x}, \quad \mathbf{r} \times \dot{\mathbf{v}} = x\ddot{y} - y\ddot{x}.$$

Between the moments of velocity and acceleration, there exists the relation

(14) $$\mathbf{r} \times \dot{\mathbf{v}} = \frac{d}{dt}(\mathbf{r} \times \mathbf{v}).$$

It derives from the fact that $\frac{d\mathbf{r}}{dt} = \mathbf{v}$ and $\mathbf{v} \times \mathbf{v} = 0$, so that

(14a) $$\frac{d}{dt}(\mathbf{r} \times \mathbf{v}) = \mathbf{r} \times \frac{d\mathbf{v}}{dt} + \mathbf{v} \times \mathbf{v} = \mathbf{r} \times \dot{\mathbf{v}}.$$

The customary proof by means of decomposition into coordinates runs exactly parallel to Eq. (14a):

(14b) $\dfrac{d}{dt}(x\dot{y}-y\dot{x}) = x\ddot{y}+\dot{x}\dot{y}-y\ddot{x}-\dot{y}\dot{x} = x\ddot{y}-y\ddot{x}.$

If in Fig. 5 one thinks of the velocity $\mathbf{v}$ of point P replacing the arbitrary vector $\mathbf{E}$, with P describing an arbitrary path, one can read off another simple relation, now between angular momentum and the so-called *areal velocity*. Indeed the infinitesimal element of area dS swept out by the radius vector $\mathbf{r}$ with origin at O is equal to one half the area of the parallelogram $\mathbf{r} \times \mathbf{ds}$, so that the areal velocity

$$\frac{d\mathbf{S}}{dt} = \frac{1}{2}(\mathbf{r} \times \mathbf{v}).$$

We therefore obtain the relation between areal velocity and angular momentum

(15) $\mathbf{r} \times \mathbf{p} = 2m\dfrac{d\mathbf{S}}{dt}.$

(3) Kinematics in Space

We decompose vectors along the three directions s (tangent), n (principal normal) and b (binormal) of the three-dimensional trajectory to obtain the following components:

$$\mathbf{v} = (v,0,0),$$
$$\dot{\mathbf{v}} = \left(\dot{v},\frac{v^2}{\rho},0\right).$$

ρ is the radius of curvature introduced in (9) or (10), now constructed so as to lie in the osculating plane of the trajectory.

If we pass to the moments of velocity and acceleration, we keep the definitions $\mathbf{r} \times \mathbf{v}$ and $\mathbf{r} \times \dot{\mathbf{v}}$, but note that Fig. 5 must now be thought of as three-dimensional. In addition to magnitude and sense of rotation the parallelogram drawn there also has position in space. Because it is helpful in visualizing this point it has become customary to indicate the position by a normal to the plane of the parallelogram. By convention that side of the normal is chosen which points in the direction of advance of a right-handed screw rotated in the sense of rotation of the moment (from $\mathbf{r}$ to $\mathbf{v}$ or $\dot{\mathbf{v}}$ through an angle less than π). The vector picture of the moment then becomes an arrow pointing along this normal, its length being equal to the magnitude of the moment. In Fig. 5 one should, therefore, think of the moment as directed out of and perpendicular to the plane of the paper. We shall postpone a thorough investigation of this procedure and of the difference between axial and polar vectors to Chapter IV, § 23.

So far we have described the moment about an arbitrarily chosen point of reference O. In the following subsection we shall explain what is meant by the moment about a given axis.

(4) Statics in Space; Moment of Force About a Point and About an Axis

The moment of a force $\mathbf{F}$ about a reference point O is completely defined by

(16) $$\mathbf{L} = \mathbf{r} \times \mathbf{F},$$

where $\mathbf{r}$ is the radius vector from O to the point of application P of $\mathbf{F}$,

(16a) $$\mathbf{r} = (x, y, z)$$

if O is taken as origin of coordinates $\mathbf{L}$ can be represented as a vector by the rule just given for moments (rule of the right-hand screw, with length of vector equal to $|\mathbf{L}|$). We now ask: what are the components of $\mathbf{L}$ along the coordinate axes ? We can define them as the projections of the moment vector on these three axes; for instance,

(17) $$L_z = |\mathbf{L}| \cos(\mathbf{L}, z)$$

But $|\mathbf{L}|$ is the area of the parallelogram having sides $\mathbf{r}$ and $\mathbf{F}$. The right member of (17) is therefore at the same time the projection of the area of the parallelogram on the x-y-plane. The latter has sides

$$\mathbf{r}_{\text{proj}} = (x, y); \quad \mathbf{F}_{\text{proj}} = (X, Y),$$

so that, with the help of (17), we obtain as in (12a)

(17a) $$L_z = xY - yX,$$

and similarly

(17b) $$L_x = yZ - zY, \quad L_y = zX - xZ.$$

The components L_x, L_y, L_z of $\mathbf{L}$ can be called the moments of the force $\mathbf{F}$ about axes x, y, z. Cf. problem I.10.

What has been said of the coordinate axes also applies to any arbitrary axis a. The moment of a force $\mathbf{F}$ about an axis a is defined, just as in (17), by taking the moment about a point O located on a and projecting the corresponding moment vector on a. Or it can be formed as in (17a, b) by projecting the area of the moment about O on a plane perpendicular to a. A third method consists in finding the shortest distance from the point of application of the force to a, which we shall call the lever arm l. In this case F is decomposed into three components, F_a parallel to a, F_l in the direction of l, and F_n in a direction perpendicular to both a and l. We then have

(18) $$L_a(\mathbf{F}) = L_a(F_a) + L_a(F_l) + L_a(F_n).$$

The first two terms on the right must vanish; for there can arise no moment of force about an axis a if the force is either parallel to a or if it intersects a.

There remains the third term which results from a force perpendicular to a acting with a lever arm l. Instead of (18) we have, therefore,

(18a) $$L_a(\mathbf{F}) = L_a(F_n) = F_n \cdot l.$$

At this point it may be well to say a few words about the different notations for the products of two vectors. The following list shows that unfortunately these notations vary widely, both in historical and national usage.

Name of product	This book	German Ed. SOMMERFELD	GIBBS	HEAVISIDE	ITALIANS	GRASSMANN
Scalar or inner ..	$\mathbf{A} \cdot \mathbf{B}$	$(\mathfrak{A}\,\mathfrak{B})$	$\mathfrak{A}\,\mathfrak{B}$	$\mathfrak{A}\,\mathfrak{B}$	$\mathfrak{A} \times \mathfrak{B}$	$\mathfrak{A}\,|\,\mathfrak{B}$
Vector or outer ..	$\mathbf{A} \times \mathbf{B}$	$[\mathfrak{A}\,\mathfrak{B}]$	$\mathfrak{A} \times \mathfrak{B}$	$V\,\mathfrak{A}\,\mathfrak{B}$	$\mathfrak{A} \wedge \mathfrak{B}$	$\mathfrak{A}\,\mathfrak{B}$ or $|\mathfrak{A}\,\mathfrak{B}$

Some explanatory remarks follow. The great thermodynamicist Willard Gibbs made a short summary of vector analysis, then still little known, for the use of his students. His notation is still followed (with slight variations) by many American and British authors. Heaviside's notation for the vector product, in which V stands for vector, was thereupon generally abandoned. The Italian notation originated with Marcolongo. Hermann Grassmann, in his " Ausdehnungslehre " (Extension Analysis, 1844 and 1862), had developed a logical system of calculation with segments and points. According to him the simplest relation between two directed segments a and b, is the " planar magnitude " (*Plangrösse*), i.e., the parallelogram formed by a and b, which he therefore denotes by ab (though occasionally also by [ab]). The vertical line in Grassmann's notation for the vector product means " complement " (*Ergänzung*), that is, denotes passing to the vector arrow perpendicular to the planar magnitude.

§ 6. Dynamics (Kinetics) of the Freely Moving Mass Point ; Kepler Problem ; Concept of Potential Energy

(1) Kepler Problem with Fixed Sun

The simplest example we can think of in connection with a freely moving mass point is, at the same time, the most important for our picture of the universe, namely, the motion of the planets. It is a two-dimensional problem, and motion takes place in the ecliptic if the planet in question is the earth. We assume the sun to be fixed in position and justify this by its large relative mass,

$$\text{Sun } 330000, \quad \text{Jupiter } 320, \quad \text{Earth } 1, \quad \text{Moon } \frac{1}{81}.$$

We shall deal with the problem including the sun's motion in Part 2 of this section. Let M be the mass of the sun, m that of the planet. The Newtonian attraction is

$$|\mathbf{F}| = G\frac{mM}{r^2}, \qquad G = \text{gravitational constant,}$$

or, vectorially,

(1) $$\mathbf{F} = -G\frac{mM}{r^2}\frac{\mathbf{r}}{r}.$$

It passes through the fixed point O at the center of the sun, which serves as origin for the radius vector $\mathbf{r}$.

It follows that $\mathbf{r} \times \mathbf{F} = 0$ and therefore, by the Second Law,

$$\mathbf{r} \times \dot{\mathbf{p}} = 0 \text{ and in view of (5.14), } \mathbf{r} \times \mathbf{p} = \text{const.}$$

The angular momentum about the sun is constant, therefore also the areal velocity of Eq. (5.15) is constant. This is the second Kepler law:

The radius vector from the sun to the planet sweeps over equal areas in equal times.

Let the constant areal velocity multiplied by two be called the "areal velocity constant" C,

(2) $$2\frac{dS}{dt} = C.$$

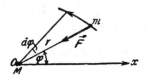

F<small>IG</small>. 6.   Polar coordinates for the Kepler problem with sun as origin; area swept out by the radius vector.

We now introduce the polar angle ϕ, the *true anomaly*[17] of the astronomers (cf. Fig. 6), and obtain

$$dS = \frac{1}{2}r^2 d\phi, \quad 2\frac{dS}{dt} = r^2 \dot\phi = C$$

---

[17] True anomaly is here defined as the angular distance of a planet from its aphelion, as seen from the sun.—T<small>RANSLATOR</small>.

so that

(3) $$\dot{\phi} = \frac{C}{r^2}.$$

In order to derive the first Kepler law, the equation of the trajectory, we decompose the forces along Cartesian coordinates. After division by m the equation of motion becomes

(4)
$$\frac{d\dot{x}}{dt} = -\frac{GM}{r^2}\cos\phi$$

$$\frac{d\dot{y}}{dt} = -\frac{GM}{r^2}\sin\phi.$$

If we multiply both sides of both equations by $\frac{1}{\dot{\phi}}$ and recall (3), we obtain

$$\frac{d\dot{x}}{d\phi} = -\frac{GM}{C}\cos\phi$$

$$\frac{d\dot{y}}{d\phi} = -\frac{GM}{C}\sin\phi.$$

These can now be integrated. Let A and B be constants of integration. The result is

(5)
$$\dot{x} = -\frac{GM}{C}\sin\phi + A$$

$$\dot{y} = \frac{GM}{C}\cos\phi + B.$$

This means that the hodograph of planetary motion is a circle,

(5a) $$(\dot{x} - A)^2 + (\dot{y} - B)^2 = \left(\frac{GM}{C}\right)^2.$$

We shall return to this point in problem I.11. Let us transform the left members of (5) into polar coordinates,

$$x = r\cos\phi, \quad y = r\sin\phi,$$

so that

$$\dot{x} = \dot{r}\cos\phi - r\dot{\phi}\sin\phi = -\frac{GM}{C}\sin\phi + A$$

$$\dot{y} = \dot{r}\sin\phi + r\dot{\phi}\cos\phi = \frac{GM}{C}\cos\phi + B.$$

We now eliminate $\dot{r}$ by multiplication of the first equation by $-\sin\phi$, the second by $\cos\phi$, and subsequent addition. We obtain

$$r\dot{\phi} = \frac{GM}{C} - A\sin\phi + B\cos\phi$$

or, recalling (3),

(6)
$$\frac{1}{r} = \frac{GM}{C^2} - \frac{A}{C} \sin \phi + \frac{B}{C} \cos \phi.$$

This is the equation of a conic section in polar coordinates whose origin coincides with one of the foci of the conic section. We therefore obtain the first Kepler law: *The planet describes an ellipse with the sun at one focus.* In this connection we note that two equally possible trajectories, the hyperbola and the parabola, evidently do not apply to the planets, but only to comets. We shall not discuss them here, but refer the reader to problem I.12.

The derivation of the first Kepler law given here differs from that offered in most texts. The latter starts out with the equation of energy, which we shall now derive. We turn back to Eq. (4), where we replace $\cos \phi$ by $\frac{x}{r}$, $\sin \phi$ by $\frac{y}{r}$ in the right members. We then multiply the first of Eqs. (4) by $\dot{x}$, the second by $\dot{y}$, add the two, and get

$$\frac{d}{dt} \frac{1}{2} (\dot{x}^2 + \dot{y}^2) = -\frac{1}{2} \frac{GM}{r^3} \frac{d}{dt} (x^2 + y^2) = -\frac{GM}{r^2} \frac{dr}{dt}.$$

An integration with respect to t yields

(7)
$$\frac{1}{2} (\dot{x}^2 + \dot{y}^2) = \frac{GM}{r} + E.$$

The left member is the kinetic energy divided by m; the first term on the right is, apart from sign, the potential energy divided by m (cf. Part 3 of this section); E is therefore the total energy divided by m. Our Eq. (7) has the same form as the equation of energy of one-dimensional motion, (3.8).

In order to pass from (7) to the equation of the path (6) in the simplest possible manner, we recall that in polar coordinates the square of the element of line is

$$dx^2 + dy^2 = dr^2 + r^2 d\phi^2.$$

We therefore have

$$\dot{x}^2 + \dot{y}^2 = \left(\frac{dr}{dt}\right)^2 + r^2 \left(\frac{d\phi}{dt}\right)^2 = \left(\frac{d\phi}{dt}\right)^2 \left\{ \left(\frac{dr}{d\phi}\right)^2 + r^2 \right\},$$

or, in view of (3),

$$C^2 \left\{ \left(\frac{1}{r^2} \frac{dr}{d\phi}\right)^2 + \frac{1}{r^2} \right\}$$

If we put $s = \frac{1}{r}$, this becomes

$$C^2 \left\{ \left(\frac{ds}{d\phi}\right)^2 + s^2 \right\},$$

so that our equation of energy (7) is transformed into

$$\frac{1}{2}C^2\left\{\left(\frac{ds}{d\phi}\right)^2+s^2\right\}-GMs=E.$$

Differentiation with respect to ϕ gives

$$\frac{ds}{d\phi}\left\{C^2\left(\frac{d^2s}{d\phi^2}+s\right)-GM\right\}=0.$$

Since $\frac{ds}{d\phi}\neq0$, the bracket must vanish. Thus we obtain a linear homogeneous differential equation with constant coefficients of second order in s,

$$\frac{d^2s}{d\phi^2}+s=\frac{GM}{C^2}.$$

The general solution of such an equation consists of a particular solution of the inhomogeneous equation plus the general solution of the homogeneous equation. Evidently

$$s=\text{constant}=\frac{GM}{C^2}$$

is a particular integral of the inhomogeneous equation. The general solution of the homogeneous equation is the sum of $\sin\phi$ and $\cos\phi$. We can now take A/C and B/C as our constants of integration and finally obtain

$$s=\frac{GM}{C^2}-\frac{A}{C}\sin\phi+\frac{B}{C}\cos\phi,$$

which is precisely the previously obtained Eq. (6).

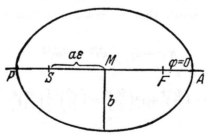

FIG. 7. Kepler ellipse with major and minor axes; perihelion, aphelion, eccentricity.

We shall now specialize this equation in such a way that the line $\phi=0$, which starts out from one focus, passes through the other focus as well, i.e., that it forms, together with the line $\phi=\pi$, the major axis of the ellipse (cf. Fig. 7). On this axis are located the points P, " perihelion " (closest to the sun) and A, " aphelion " (farthest from the sun), at which r must be a minimum and a maximum respectively. We therefore impose the condition

$$\frac{dr}{d\phi} = 0 \text{ for } \phi = \left\{ \begin{array}{c} 0 \\ \pi \end{array} \right.$$

which, from (6), requires $A = 0$.

If, in addition, ϵ is the eccentricity of the ellipse, Fig. 7 shows that at

$$\text{perihelion,} \quad r = SP = a\,(1-\epsilon),\ \phi = \pi,$$

$$\text{aphelion,} \quad r = SA = a\,(1+\epsilon),\ \phi = 0.$$

According to (6) we then have at

$$\text{perihelion,} \quad \frac{1}{a(1-\epsilon)} = \frac{GM}{C^2} - \frac{B}{C},$$

$$\text{aphelion,} \quad \frac{1}{a(1+\epsilon)} = \frac{GM}{C^2} + \frac{B}{C}.$$

From these we obtain by addition and subtraction

(8) $$\frac{GM}{C^2} = \frac{1}{a(1-\epsilon^2)}, \qquad \frac{B}{C} = -\frac{\epsilon}{a(1-\epsilon^2)},$$

respectively.

We shall finally express the areal velocity constant C in terms of the period T. From (2) we immediately obtain

$$C = \frac{2S}{T} \text{ with } S = \pi\,ab = \pi a^2 (1-\epsilon^2)^{\frac{1}{2}}$$

as the total area swept out by the radius vector. It follows that

(9) $$C^2 = \frac{4\pi^2 a^4 (1-\epsilon^2)}{T^2}.$$

If we introduce this in the first of Eqs. (8), we have

(10) $$\frac{T^2}{a^3} = \frac{4\pi^2}{GM}.$$

Since G and M are the same for all planetary trajectories, (10) is the expression of the third Kepler law: *the squares of the periodic times are proportional to the cubes of the major axes.*

Kepler greeted the discovery of this law with the enthusiastic statement[18]: " Finally I have brought to light and verified beyond all my hopes and

---

[18] *Harmonice mundi*, 1619. The two first Kepler laws had been published in the *Astronomia Nova*, 1609.

expectations that the whole Nature of Harmonies permeates to the fullest extent, and in all its details, the motion of the heavenly bodies; not, it is true, in the manner in which I had earlier thought, but in a totally different, altogether complete way."

Actually the third Kepler law in the form (10) is not yet quite exact. It is valid only as long as one can neglect the planetary mass m in comparison with the mass M of the sun. We shall now drop this assumption, thereby passing to the two-body problem proper of astronomy. This problem is not significantly more difficult than the one-body problem treated so far.

(2) Kepler Problem Including Motion of the Sun

Let x_1, y_1 be the coordinates of the sun S; x_2, y_2 those of the planet P.

According to Newton's third law the force on S must be equal and opposite to that on P, so that the complete equations of motion are

for the sun,   for the planet,

$$M\,\frac{d^2x_1}{dt^2} = \frac{m\,MG}{r^2}\cos\phi; \qquad\qquad m\,\frac{d^2x_2}{dt^2} = -\frac{m\,MG}{r^2}\cos\phi;$$

$$M\,\frac{d^2y_1}{dt^2} = \frac{m\,MG}{r^2}\sin\phi; \qquad\qquad m\,\frac{d^2y_2}{dt^2} = -\frac{m\,MG}{r^2}\sin\phi.$$

We now introduce the relative position coordinates

(11a) $$x_2 - x_1 = x,\ \ y_2 - y_1 = y,$$

and furthermore the center of mass coordinates

(11b) $$\frac{m\,x_2 + M\,x_1}{m+M} = \xi,\quad \frac{m\,y_2 + M\,y_1}{m+M} = \eta.$$

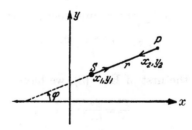

FIG. 8.   Kepler problem with motion of sun taken into account.

Subtraction of the equations of motion gives

(12)
$$\frac{d^2x}{dt^2} = -\frac{(M+m)\,G}{r^2}\cos\phi,$$

$$\frac{d^2y}{dt^2} = -\frac{(M+m)\,G}{r^3}\sin\phi;$$

whereas addition yields

(13) $$\frac{d^2\xi}{dt^2} = 0, \qquad \frac{d^2\eta}{dt^2} = 0.$$

Comparison of (12) with the earlier Eqs. (4) shows at once that the first two laws of Kepler remain intact, i.e., are valid for the relative motion as well. The third law takes the form

(14) $$\frac{T^2}{a^3} = \frac{4\pi^2}{G(M+m)}.$$

The ratio T^2/a^3 is therefore no longer a universal constant, but in principle is somewhat different for every planet. Because of the preponderance of the sun's mass the deviations from (10) are, however, exceedingly slight.

Eqs. (13) further show that the mass center of sun and planet moves with constant velocity. If we do our calculations in terms of a coordinate system in which the mass center is fixed at the origin, this velocity must be put equal to 0; the same applies to the coordinates ξ, η of the center of mass themselves.

Eqs. (11b) are simplified accordingly. With their help and that of Eqs. (11a) the coordinates x_1, y_1 of the sun and the coordinates x_2, y_2 of the planet can now be expressed separately in terms of the relative position coordinates x, y:

$$x_1, y_1 = -\frac{m}{M+m}(x, y),$$

$$x_2, y_2 = \frac{M}{M+m}(x, y).$$

It follows that in the center of mass system the trajectories of the sun and planet are also ellipses; that of the planet is almost identical to the ellipse considered in Part 1 of this section. That of the sun is a very dwarfed ellipse, traversed in the same sense, but π out of phase with the planet's trajectory.

If we change the law of gravitation to

(15) $$\mathbf{F} = k r^n \cdot \frac{\mathbf{r}}{r}, \ n \text{ arbitrary,}$$

the second Kepler law will hold unchanged; the trajectories, however, become transcendental curves which are, in general, not closed. It is only in the case $n = 1$ that we obtain ellipses just as in the case of gravitation, $n = -2$. (Cf. problem I.13).

(3) When Does a Force Field Have a Potential?

In one-dimensional motion we were able to define a potential energy V connected with a force X without any difficulty—see Eq. (3.7). As mentioned at that time, this is possible for two- or three-dimensional motions only if certain conditions are met. If X, Y, Z are the Cartesian components of the force $\mathbf{F}$ the definition of potential energy for the three-dimensional case analogous to (3.7) would be

$$(16) \qquad V = - \int^{xyz} (X\,dx + Y\,dy + Z\,dz).$$

If V is to be a quantity independent of the path of integration and dependent only on its endpoint (the choice of the initial point merely gives rise to an additive constant which remains arbitrary in any case), the expression

$$X\,dx + Y\,dy + Z\,dz$$

must be a *perfect differential*; i.e., X, Y, Z must be the derivatives with respect to x, y, z of a " field function." In our case the function is just $-V$, and we say that $\mathbf{F}$ is " derivable from the potential V." The well-known conditions for this are that

$$(17) \qquad \frac{\partial Y}{\partial x} = \frac{\partial X}{\partial y}, \quad \frac{\partial Z}{\partial y} = \frac{\partial Y}{\partial z}, \quad \frac{\partial X}{\partial z} = \frac{\partial Z}{\partial x}.$$

It is only if these conditions are fulfilled that a field function $V(x, y, z)$ can be defined for each point x, y, z; V is called the potential energy or simply the potential.

In the two-dimensional case, where $Z = 0$ and X, Y are independent of z, the three Eqs. (17) are reduced to the first one of these.

Vector analysis (which has been relegated to Vol. II of this series since we need only vector algebra in this volume) shows that the conditions (17) have an invariant meaning, i.e. are independent of the choice of coordinates. In Vol. II these conditions will be summarized in the vector equation, curl $\mathbf{F} = 0$ (this is often expressed by saying that the vector field $\mathbf{F}$ is *irrotational*).

Evidently one can without difficulty write down expressions for X, Y, Z in terms of x, y, z which do not satisfy conditions (17). On the other hand we see that these conditions are satisfied for the gravitational field

$$X = Y = 0, Z = -mg$$

and lead to

$$(18) \qquad V = mgz.$$

The same is true of the general gravitational fields based on Newton's law and the mathematically similar fields of electrostatics and magnetostatics. As a matter of fact fields that are irrotational and simultaneously time-independent (" potential fields ") occupy a unique position in nature.

They will play a special role in the general developments of Chapters VI and VIII.

A mechanical system in which only forces derivable from potentials act is called a *conservative system* because its energy is conserved; otherwise we speak of non-conservative or *dissipative systems*.

CHAPTER II

MECHANICS OF SYSTEMS, PRINCIPLE OF VIRTUAL WORK, AND D'ALEMBERT'S PRINCIPLE

§ 7. Degrees of Freedom and Virtual Displacements of a Mechanical System ; Holonomic and Non-holonomic Constraints

The single mass point has one degree of freedom if its motion is restricted to a straight line or a curve, two degrees of freedom if it is made to move in a plane or on a curved surface; the mass point moving freely in space has three degrees of freedom.

Two mass points connected by a weightless, rigid rod have five degrees of freedom; for the first point can be regarded as freely moving, in which case the second is restricted to the surface of a sphere described about the first, its radius equal to the length of the rod.

The number of degrees of freedom for n mass points which are coupled by r relations between their coordinates is

$$(1) \qquad f = 3n - r.$$

If there is an infinity of mass points connected by infinitely many conditions such an enumeration is of course not feasible. The procedure to be used in that case will now be shown, the rigid body serving as example.

(a) Freely Moving Rigid Body

We single out a point of the rigid body. It has three degrees of freedom. A second point, at a constant distance from the first (definition of "rigid"!), can move only on a spherical surface about the first point as center. This gives two more degrees of freedom. Finally a third point can describe a circle about the axis connecting the first two points, thus contributing one more degree of freedom. Once the motions of these three points have been specified, the paths of all other points of the rigid body are uniquely determined. It follows that

$$f = 3 + 2 + 1 = 6.$$

(b) Top on a Plane Surface

We assume that the bottom of the spinning top terminates in a point, and take this as the first point of our enumeration; it has two degrees of freedom. A second point can move on a hemisphere about the first, and a third one on a circle about a line connecting the first two. Thus

$$f=2+2+1=5.$$

(c) Top with Fixed Point

Now the two degrees of freedom of the first point are lost, so that

$$f=2+1=3.$$

(d) Rigid Body with Fixed Axis—Pendulum

Here

$$f=1.$$

If the center of mass of the body does not lie on the axis we speak of a physical or *compound pendulum*. From this we obtain a mathematical or *simple pendulum* if the body shrinks to a point. The *spherical pendulum* — a mass point restricted in its motion to the surface of a sphere — has

$$f=2.$$

(e) Infinitely Many Degrees of Freedom

For a deformable solid body or a liquid

$$f=\infty.$$

In that case the equations of motion become partial differential equations. By contrast a system with a finite number of degrees of freedom n is determined by an equal number n of ordinary differential equations of second order.

(f) Machine with One Degree of Freedom

Such a machine consists of a series of nearly rigid bodies coupled to each other either by links or by means of guides of various types. The classical example of such a machine is the drive mechanism of a piston engine (Fig. 9). If the machine is provided with a centrifugal governor (also called Watt governor because it was first proposed by the inventor of the steam engine), it acquires a second degree of freedom.

In the aforementioned examples the number of degrees of freedom equals the number of independent coordinates which are necessary to determine the position of the system. The coordinates need not be Cartesian. In case of the drive mechanism we can equally well specify either the coordinate x determining the position of the piston or the angle ϕ giving the position of the crank pin on the shaft. In general we shall call the independent coordinates of a system of f degrees of freedom

(2) $q_1, q_2 \ldots q_f.$

They can, within certain limits, be chosen arbitrarily. The r conditions among the coordinates referred to in Eq. (1) can be satisfied identically by suitable choice of the q, so that they drop out of the subsequent treatment of our system.

The mechanics of Hertz mentioned on p. 5 has the important merit of having called attention to *conditions of differential form*, to which the foregoing cannot be applied. Such a condition can be written as

$$(3) \qquad \sum_{k=1}^{f} F_k(q_1 \ldots q_f) dq_k = 0.$$

Here we assume that the F_k do not all have the form $\dfrac{\partial \Phi}{\partial q_k}$, so that (3) is not the total differential of some function Φ $(q_1 \ldots q_f)$, and we assume, moreover, that it cannot be converted into a total differential by means of an integrating factor.

In agreement with Hertz we shall call conditions of the form $\Phi(q_1 \ldots q_f)$ = const. *holonomic* (*holos* in Greek = *integer* in Latin = whole = integrable); conditions of the form (3) which cannot formally be integrated will be called *non-holonomic*.[1] The simplest example of a non-holonomic condition is furnished by a sharp-edged wheel rolling on a horizontal plane, cf. problem II.1 (the sleigh and the flexible coupling mechanism of a bicycle also belong in this category). Such a wheel is restricted to move always in the direction it may have at any given instant. Nevertheless it is able to reach all points of its supporting plane, even if at times only by pivoting about its sharp point of contact. It therefore possesses more degrees of freedom in finite than in infinitesimal motion. In general, *if a system subject to r non-holonomic conditions has f degrees of freedom in finite motion, it has only f−r degrees of freedom in infinitesimal motion.* This point will be investigated in problem II.1.

The foregoing distinction is important for the concept of virtual displacements. *A virtual displacement is an arbitrary, instantaneous, infinitesimal change of the position of the system compatible with the conditions of constraint.* Whereas we shall denote real displacements due to given forces under given conditions by

$$dq_1, dq_2 \ldots dq_f,$$

the symbols

$$\delta q_1, \delta q_2 \ldots \delta q_f$$

will be used to denote virtual displacements. The δq have nothing to do

---

[1] A. Voss made a general study of such conditions in 1884, long before Hertz; cf. *Math. Ann.* **25**.

with the actual motion. They are introduced, so to speak, as test quantities, whose function it is to make the system reveal something about its internal connections and about the forces acting on it.

For purely holonomic constraints the δq are independent of each other, *one* δq corresponding to each degree of freedom. A larger number of δq must be introduced for non-holonomic constraints; in that case the δq are related by differential conditions of the form (3), or, for virtual displacements,

$$(4) \qquad \sum_{k=1}^{f} F_k(q_1 \ldots q_f)\delta q_k = 0.$$

Here f is the number of degrees of freedom for finite motion. As previously emphasized, this number is greater than that for infinitesimal motion.

§ 8. The Principle of Virtual Work

Let us consider a mechanical system in equilibrium under applied forces. The forces may have any desired direction, may act on various parts of the system, and need not have the positions required for the equilibrium of a simple rigid body. Whether the forces lead to the equilibrium of the system under investigation depends as much on the system as on the forces.

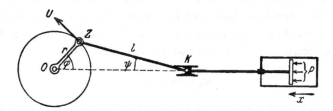

Fig. 9. Schematic diagram of the drive mechanism of a piston engine.

In the spirit of elementary particle mechanics we would ask for the *reactions* which are exerted by one part of the system on another due to the applied forces. This procedure would, for instance, be used by a mechanical engineer in the analysis of the crank mechanism (Fig. 9). The steam pressure P acting on the piston is transmitted by the piston rod to the crosshead K, whence it is passed on as longitudinal compression to the connecting rod of length l. The connecting rod acts on the crank pin Z with a thrust which has the direction of the rod. In order that the system be in equilibrium only that part U of the thrust which is perpendicular to the crank, therefore tangential to the crank circle, need be opposed by an equal applied force. The component in the direction of the crank, i.e., toward the center of the crank shaft, is absorbed by the rigidly fixed shaft bearing, O. It only puts a stress on the bearing and is irrelevant to the question of the equilibrium of the system.

It is therefore the reactions within the system which make equilibrium possible. To investigate them individually is possible in simple cases, but tedious in general. We can, however, assert without knowing them in detail that *they do no work on the system*. In our case the guide pressure at the guide rails is perpendicular to the motion of the crosshead, and that part of the force acting on the crank pin which is transmitted to the crank shaft acts through the fixed point O of the crank shaft bearing. We establish this assertion in the general case by giving the system a tentative virtual displacement from its position of equilibrium. The " virtual work " of the reactions in such a displacement is found to be zero.

Let us verify the principle in detail on the simple rigid body. We must imagine that every point i is related to every point k of the body by means of reactions $\mathbf{R}_{ki}$ and $\mathbf{R}_{ik}$ acting on i and k respectively. If we single out two such points, we have the system of two mass points mentioned at the beginning of § 7, the two masses being connected by a weightless, rigid rod. The reactions acting in this rod must satisfy Newton's third law,

(1) $$\mathbf{R}_{ik} = -\mathbf{R}_{ki}.$$

Just as in § 7, in the enumeration of the degrees of freedom, we shall now decompose the virtual displacement into a translation $\delta\mathbf{s}_i$ common to both points and a rotation $\delta\mathbf{s}_n$ of point k about the already displaced point i, this rotation being a motion normal to the rod. Then

$$\delta\mathbf{s}_k = \delta\mathbf{s}_i + \delta\mathbf{s}_n$$

For the virtual work of translation we therefore obtain, in view of (1),

$$\delta W_{tr} = \mathbf{R}_{ik} \cdot \delta\mathbf{s}_i + \mathbf{R}_{ki} \cdot \delta\mathbf{s}_i = 0;$$

for that of rotation, for which i remains fixed and k is displaced normal to the rod,

$$\delta W_{rot} = \mathbf{R}_{ik} \cdot \delta\mathbf{s}_n = 0.$$

This example illustrates that Newton's law of action and reaction is the salient point in the transition from particle mechanics to the mechanics of systems.

We shall now expand what we have learned with the help of the foregoing examples into a general postulate: *in any mechanical system the virtual work of the reactions equals zero*. Far be it from us to want to give a general proof of this postulate.[2] Rather we regard it practically as definition of a " mechanical system."

It is now only a small step to the general formulation of the principle of virtual work. We argue as follows: every physically given force acting on a system in equilibrium is in equilibrium with the reactions induced at its point of application; the work done by such an applied force plus that done by its reactions in any virtual displacement of the point of application is therefore zero. The same is true of the sum of all applied forces and the sum of all the reactions induced by them. Now the reactions, taken by themselves, do no virtual work (by the previous paragraph). *Therefore the virtual work done by the applied forces keeping a system in equilibrium must equal zero* as well. The tedious investigation of the reactions is thereby eliminated.

This is the principle of virtual work, often called *Prinzip der virtuellen Verrückungen oder Verschiebungen* (principle of virtual displacements) in the German literature. This name is not as fortunate as the one used in English-speaking countries, which was taken over from the Italian *principio dei lavori virtuali*. The term, principle of virtual velocities, which is often used in the mathematical literature and was first proposed by Jean Bernoulli, seems unsuited to us.

Historically the principle was already sketched by Galileo. It was further developed by Stevin, Jacques and Jean Bernoulli and d'Alembert. It achieved its dominating position as the most general equilibrium principle only with the " Mécanique analytique " of Lagrange.

Whether the constraints of the system are of the holonomic or the non-holonomic variety affects the application of the principle of virtual work but little. Indeed, a condition of the form (7.4) can be introduced in the expression for the virtual work by elimination of one of the δq, regardless of whether this condition is integrable or not.

Instead of forces of reaction we can use the more descriptive term, forces of geometric origin. For they are given by the geometric relations between the various parts of the system, or, as in the case of the rigid body, between its individual mass points.

Antonymous to forces of geometric origin are the " forces of physical origin " or *applied forces*. The commonly used term "external forces " is less clear and will not be used here in this sense. Applied forces are caused by physical effects, such as gravity, steam pressure, cable tensions acting on the system from the outside, etc. They betray their physical origin by the fact that their mathematical expressions contain specific

---

[2] Lagrange attempts this in the introduction of his *Mécanique Analytique* (cf. p. 1) by means of certain block and tackle constructions.

constants (gravitational constant, readings of the scale of a manometer or barometer, etc.) which can be determined only experimentally. In § 14 we shall talk about the force of friction, which must sometimes be counted among the forces of reaction, sometimes among the applied forces. It is a force of reaction if it occurs as static friction; an applied force if it occurs as sliding or kinetic friction. Static friction is automatically eliminated by the principle of virtual work; kinetic friction must be introduced as an applied force. An external indication of this is the occurrence of the experimental constant μ in the law of sliding friction (14.4).

§ 9. *Illustrations of the Principle of Virtual Work*

(1) The Lever (Archimedes)

The lever possesses one degree of freedom, $f=1$, therefore only one displacement δq which corresponds to the virtual angular displacement $\delta\phi$.

Equilibrium exists if, and only if, the virtual work done in a rotation $\delta\phi$ of the lever is zero. Let δs_A, δs_B be the virtual displacements of the points of application P and Q of the forces A and B respectively. We then demand that

$$A\ \delta s_A + B\ \delta s_B = 0.$$

But from Fig. 10a $\delta s_A = a\ \delta\phi, \quad \delta s_B = -b\ \delta\phi.$ Therefore

$$(Aa - Bb)\ \delta\phi = 0$$

and consequently

$$Aa = Bb.$$

The moments of the forces about the fulcrum O are equal, i.e., their algebraic sum is zero.

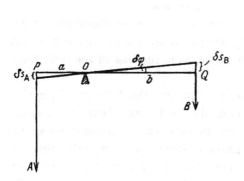

FIG. 10a. Lever with arms a and b under vertical loads A and B.

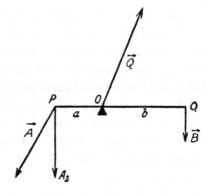

FIG. 10b. Lever under oblique load, showing the reaction of the fulcrum on the beam.

If, as in Fig. 10b, the force **A** is not perpendicular to the lever arm, we can decompose it into a component A_1 in the direction of the arm, and A_2 perpendicular to it. With point O fixed, A_1 has no effect, so that we have

$$A_2 a = |\mathbf{B}| b.$$

In order to obtain the load at O, we must introduce an opposing force acting on the arm; in Fig. 10a it is directed vertically upward and has magnitude $Q = A + B$; the load on the fulcrum is equal and opposite to this force Q. In the case of Fig. 10b we have the vector equation $\mathbf{Q} = \mathbf{A} + \mathbf{B}$; here, too, the force on O is the opposite (i.e., " equilibrant ") of $\mathbf{Q}$. In posing these questions, we actually transgress the limits of the principle of virtual work. The fixed position of the pivot O is characteristic of the mechanical system of the lever. Its virtual displacement, and the virtual work done on it, are therefore zero. In order to obtain Q or $\mathbf{Q}$ by means of our principle we should have to consider an altogether different mechanical system: we should have to provide O with two degrees of freedom and ask for the condition of equilibrium when we add a virtual translation of the whole lever parallel to itself to the rotation so far considered.

(2) Inverse of the Lever: Cyclist, Bridge

Consider the bicycle of Fig. 11a. The earth opposes the weight in the two points R (rear wheel) and F (front wheel). The rear wheel is exposed to the greater pressure, since the weight Q of bicycle and rider lies closer to R than to F. Accordingly a cyclist pumps his rear wheel to a higher

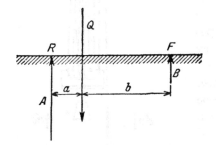

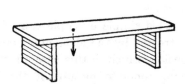

FIG. 11a. Distribution of weight on front and rear wheels of a bicycle with rider.

FIG. 11b. Distribution of load on the two supports of a schematic bridge.

pressure than his front wheel. The load on the rear wheel is $A = \dfrac{b}{a+b} Q$, that on the front wheel, $B = \dfrac{a}{a+b} Q$.

The same situation obtains with a bridge loaded off center (Fig. 11b).

(3) The Block and Tackle (also known to the Greeks)

Let n be the number of pulleys at both the upper and the lower end of the tackle. Q is the load to be lifted, P the force required at the loose end of the rope. In a virtual displacement of the system let

$$P \text{ move a distance } \delta p,$$

$$Q \text{ move a distance } \delta q,$$

the positive direction of motion being indicated by the arrows of Fig. 12. Equilibrium exists if

(1) $P\, \delta p - Q\, \delta q = 0.$

If now Q is lifted an amount δq, the $2n$ rope lengths between the upper and lower pulleys are shortened by δq each, the total shortening therefore being $2n\,\delta q$. The loosely hanging rope at P must lengthen by precisely the same amount. Thus

$$\delta p = 2n\,\delta q$$

and, in view of (1),

$$(Q - 2\,n\,P)\,\delta q = 0.$$

We then obtain

(2) $P = \dfrac{Q}{2n}.$

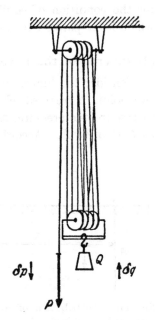

We have here treated the block and tackle as an "ideal" mechanical system, i.e., we have neglected the friction between ropes and pulleys and the friction in the pulley bearings.

This simple example can of course also be treated by the elementary method of rope tension, which in this case affords perhaps a more concrete picture of the interplay of forces.

Let S be the tension in the rope, taken over its total cross-section. If we neglect all frictional effects, the tension must be the same at every point of the rope; no matter

Fig. 12. Block and tackle. Virtual displacements of load and force.

where the rope is cut, one encounters the same tension S, which in both severed ends acts away from the point of severance. Let us cut the rope once on the left side, above P. The severed piece, in which P

acts downward and S upward, gives

$$P=S.$$

Next we cut all the ropes in the right part of the figure, thereby exposing $2n$ cross-sections on each side of the cut. The equilibrium of the forces acting on the severed lower right part demands

$$Q=2nS.$$

We therefore have again

$$P=\frac{Q}{2n}.$$

In addition, a consideration of the upper part of the system yields the loading of the beam from which the block is suspended. Evidently it amounts to $P+Q$.

(4) The Drive Mechanism of a Piston Engine

As in Fig. 9, P is the total force due to the steam pressure exerted on the piston, so that the virtual work done on the piston is $P\delta x$. Let Q be the equilibrant of the peripheral force U on the crank, i.e., the force causing P to be in equilibrium. The virtual work done by Q is $-Qr\,\delta\phi$. Our principle requires

$$(3) \qquad Q r\,\delta\phi=P\,\delta x, \quad Q=P\frac{\delta x}{r\delta\phi}.$$

The calculation of Q therefore reduces to the purely kinematic task of determining the relation between δx and $\delta\phi$.

According to Fig. 9 (projection on the x-direction),

$$(4) \qquad r\cos\phi+l\cos\psi=\text{const}-x,$$

so that, differentiating,

$$(4a) \qquad r\sin\phi\,\delta\phi+l\sin\psi\,\delta\psi=\delta x.$$

The triangle OZK gives

$$(4b) \qquad \sin\psi=\frac{r}{l}\sin\phi, \quad \delta\psi=\frac{r\cos\phi}{l\cos\psi}\delta\phi=\frac{r}{l}\frac{\cos\phi}{\left[1-\left(\frac{r}{l}\right)^2\sin^2\phi\right]^{\frac{1}{2}}}\delta\phi$$

If we introduce this in (4a), we obtain

$$(4c) \qquad r\sin\phi\,\delta\phi\left(1+\frac{r}{l}\frac{\cos\phi}{\left[1-\left(\frac{r}{l}\right)^2\sin^2\phi\right]^{\frac{1}{2}}}\right)=\delta x.$$

This relation furnishes the kinematic quantity $\frac{\delta x}{r\,\delta\phi}$. Substitution in (3) now gives

(5) $$Q = P \sin \phi \left(1 + \frac{r}{l} \frac{\cos \phi}{\left[1 - \left(\frac{r}{l} \right)^2 \sin^2 \phi \right]^{\frac{1}{2}}} \right).$$

Thus the peripheral force $U = Q$ transmitted by the crank pin Z is determined for every crank position ϕ. Its precise knowledge is essential for an evaluation of the amount of cyclic fluctuation of the machine, and hence for the determination of the flywheel required. Since r/l is a small proper fraction, (5) can be expanded into a rapidly converging series in r/l. Cf. also problem II.2.

Finally, for the sake of a later application we shall calculate the piston position x as a power series in r/l. According to (4) and (4b) we have

(6) $$x + r \left(\cos \phi - \frac{1}{2} \frac{r}{l} \sin^2 \phi + \ldots \right) = \text{const.}$$

(5) Moment of a Force About an Axis and Work in a Virtual Rotation

Let a point P be at a distance l from an axis a. Let a force $\mathbf{F}$ of arbitrary direction act at P. In a virtual rotation $\delta\phi$ about the axis a, P is displaced by

$$\delta s_P = l \, \delta\phi.$$

What is the work δW done by $\mathbf{F}$ in this displacement ?

We decompose $\mathbf{F}$ into the mutually perpendicular components F_a, F_l, F_n, just as for Eq. (5.18). The work done depends only on F_n, for

$$\delta W = F_n \, \delta s_P = F_n l \, \delta\phi.$$

A comparison with (5.18a) will allow us to make a general statement:

The moment of a force about an axis can be regarded as the virtual work of the force in a rotation $\delta\phi$ of its point of application about the axis, divided by $\delta\phi$,

(7) $$L_a(\mathbf{F}) = \frac{\delta W}{\delta\phi} = l F_n.$$

The concept of moment, basic to statics, is thereby brought into relation with the concept of virtual work basic to all questions of equilibrium.

Let us remark in this connection that the dimensions of moment (force· lever arm) are the same as those of work (force · distance). This is in agreement with (7) if, as is customary, we regard the angle measured in radians as dimensionless.

§ 10. D'Alembert's Principle ; Introduction of Inertial Forces

As we have seen, all bodies have the tendency to remain in a state of rest or of uniform rectilinear motion. We can think of this tendency as a resistance to changes in the motion, an inertial resistance, or, for brevity, as an *inertial force*. The definition of inertial force $\mathbf{F}^*$ for the single mass point is therefore

$$(1) \qquad\qquad \mathbf{F}^* \equiv -\dot{\mathbf{p}}$$

and the fundamental law $\dot{\mathbf{p}} = \mathbf{F}$ takes on the form

$$(2) \qquad\qquad \mathbf{F}^* + \mathbf{F} = 0.$$

The inertial force is in vectorial equilibrium with the applied force.

While $\mathbf{F}$ is a force given by the physical situation, $\mathbf{F}^*$ is a fictitious force. We introduce it in order to reduce problems of motion to problems involving equilibrium, a procedure that is often convenient.

Inertial forces are familiar to us from everyday life. When we set the heavy revolving door of an hotel in motion, it is not the force of gravity or friction, but the inertia of the door that has to be overcome. A similar example is that of the sliding doors of street cars and trolleys.[3] On the forward platform the door opens in the direction of travel. When the car brakes, the door tends to move forward and can therefore be opened easily. When the car accelerates after a stop, the open door seeks to retain its position of rest; it therefore tends to move to the rear and can be closed without effort. It is easier to get on and off at the front platform than at the rear, where the door opens in the reverse manner.

The best-known form of an inertial force is the *centrifugal force*, which is noticeable in any curved motion. It, too, is a fictitious force. It corresponds to the acceleration $\dot{v}_n$ normal to the curve, which is a *centripetal* acceleration, i.e., directed toward the center of curvature. According to (5.9) the centrifugal force is given by

$$(3) \qquad\qquad \mathbf{C} = -m\dot{\mathbf{v}}_n, \quad |\mathbf{C}| = m|-\dot{\mathbf{v}}_n| = m\frac{v^2}{\rho},$$

where the minus sign refers to the outward direction.

The *Coriolis force* (cf. § 28) and the various gyroscopic effects (cf. § 27) also come under the heading of inertial forces.

Incidentally the operation of railroads furnishes a very vivid example of the fact that the " fictitious " centrifugal force has a very real existence.

---

[3] The translator does not guarantee that the following is applicable to trolleys in the United States. It applies at least in part to the streetcars of San Francisco, which belong, however, to a breed rapidly approaching extinction.

On a curve the rail bed is banked in such a way that the outer rail is higher than the inner. The difference in height is always such that for some mean velocity of the train the resultant of gravity and centrifugal force is perpendicular to the rail bed. This procedure eliminates not only the danger of overturning about the outer rail, but also a harmful unequal loading of the rails.

Strangely enough, the great Heinrich Hertz raises objections to the introduction of the centrifugal force in the unusually beautiful and beautifully written introduction to his " Mechanics " (Collected Works, Vol. III, p. 6):

" We swing a stone attached to a string in a circle; we thereby consciously exert a force on the stone; this force constantly deviates the stone from a straight path, and if we alter this force, the mass of the stone or the length of the string, we discover that indeed the motion of the stone occurs at all times in agreement with Newton's second law. Now the third law demands a force opposing that which is exerted by our hand on the stone. If we ask for this force, we obtain the answer familiar to everybody, that the stone reacts on the hand by virtue of the centrifugal force, and that this centrifugal force is indeed equal and opposite to the force exerted by us on the stone. Is this mode of expression admissible ? Is that which we now call centrifugal force anything but the inertia of the stone ? "

We answer this question with a flat no; indeed the centrifugal force, by virtue of our definition (3), is identical to the inertia of the stone. But the force opposing that which we exert on the stone, i.e., really on the string, is the pull which the string exerts on our hand. Hertz further remarks that " we are forced to the conclusion that the classification of the centrifugal force as a force is not suitable; its name, just like that of live force, is to be regarded as a heritage passed down from former times; and from the point of view of usefulness the retention of this name is easier to excuse than to justify." In regard to this we would like to say that the name centrifugal force needs no justification, for it rests, like the more general term, inertial force, on a clear definition.

Incidentally, it is precisely this alleged lack of clarity of the force concept which induced Hertz, in an interesting but not very fruitful attempt, to construct his mechanics entirely without the notion of force (cf. § 1, p. 5).

We now come to the achievement of d'Alembert (mathematician, philosopher, astronomer, physicist, encyclopedist; " Traité de Dynamique," 1758).

If a mass point k, part of an arbitrary mechanical system, is acted on by an applied force $\mathbf{F}$, Eq. (2) must be changed to read

(4) $$\mathbf{F}_k{}^* + \mathbf{F}_k + \sum_i \mathbf{R}_{ik} = 0.$$

Here $\mathbf{R}_{ik}$ is the reaction which the mass point i connected with k exerts on k. According to our general postulate of p. 52, the $\mathbf{R}_{ik}$, taken together, do no work in an arbitrary virtual displacement compatible with the (here internal) constraints. It follows that the virtual work of the sum of all the $\mathbf{F}^* + \mathbf{F}$ is zero as well,

(5) $$\sum_k (\mathbf{F}_k{}^* + \mathbf{F}_k) \cdot \delta \mathbf{s}_k = 0.$$

Recalling now the principle of virtual work, we can express Eq. (5) by saying that *the inertial forces of a system are in equilibrium with the forces applied to the system.* A knowledge of the reactions is not required.

This is *d'Alembert's principle* in its simplest and most natural form. In order to obtain another interesting formulation of the principle, let us look at the quantity

$$\mathbf{F}_k + \mathbf{F}_k{}^* = \mathbf{F}_k - \dot{\mathbf{p}}_k.$$

It is that part of the force $\mathbf{F}_k$ that cannot be converted into motion of the point k. We can call this part the "lost force" and can therefore re-frame (5) by stating that *the lost forces of a system are in equilibrium.*

A formulation of d'Alembert's principle widely used in textbooks is that expressed in Cartesian coordinates. We call the components of $\mathbf{F}_k$, X_k, Y_k, Z_k and those of $\delta \mathbf{s}_k$, $\delta x_k, \delta y_k, \delta z_k$. Furthermore, we stipulate that the masses m_k involved are constant; for a system consisting of n mass points we can then replace (5) by

(6) $$\sum_{k=1}^{n} \{(X_k - m_k \ddot{x}_k)\delta x_k + (Y_k - m_k \ddot{y}_k)\delta y_k + (Z_k - m_k \ddot{z}_k)\delta z_k\} = 0.$$

It is here required that the δx_k, δy_k, δz_k be compatible with the constraints of the system. Let us at once consider the general case of non-holonomic constraints. There relations of type (7.4) exist; if we replace the general coordinates q of (7.4) by Cartesian coordinates, these relations become

(6a) $$\sum_{\mu=1}^{n} \left[F_\mu(x_1 \ldots z_n)\delta x_\mu + G_\mu(x_1 \ldots z_n)\delta y_\mu + H_\mu(x_1 \ldots z_n)\delta z_\mu \right] = 0.$$

If f is the number of degrees of freedom for infinitesimal motion, there must be $3n - f$ such relations for the δx, δy, δz (cf. p. 50). In the case of holonomic constraints the F_μ, G_μ, H_μ are derivatives of one and the same function with respect to x_μ, y_μ, z_μ.

Let the reader be warned emphatically not to look for the true content of d'Alembert's principle in the clumsy formulation (6), (6a). Equation (5) or the statement of equilibrium equivalent to it is not only more readily useful, but also, by virtue of its invariant form, more natural.

§ 11. Application of d'Alembert's Principle to the Simplest Problems

(1) Rotation of a Rigid Body About a Fixed Axis

Here we are dealing with a single degree of freedom, viz., the angle of rotation ϕ. We let $\dot\phi = \omega$ be the angular velocity, $\ddot\phi = \dot\omega$ the angular acceleration. For the present we are not interested in the axle bearings.

We suppose that arbitrary applied forces $\mathbf{F}$ act on the body. According to § 9, Eq. (7), their virtual work is given by the sum of their moments about the axis of rotation, i.e., by

$$(1) \qquad \delta W = \mathbf{L} \cdot \delta \boldsymbol{\phi} = L_a \delta \phi$$

where L_a is the sum of the moments of the $\mathbf{F}$ about the axis of rotation a. We also wish to know the work done by the inertial forces $\mathbf{F}^*$. For this purpose we subdivide the body into mass elements dm. In view of (10.3) the inertial force acting on dm directed normal to the path is the centrifugal force $dm \dfrac{v^2}{r} = dm \omega v$. (In circular motion the radius of curvature ρ is of course equal to the distance r from the rotation axis, the velocity v of each element of mass therefore becomes $r\omega$, and its acceleration $\dot v$ along the path is $r \cdot \dot\omega$). But the centrifugal force does no work. Along the path direction, on the other hand, the inertial force is

$$-dm\,\dot v = -dm\,r\dot\omega.$$

The total virtual work of the inertial forces is therefore

$$(2) \qquad \sum(-dm\,\dot v)\delta s = \sum -dm\,r\dot\omega\,r\,\delta\phi = -\delta\phi\dot\omega\int r^2\,dm = -\delta\phi\dot\omega I,$$

where

$$(3) \qquad I = \int r^2\,dm$$

is the *moment of inertia* of the body. The dimensions of I are ML^2, therefore g cm^2 in the absolute system, g cm sec^2 in the gravitational system.

By virtue of (1) and (2) d'Alembert's principle takes the form

$$\delta\phi(L_a - I\dot\omega) = 0$$

so that we obtain the basic equation of rotational motion

(4) $$I\dot\omega = L_a.$$

Let us compare this equation with the basic equation of translational motion of one degree of freedom, say in the x-direction,

$$m\ddot x = F_x.$$

We see that in rotational motion I takes the place of m.

The same substitution holds in the expression for the kinetic energy. The kinetic energy of rotation of a rigid body is

(5) $$E_{\text{kin}} = T = \int \frac{dm}{2} v^2 = \int \frac{dm}{2} r^2 \omega^2 = \frac{\omega^2}{2} \int r^2\, dm = \frac{\omega^2}{2} I$$

and therefore corresponds exactly to the elementary expression of particle mechanics,

(5a) $$E_{\text{kin}} = T = \frac{\dot x^2}{2} m.$$

In the case of a rigid body with fixed axis, I is time-independent; in mechanisms with flexible joints and in living beings it is, however, variable in a characteristic manner. In § 13 we shall see that all athletic activities, in particular apparatus gymnastics, are based primarily on the ability of the human body to change its moment of inertia.

An investigation of the manner in which the moment of inertia of a rigid body depends on the position of the axis of rotation will be deferred to § 22.

Finally we shall turn to the connection of the kinetic energy with the basic equation of motion. Just as, in the case of constant mass, we can obtain the equation of motion $m\ddot x = F_x$ from the law of kinetic energy in particle mechanics, i.e.,

$$\frac{dT}{dt} = \frac{dW}{dt} \text{ with } dW = F_x\, dx,$$

we obtain, in the case of constant I, the equation of motion (4) for rotation. We need merely make use of (5) in

$$\frac{dT}{dt} = \frac{dW}{dt} \text{ with } dW = L_a\, d\phi \ [\text{Eq. (9.7)}].$$

The moment of inertia occurs also in the expression for the *moment of momentum* or *angular momentum* of the rotating body. If we let M be the angular momentum of the body, we evidently have

(6) $$M = \sum dm\, vr = \omega \sum dm\, r^2 = \omega I.$$

(2) Coupling of Rotational and Translational Motion

Think of the coal basket in a mine, or of an elevator. The cable carrying the elevator is wound around a drum and driven by a force P. Let r be the drum radius. The two virtual displacements that take place (cf. Fig. 13) are related by

(7) $\delta z = r\,\delta\phi$.

d'Alembert's principle requires

(7a) $(-Q - M\ddot{z})\delta z + (rP - I\dot{\omega})\delta\phi = 0$.

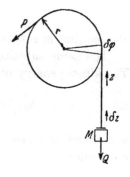

It is convenient to " reduce " the mass of the drum, so to speak, to the periphery of the drum, i.e., to replace I by a " reduced mass " defined by

(8) $I = M_{\text{red}}r^2$.

By virtue of (7), Eq. (7a) can then be rewritten in the form

$$(P - Q - M\ddot{z} - M_{\text{red}}r\dot{\omega})\delta z = 0.$$

FIG. 13. Coupling of translational and rotational motion (elevator, coal basket).

Since $r\omega = \dot{z}$, $r\dot{\omega} = \ddot{z}$, we then obtain the equation of motion

(9) $(M + M_{\text{red}})\ddot{z} = P - Q$.

The inertia of the drum therefore adds a term M_{red} to the mass of the elevator.

(3) Sphere Rolling on Inclined Plane

Here again we are dealing with the coupling of translation (motion down the incline) and rotation (about an axis through the center of the sphere perpendicular to the plane of the paper in Fig. 14). The component of gravity effective in this case is $P = Mg \sin \alpha$; the static friction F indicated on the diagram does not enter d'Alembert's principle, since it acts at the point of contact which is instantaneously at rest. The condition for pure rolling motion is

(10) $\dot{z} = r\omega$, or, written for virtual motion, $\delta z = r\delta\phi$.

With d'Alembert we now require that

(11) $\delta z(Mg \sin \alpha - M\ddot{z}) + \delta\phi(-I\dot{\omega}) = 0$.

The calculation of I is a problem of integral calculus. We shall state without proof that the moment of inertia of a homogeneous ellipsoid of semi-axes a, b, c about axis c (and correspondingly about a and b) is

(12) $$I_c = \frac{M}{5}(a^2 + b^2).$$

As a special case we obtain for the moment of inertia of a sphere

(12a) $$I = \frac{2}{5}Mr^2.$$

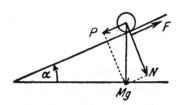

As in (8) we introduce a mass reduced to the distance r, which by virtue of (12a) becomes

FIG. 14. Sphere on inclined plane. The static friction F causes pure rolling, but does not enter d'Alembert's principle.

(12b) $$M_{\text{red}} = \frac{2}{5}M.$$

If we substitute this in (11) and also take (10) into account, we easily obtain

(13) $$\ddot{z} = \frac{5}{7}g \sin \alpha.$$

The factor $\frac{5}{7}$ shows how the " fall " on an inclined plane is delayed by the angular acceleration of the sphere and the increased inertia due to it.

Whereas from (3.13) the final velocity in a free fall was found to be

$$v = (2gh)^{\frac{1}{2}}, \quad h = \text{height of fall},$$

equation (13) now gives the final velocity

$$v = (2 \cdot \frac{5}{7}gh)^{\frac{1}{2}}.$$

The difference is due to the fact that now the gravitational potential energy is converted not only into kinetic energy of descent, but also into rotational energy of the rolling sphere.

(4) Mass Guided Along Prescribed Trajectory

If we assume the guide ways to be frictionless, d'Alembert's principle applied to the one degree of freedom here present (displacement along the guide) simply says that

$$\delta s (F_s{}^* + F_s) = 0,$$

i.e., according to (5.8),

(14) $$m\dot{v}_s = m|\dot{v}| = F_s$$

with arbitrary direction of the applied force **F**. The component F_n of **F** normal to the guide, which we may take positive in the centripetal direction, must then add to the reaction R_n (counted positive in the same direction) to give the equilibrant of the centrifugal force C; i.e.,

$$(15) \qquad R_n + F_n = C = m \frac{v^2}{\rho}.$$

In general, especially if the guiding action is achieved by a material device such as a rail, we are compelled to take into account also a tangential component R_s of the reaction, the *friction*. If we count the friction positive in the negative direction of δs, Eq. (14) is therefore enlarged to

$$(16) \qquad m\dot{v} = F_s - R_s.$$

Whereas R_n is determined by Eq. (15), R_s in (16), on the other hand, remains " statically and dynamically indeterminate " and can be determined only from experiment. In § 14 we shall discuss how such experiments are carried out.

§ 12. Lagrange's Equations of the First Kind

Let us consider a system of discrete mass points $m_1, m_2, \ldots m_n$, connected with each other by r holonomic conditions

$$(1) \qquad F_1 = 0, F_2 = 0, \ldots F_r = 0.$$

The number of degrees of freedom is then $f = 3n - r$. We operate in Cartesian coordinates and make use of the formulation (10.6) of d'Alembert's principle. In order to write the clumsy sums occurring there in a more convenient way, we number the coordinates $x_1, y_1, z_1, \ldots, x_n, y_n, z_n$ consecutively as

$$x_1, x_2, x_3, x_4, \ldots x_{3n-1}, x_{3n},$$

and likewise the components of force X, Y, Z. The mass belonging to x_k, X_k will be denoted by m_k; evidently the m_k will be equal in groups of three. Eq. (10.6) now becomes

$$(2) \qquad \sum_{k=1}^{3n} (X_k - m_k \ddot{x}_k)\delta x_k = 0.$$

By virtue of the r conditions of constraint (1), the δx_k are subject to the restrictions

$$(3) \qquad \delta F_i = 0, \; i = 1, 2, \ldots r,$$

which can also be written

$$(4) \qquad \sum_{k=1}^{3n} \frac{\partial F_i}{\partial x_k}\delta x_k = 0, \; i = 1, 2, \ldots r.$$

Let us multiply each of the δF_i by an arbitrary numerical factor λ_i (Langrange multiplier) and add it to the d'Alembert equation (2), giving

$$(5) \qquad \sum_{k=1}^{3n} \left(X_k - m_k \ddot{x}_k + \sum_{i=1}^{r} \lambda_i \frac{\partial F_i}{\partial x_k} \right) \delta x_k = 0.$$

Only f of the $3n$ displacements δx are independent of each other. The remaining r are functions of these independent ones. Let these r displacements be given by the quantities $\delta x_1, \delta x_2, \ldots \delta x_r$. Now we have precisely r quantities $\lambda_1, \lambda_2, \ldots \lambda_r$ over which we can dispose freely. We choose them so as to make

$$(6) \qquad X_k - m_k \ddot{x}_k + \sum_{i=1}^{r} \lambda_i \frac{\partial F_i}{\partial x_k} = 0; \quad k = 1, 2, \ldots r.$$

Eq. (5), with the numbers λ_i now determined, reduces to

$$(7) \qquad \sum_{k=r+1}^{3n} \left(X_k - m_k \ddot{x}_k + \sum_{i=1}^{r} \lambda_i \frac{\partial F_i}{\partial x_k} \right) \delta x_k = 0$$

where the δx_k are completely independent, there being indeed $f = 3n - r$ of these. If, for example, we choose

$$(8) \qquad \delta x_{r+v} \neq 0; \; \delta x_{r+1} = \delta x_{r+2} = \cdots = \delta x_{r+v-1} = \delta x_{r+v+1} = \cdots = \delta x_{3n} = 0,$$

we see that the factor of δx_{r+v} must vanish. Letting v run through all values $1, 2, \ldots f$, we conclude that all expressions in parentheses have to be $= 0$,

$$X_k - m_k \ddot{x}_k + \sum_{i=1}^{r} \lambda_i \frac{\partial F_i}{\partial x_k} = 0; \quad k = r+1, r+2, \ldots 3n.$$

Together with the Eqs. (6) these form $3n$ differential equations

$$(9) \qquad m_k \ddot{x}_k = X_k + \sum_{i=1}^{i=r} \lambda_i \frac{\partial F_i}{\partial x_k}; \quad k = 1, 2, \ldots 3n,$$

which are called the *Lagrange equations of the first kind*. Of course the m_k are equal in groups of three; thus $m_1 = m_2 = m_3$, since we are dealing with the same mass point m_1 having the three coordinates $x_1 = x_1, x_2 = y_1, x_3 = z_1$.

So far we have assumed that the conditions (1) are holonomic; we can easily convince ourselves that all of the preceding can be carried over to the case of non-holonomic constraints with only slight modification. The only difference is that the factors $\dfrac{\partial F_i}{\partial x_k}$ in (4) must be replaced by general functions of the coordinates, F_{ik}, which cannot be written in the form of partial derivatives of a function. If we make this replacement in Eqs. (9), we at once obtain Lagrange's equations of the first kind for non-holonomic systems,

(9a)
$$m_k \ddot{x}_k = X_k + \sum_{i=1}^{i=r} \lambda_i F_{ik}.$$

Let us make a more interesting generalization by assuming that the conditions (1) vary with time. Then the F_i depend explicitly not only on the x_k but also on t. We must now demand that in forming (4) the time be held constant, a stipulation which is not only permissible but also plausible since our virtual displacement has nothing to do with the passage of time. The derivation of (9) is not affected by this requirement. But we obtain an important consequence regarding the form of the equation of energy.

If we want to derive this equation in the case of time-independent constraints, we proceed as follows: we multiply (9) by dx_k and sum over k. On the left we obtain

(9b)
$$dt \sum m_k \dot{x}_k \ddot{x}_k = dt \frac{d}{dt} \sum \frac{m_k}{2} \dot{x}_k^2 = dt \frac{dT}{dt} = dT.$$

The first term of the right member gives the work done by the applied forces in time dt,

(9c)
$$\sum dx_k X_k = dW.$$

The second term on the right vanishes. For

(9d)
$$\sum_{i=1}^{r} \lambda_i \sum_{k=1}^{3n} \frac{\partial F}{\partial x_k} dx_k = \sum_{i=1}^{r} \lambda_i dF_i = 0$$

by virtue of the fact that the F_i depend only on the x_k, so that $F_i = 0$ implies

(9e)
$$dF_i = \sum \frac{\partial F_i}{\partial x_k} dx_k = 0.$$

From (9b, c) we then have

(10)
$$dT = dW.$$

This is no longer so if the F_i also depend on t. Then the zero in (9d, e) is to be replaced by

$$- \sum_{i=1}^{r} \lambda_i \frac{\partial F_i}{\partial t} dt \quad \text{and} \quad - \frac{\partial F_i}{\partial t} dt$$

respectively. For time-dependent constraints the equation of energy is then

(10a)
$$dT = dW - dt \sum_{i=1}^{r} \lambda_i \frac{\partial F_i}{\partial t}.$$

This means that *time-dependent constraints do work on the system.*

To make this principle more concrete, let us think of a tennis racquet. If the racquet is kept fixed, it reflects the ball with unchanged energy. If instead it yields backward or swings toward the ball, it takes energy away from or gives energy to the ball.

In non-holonomic systems an explicit dependence on t of the F_{ik} occurring in (9a) would be compatible with an equation of energy of the form (10). If, however, the non-holonomic conditions had the form

$$\sum F_{ik} dx_k + G_i dt = 0$$

instead of (7.4), it would be necessary to add members in G_i to (10) which would then take on a form analogous to (10a), i.e.,

(10b)
$$dT = dW - dt \sum_{i=1}^{r} \lambda_i G_i.$$

We shall learn from the example of the spherical pendulum in the following chapter that the λ_i can be regarded as the reactions of the system against the constraint exerted by the holonomic or non-holonomic conditions. There we shall also see that the determination of the λ cannot be effected by means of r Lagrange equations arbitrarily singled out, even though this was a permissible assumption for purposes of our derivation. Instead the λ must be determined from all $3n$ of the Lagrange equations taken together. It should be emphasized that the method of Lagrange multipliers plays an important role not only in the Lagrange equations of the first kind, but also (cf. Ch. VI, § 34) in types of equations of a much more general nature. Apart from their use in mechanics, the Lagrange multipliers are encountered in the elementary theory of maxima and minima.

§ 13. Equations of Momentum and of Angular Momentum

We derive these equations for a system of discrete mass points which can be translated and rotated as a whole in space. Through a limiting process they can, however, be applied equally well to a freely moving rigid body or to an arbitrary mechanical system whose motion is not restricted by external constraints.

We divide the forces acting into *external* and *internal forces*. This classification says nothing about the origin of the forces and is therefore by no means identical with the classification of p. 53 into applied forces and forces of reaction. Our present distinction is strictly based on the criterion of whether the law of action and reaction is or is not satisfied within the system itself. In the first case we speak of internal forces, in the second of external forces. The internal forces of the solar system, for instance, are applied forces because they are gravitational, whereas the external force which drives a railroad train forward is a force of reaction (as we shall see on p. 84), viz., the static friction at the rolling wheels.

We call $\mathbf{F}_k$ the external force acting at the point k; the internal forces will be called $\mathbf{F}_{ik}$ to remind us that they act between two points contained in the system and therefore within the system satisfy Newton's third law,

$$(1) \qquad \mathbf{F}_{ik} = -\mathbf{F}_{ki}.$$

(1) Equation of Momentum

Let us now make use of d'Alembert's principle in the form (10.5). We replace $\mathbf{F}_k$ by $\mathbf{F}_k + \sum_i \mathbf{F}_{ik}$, $\mathbf{F}_k^*$ by $-\dot{\mathbf{p}}_k$ in agreement with definition, and make all the $\delta\mathbf{s}_k$ equal to each other. We therefore impart the same virtual displacement to all the mass points of the system. The $\mathbf{F}_{ik}$ drop out because of (1) once we sum over i and k, and we are left with

$$(2) \qquad \delta\mathbf{s} \cdot \left(\sum_k \mathbf{F}_k - \sum_k \dot{\mathbf{p}}_k \right) = 0.$$

Let us indicate the summation over k by means of a bar. From (2) we conclude that

$$(3) \qquad \dot{\bar{\mathbf{p}}} = \bar{\mathbf{F}}.$$

$\bar{\mathbf{p}}$ is the total momentum of the system, equal to the vector sum of the individual momenta. We define the center of mass velocity $\mathbf{V}$ by

$$M\mathbf{V} = \overline{m\mathbf{v}} = \bar{\mathbf{p}}, \quad M = \overline{m}$$

and have, in lieu of (3),

$$(3a) \qquad M\dot{\mathbf{V}} = \bar{\mathbf{F}}.$$

We now choose an arbitrary but fixed point of reference O. We measure the distance $\mathbf{r}_k$ of the points of the system from O and define the position $\mathbf{R}$ of the center of mass with respect to O by the equation

$$(3b) \qquad M\mathbf{R} = \overline{m\mathbf{r}}.$$

The content of equations (3a, b) can be summed thus: *the center of mass of a freely moving mechanical system moves like a single mass point, having a mass M equal to the total mass of the system, and acted on by the resultant* $\bar{\mathbf{F}}$ *of all the external forces acting on the system.*

(2) Equation of Angular Momentum

Suppose we impart to the system a virtual rotation $\delta\phi$ about an arbitrary axis passing through a point O. The displacements $\delta\mathbf{s}_k$ of the various points m_k of the system are then unequal; for

(4) $$\delta\mathbf{s}_k = \delta\phi \times \mathbf{r}_k.$$

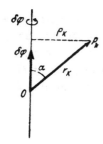

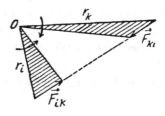

FIG. 15. The virtual displacement δs resulting from a virtual rotation $\delta\phi$.

FIG. 16. The moments of internal forces cancel in pairs.

To prove this, let us look at Fig. 15. $\delta\phi$ is there drawn as a vector along the axis of rotation and at the same time as a curved arrow about this axis in agreement with the rule of the right-handed screw. By virtue of the definition of the vector product the magnitude δs_k of $\delta\mathbf{s}_k$ is

$$\delta s_k = \delta\phi \, |\mathbf{r}_k| \sin\alpha = \delta\phi\rho_k,$$

as must be the case for the rotation in question. The direction and sense of $\delta\mathbf{s}_k$ are likewise correctly given by (4). $\delta\mathbf{s}_k$ is directed normal to the plane of the drawing, into the paper.

We introduce (4) in (10.5), while replacing $\mathbf{F}^*$ and $\mathbf{F}$ as in subsec. 1, and immediately obtain

(5) $$\sum_k \left\{ \left(\mathbf{F}_k + \sum_i \mathbf{F}_{ik} - \dot{\mathbf{p}}_k\right) \cdot (\delta\phi \times \mathbf{r}_k) \right\} = 0.$$

Next we use a rule of elementary vector algebra,

(6) $$\mathbf{A} \cdot \mathbf{B} \times \mathbf{C} = \mathbf{B} \cdot \mathbf{C} \times \mathbf{A} = \mathbf{C} \cdot \mathbf{A} \times \mathbf{B}$$

which says that the parallelopiped formed by any three vectors $\mathbf{A}$, $\mathbf{B}$, $\mathbf{C}$ has a volume which is independent of cyclic permutation of the labels of its three edges.

Instead of (5) we can therefore write

(7) $$\delta\phi \cdot \left\{ \sum_{k}(\mathbf{r}_k \times \mathbf{F}_k) + \sum_{i}\sum_{k}(\mathbf{r}_k \times \mathbf{F}_{ik}) - \sum_{k}(\mathbf{r}_k \times \dot{\mathbf{p}}_k) \right\} = 0.$$

In this fashion the connection between $\delta\phi$ and $\mathbf{r}$ is severed, so that, with $\delta\phi$ arbitrary, the factor in brackets $\{\ \}$ must itself vanish. In order to write this factor more simply we introduce the following notation:

(7a) $$\mathbf{L}_k = \mathbf{r}_k \times \mathbf{F}_k \text{ as in (5.12), } \overline{\mathbf{L}} = \sum \mathbf{L}_k;$$

(7b) $$\mathbf{M}_k = \mathbf{r}_k \times \mathbf{p}_k, \quad \mathbf{r}_k \times \dot{\mathbf{p}}_k = \frac{d}{dt}(\mathbf{r}_k \times \mathbf{p}_k) = \dot{\mathbf{M}}_k \text{ as in (5.14);}$$

(7c) $$\overline{\mathbf{M}} = \sum \mathbf{M}_k, \quad \dot{\overline{\mathbf{M}}} = \sum \dot{\mathbf{M}}_k.$$

L is therefore the vector sum of all moments of the external forces about the common point of reference O, $\overline{\mathbf{M}}$ is the vector sum of the angular momenta of all the mass points of the system about the same point of reference, or, more briefly, the total angular momentum of the system about O.

Moreover we show with the help of Fig. 16 that in the double sum of Eq. (7) all the terms cancel in pairs, viz., that

(8) $$\mathbf{r}_k \times \mathbf{F}_{ik} + \mathbf{r}_i \times \mathbf{F}_{ki} = 0.$$

We see that in this expression the Third Law, Eq. (1), acts essentially as the definition of internal force.

From (8) it follows that the double sum in (7) vanishes. Recalling (7a, b, c) we therefore conclude from (7) that

(9) $$\dot{\overline{\mathbf{M}}} = \overline{\mathbf{L}}.$$

This equation is the exact counterpart of Eq. (3). It states that *the time rate of change of the total angular momentum of the system is equal to the resultant moment of the external forces*, just as Eq. (3) stated that *the time rate of change of the total momentum of the system is equal to the resultant of all the external forces*.

These two laws will be called the equations (or principles) of angular momentum and (linear) momentum respectively.

Formerly it was the custom in the German literature to call basic equation (9) the principle of areas (Flächensatz). This name had its origin in the Kepler problem. There we found that in the case of *one* planet the areal velocity was proportional to the angular momentum, and the direction of the angular momentum was normal to the orbital plane of the planet. This is no longer the case for the planetary many-body problem, where we have instead

(10) $$\overline{\mathbf{M}} = \sum 2m_k \frac{d\mathbf{A}_k}{dt},$$

so that not only the different planetary masses occur as factors, but the individual areal velocities corresponding to the planets must be added vectorially. The areal velocity thus arising for a complete planetary system is defined, as is well known, by the *invariable plane* (plane normal to $\mathbf{\overline{M}}$). It is invariable because in a planetary system external forces are absent, so that $\mathbf{\overline{L}} = 0$ and, according to (9),

(10a) $$\mathbf{\overline{M}} = \text{const.}$$

In general for $\mathbf{\overline{L}} = 0$ we obtain the special *principle of the conservation of angular momentum*. The notion of areal velocity is even more difficult to visualize, hence less useful, for a system of infinitely many particles such as a rigid body, so that the term " Flächensatz " should be abandoned for general use.

(3) Proof Using the Coordinate Method

We shall now sketch the proof of our principles by an alternate method, that of decomposition into Cartesian coordinates, because the use of these coordinates is so widespread and has been so greatly favored by older texts that we wish to defer to usage in some measure.

We begin with the equations

(11) $$m_k \ddot{x}_k = X_k + \sum_i X_{ik}$$
$$m_k \ddot{y}_k = Y_k + \sum_i Y_{ik}$$

which are written in easily understandable form. Summation of the first of these equations over k, with $X_{ik} = -X_{ki}$, at once yields the x-component of the equation of momentum,

(12) $$\frac{d^2}{dt^2} \sum_k m_k x_k = \sum_k X_k.$$

Multiplication of the first equation by $-y_k$, the second by x_k, yields as their sum

(13) $$\sum_k m_k (x_k \ddot{y}_k - y_k \ddot{x}_k) = \sum_k (x_k Y_k - y_k X_k) + \cdots$$

We group together in pairs ik and ki the terms ... not written down, thereby bringing out the direction of the internal forces, $i \rightarrow k$ and $k \rightarrow i$. We then obtain

$$x_k Y_{ki} - y_k X_{ki} + x_i Y_{ik} - y_i X_{ik}$$
$$= \frac{|F_{ik}|}{r_{ik}} \left[x_k (y_i - y_k) - y_k (x_i - x_k) + x_i (y_k - y_i) - y_i (x_k - x_i) \right].$$

Simplification shows this to equal zero, in agreement with Fig. 16. With the help of (5.17a) the right member of (13) reduces to

$$\sum_k L_{kz} = \bar{L}_z.$$

The left member of (13) is, in view of (5.14b),

(13a) $$\frac{d}{dt}\sum_k m_k(x_k\dot{y}_k - y_k\dot{x}_k) = \sum_k \dot{M}_{kz} = \bar{\dot{M}}_z.$$

Equation (13) is then identical with the z-component of our equation of angular momentum (9).

(4) Examples

There exists a profound difference between the principles of linear and of angular momentum which we shall explain with the help of the special case in which no external forces act on the system.

According to Eq. (3a) in this case the velocity of the mass center remains constant; for the total mass M occurring as factor is constant, even for a system with internal motion. If, then, the mass center is initially at rest, it remains at rest. *Internal forces* are *unable* to impart motion to the center of mass, even in a mechanism with flexible joints or in a living body. In order to move one's center of mass, one must be able to push against a support; therefore an *external force* is necessary.

It is evident that in the absence of external forces $\mathbf{L}=0$, so that (9) yields

(14) $$\mathbf{\bar{M}} = \text{const.}$$

If the moment of momentum is initially zero, it remains zero, even for a system with internal motion. From this it does not follow, however, that the angular position of the system is conserved permanently. Rather, this angular position can be varied *ad libitum* with the help of internal forces alone, and without a push against some outside object.

An example of this is the cat, which always manages to fall on its feet. It achieves this by suitable rotation of the anterior extremities coupled with opposite rotation of the posterior ones. This action is illustrated by the rapid exposure photographs published in the "Comptes Rendus of the Paris Academy," 1894, p. 714.

The essential points of this process can conveniently be followed by means of an experiment with a turning stool. Such a stool consists of a horizontal disc which revolves with as little friction as possible about a vertical axis. The victim of the experiment is seated on the disc, initially at rest:

$$\mathbf{M}_0 = 0.$$

He lifts his right arm forward and describes with it a backward rotation. The " area swept out " in this process must be compensated by a counter-rotation of the remainder of the body including the disc of the stool. More precisely, the moment of momentum M_1 of the moving arm induces a moment of momentum of torso and disc M_2 such that

$$\mathbf{M}_2 = -\mathbf{M}_1.$$

The experimental subject now lowers his arm; this causes no change in $\mathbf{M}$. Now the initial position of the body is restored, and the process can be repeated. With each repetition the same counter-rotation $\mathbf{M}_2$ takes place. After n repetitions the subject notices that he is facing in a direction opposite to the initial one. In contrast to the position of the center of mass, the angular position is not fixed by the initial state of rest.

One can strengthen the effect by making the subject hold a heavy weight in the right hand. The " area swept out " is thereby, so to speak, multiplied, so that the counter-rotation is also visibly increased.

Let us perform two more experiments: the subject stands on the stool with lowered arms and is given an angular momentum M_0; he now raises his arms (with weights in his hands if desired) sideways; the rotation suddenly decreases. Instead, we can set the person spinning with out-stretched arms; he next lowers his arms and usually falls off the stool because the rotation, especially when weights are used, is suddenly increased considerably.

In both foregoing cases

$$\mathbf{M}_0 = \mathbf{M}_1 \text{ and therefore } I_0 \omega_0 = I_1 \omega_1 \text{ from Eq. (11.6).}$$

In the first case, however, we have

$$I_0 \ll I_1 \text{ and hence } \omega_1 \ll \omega_0,$$

whereas in the second case

$$I_0 \gg I_1 \text{ so that } \omega_1 \gg \omega_0.$$

The changeability of the moment of inertia under conservation of angular momentum is used extensively in all athletic feats, especially in exercises on the horizontal bar. Consider, for example, the " forward upswing." In the initial act of acquiring swing the body is stretched, its moment of inertia great, and its angular velocity about the bar moderate. As he swings forward, shortly before reaching the highest point, the performer pulls in his legs, reduces his moment of inertia about the bar and his angular velocity becomes high. His mass center swings over the bar and the performer achieves an upright position on the bar. Notice that the reactions produced by the grasp of the hands on the bar do not influence the angular momentum to any noticeable degree since the bar is so thin that the forces of reaction have a vanishingly small lever arm.

The same principles are used in the " circles," (backward hip circle, knee circle, etc.). Gymnastics, ice skating and skiing are, in a way, practical lessons in experimental and theoretical mechanics.

(5) Mass Balancing of Marine Engines

Let us finally consider an illustration on a large scale, the Schlick method for balancing the reciprocating masses of marine engines.

In the transition period leading to the modern express steamers, toward the end of the last century, the shipbuilding industry went through a crisis. For technical reasons the speed of revolution of the propellor shaft is fixed at approximately 100 per min. The inertial effects of the piston engines, which have to be absorbed by the ship's body, change in this same rhythm. As the length of ships was increased more and more, the " proper frequency " of the vessel was continually depressed, so that this frequency came dangerously near to the rhythm of the inertial effects. Let us anticipate by using the word " resonance," a phenomenon with which we shall deal at great length in the next chapter. The word originated in acoustics, where resonance phenomena are most immediate and where they were studied first.

For lack of space the steam cylinders of fast steamers have to be arranged vertically. Let us assume, to make things specific, that we are dealing with four pistons (cf. Fig. 17), which are all connected to the same crank shaft oriented lengthwise, along the z-direction in our diagram. We shall see that for a smaller number of pistons a mass balance even to first order (to which we shall restrict ourselves here) is impossible. With the choice of coordinates of Fig. 17, the inertial forces are directed along the x-axis; they give rise to moments only about the y-axis. The inertial effects must be absorbed by the reactions of the body of the ship, in which they induce rhythmic countervibrations.

This is beautifully illustrated by the models which Consul Otto Schlick donated to the German Museum in Munich at the time of his invention. The ship's hull is here idealized as an elongated beam; it is suspended by spiral springs which represent the buoyancy of the water and enable the ship to oscillate. When the engine models carried by the beam are set in motion, the beam starts oscillating with slight amplitude. If the speed of revolution of the engines is increased, the vibrations of the beam grow larger the more the rotation frequency approaches the fundamental proper frequency of the beam (cf. Fig. 18). Great amplitudes of oscillation would have disastrous effects on the safety of the ship—and also on the wellbeing of the passengers. The idea of mass balancing is to bring about a cancellation of the inertial forces and torques of the reciprocating masses of the marine engine in order to protect the ship's body from their harmful effects.

If we pass at once from accelerations to position coordinates, the balancing of the inertial forces, which are all in the x-direction, demands that

(15) $$\sum M_k x_k = 0.$$

The masses M_k include not only those of the pistons and piston rods, but to first approximation also those of the connecting rods and portions of the eccentric parts of the crank shaft.

Just as important is the balancing of the moments of the inertial forces. It is mentioned above and made plausible by Fig. 17 that only the moments about the y-axis play any role here. Again we immediately pass from the accelerations to the position coordinates, which is permissible since the lever arms, i.e., the a of our Fig. 17, are constant. We then require

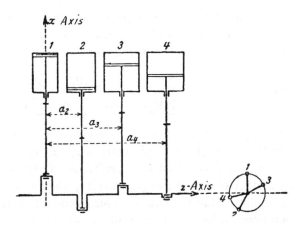

FIG. 17.   Schlick mass balance of a vertically arranged four-cylinder piston engine. Diagram at lower right shows the position of the four crank pins relative to each other.

(16) $$\sum M_k a_k x_k = 0.$$

We now express the piston coordinates x_k in terms of the crank pin

FIG. 18.   Proper frequency of a freely vibrating beam as a model for the fundamental frequency of a ship.

coordinates ϕ_k. From Fig. 9 and Eq. (9.6) we have, to a *first approximation*,

(17) $$x_k + r_k \cos \phi_k = \text{const.}$$

First approximation[4] here means that we pass to the limit of an infinitely long connecting rod, or $r/l \longrightarrow 0$. We shall not go into the calculation to second order where the first power of r/l is retained, as in Eqs. (9.5) and (9.6). Since all the pistons work on the same shaft, the ϕ_k are equal to each other apart from a phase shift α_k constant in time;

(18) $$\phi_k = \phi_1 + \alpha_k,$$

where $\alpha_1 = 0$ and α_2, α_3, α_4 can be chosen at will. By virtue of (17) and (18), the variable part of the conditions (15) and (16), which alone concerns us, gives

(19) $$\sum M_k r_k \cos(\phi_1 + \alpha_k) = 0, \qquad \sum M_k r_k a_k \cos(\phi_1 + \alpha_k) = 0.$$

If we expand the trigonometric functions, we see that with ϕ_1 arbitrary the factors of $\cos \phi_1$ and $\sin \phi_1$ must vanish separately. We then obtain four equations between the parameters a_k and α_k.

(20)
$$\sum M_k r_k \cos \alpha_k = 0, \qquad \sum M_k r_k \sin \alpha_k = 0,$$
$$\sum M_k r_k a_k \cos \alpha_k = 0, \qquad \sum M_k r_k a_k \sin \alpha_k = 0.$$

The M_k and r_k are fixed by construction. The quantities at our disposal are the three phase displacements α_2, α_3, α_4, and the two lever arm ratios $a_2 : a_3 : a_4$ [the absolute magnitudes of the a do not enter in Eq. (20)], altogether therefore five parameters; they allow a certain freedom of choice in fulfilling conditions (20). This freedom in turn makes it possible to avoid solutions which are technically objectionable. The preceding shows that the mass balancing can be carried through to first order in four-cylinder engines; it also shows that for lack of enough parameters it cannot be effected in engines with a smaller number of cylinders, as asserted above. The external characteristic of the Schlick mass balancing method is that the pistons of a four-cylinder engine are not equidistant and that their crank pins are not arranged at equal angles to each other. The latter feature is illustrated in the lower right-hand corner of Fig. 17.

The Schlick method proved its worth in the first modern steamers of the Hamburg-America Line; it eliminated the danger of resonance. It is true, however, that it had only a transient importance in the practices of ship-building, since piston engines were soon to be displaced by turbines, where there are no reciprocating masses. Even nowadays, however, mass balancing is important in automobile and airplane engines as well as in the Diesel engines of submarines.

---

[4] This first approximation defines the mass balancing to first order (i.e., the " balancing for primary forces and primary couples," as it is called). Since we want to restrict ourselves to the latter, we need not carry out the second approximation.

(6) General Rule on the Number of Integrations Feasible in a Closed System

A mechanical system is called closed if no external forces, but only internal ones, act on it.[5] In that case the equations of linear and angular momentum become principles of conservation. The conservation of momentum introduces $2 \cdot 3$ constants, that of angular momentum 3 constants of integration.[6] The equation of energy yields one additional constant. We therefore have a total number of

$$(21) \qquad 2 \cdot 3 + 3 + 1 = 10$$

integrals of the equations of motion.

So much for the three-dimensional case. In the case of two dimensions, such as the two-body problem of astronomy, we have only one component of angular momentum (directed perpendicular to the plane containing the trajectories of the two bodies), so that we obtain, together with the integral of energy,

$$(22) \qquad 2 \cdot 2 + 1 + 1 = 6$$

generally feasible integrals.

In the one-dimensional case this number evidently reduces to

$$(23) \qquad 2 \cdot 1 + 0 + 1 = 3.$$

The general expression for n dimensions is

$$(24) \qquad n + 1 + \tfrac{1}{2}n(n+1).$$

The best method of clarifying this expression is to appeal to the concepts of relativity: we put $n=3$ and add the time as the fourth coordinate. We must then form the four-vector momentum which is obtained from Eq. (2.19) by summing over all the particles of the system. The basic equations of relativistic mechanics now tell us that for a closed system this four-vector remains constant; incidentally its time component is, apart from a factor $-ic$ and an additive constant, equal to the kinetic energy. The four integrals thus obtained (conservation of momentum and energy) are represented in (24) by the term $n+1$. The second term of the expression is the result of the combination of two axes at a time in the formation of moments. Evidently the combination of two space axes yields the equations of angular momentum in the ordinary sense. The combination of the time axis with one of the space axes, on the other hand, gives the second

---

[5] Every system becomes closed, of course, if one makes it large enough, i.e., if one includes the sources of the external forces in the system.

[6] The 2·3 constants arising from the equation of the straight line described by the center of mass, and the three areal velocity constants.

integrals of motion of the mass center which express the rectilinearity of this motion. For according to (2.19), if we indicate summation over all mass points by a bar as on p. 70 and replace $(1-\beta^2)^{\frac{1}{2}}$ by unity from the start, we calculate

$$x_k p_4 - x_4 p_k = ic\,(\overline{m_k x_k} - t\,\overline{m_k \dot{x}_k}), \quad k = 1, 2, 3.$$

From the principle of conservation of angular momenta this quantity must be equal to a constant, which we may call $ic A_k$. In three-dimensional vector notation and with the symbols of (3a, b) we then have

(25) $$\mathbf{R} - t\mathbf{V} = \mathbf{A}.$$

With $\mathbf{A}$ and $\mathbf{V}$ constant this means that indeed the mass center moves in a straight line with constant speed. The foregoing should be sufficient explanation for the origin of (24); the use of the four-dimensional space-time symmetry has lent additional clarity to it.

We wish finally to make a remark concerning the enumeration of (21) and (22) pertaining to the field of astronomy. The famous three-body problem would need for its complete integration, i.e., for a determination of its $3 \cdot 3$ coordinates and $3 \cdot 3$ components of velocity,

(26) $$2 \cdot 3 \cdot 3 = 18$$

first integrals. Each of these, as exemplified by Eq. (25), would give one relation between the position and velocity coordinates involving one constant of integration. But a comparison of (26) with (21) shows that we are lacking eight integrals for the complete integration; above and beyond this the unrelenting efforts of the greatest mathematicians from Lagrange to Poincaré have shown that the missing integrals cannot be obtained in algebraic form; a conclusive proof of this was given by H. Bruns.

A similar enumeration for the two-body problem, plane by its very nature, requires only

$$2 \cdot 2 \cdot 2 = 8$$

instead of $2 \cdot 3 \cdot 3 = 18$ constants of integration for its complete integration. Thus only two constants are required beyond those which according to (22) are in all cases available for a two-dimensional problem. As a matter of fact these two integrals with their corresponding arbitrary constants can be found here, as shown by the transition from Eqs. (6.4) to (6.5). Hence the two-body problem can be solved exactly; the three-body problem is in general insoluble, i.e., it can be solved only by analytical approximation methods. It is only under very special assumptions about the type of motion that we shall be able to find a solution in closed form for the latter problem in § 32.

§ 14. The Laws of Friction

As already emphasized in § 11, subsec. 4, the guiding of a mass on a prescribed path introduces a component of reaction along the path direction which cannot be obtained from general principles of mechanics, but must be determined experimentally. Apart from some preliminary work of other investigators this determination was carried out for the first time in 1785 in the famous, and for those times very accurate, experiments of Ch. A. Coulomb, whose name, we recall, is permanently linked with the basic laws of electrostatics and magnetostatics.

With Coulomb we distinguish
    (a) Static friction
    (b) Kinetic or sliding friction.

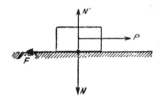

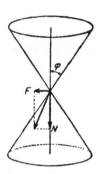

Fig. 19. Static friction on plane support.

Fig. 20. Construction of the angle of friction and the cone of friction.

(1) Static Friction

Consider a body resting on a horizontal support. If we exert a gradually increasing pull P on the body parallel to the support, no motion will occur at first. We must therefore assume that a force of friction F balances the pull P. If, however, P exceeds a very definite limit, acceleration takes place.

This limit F_{max} is, according to Coulomb (and his predecessors), proportional to the normal pressure N, which in the case of rest on a horizontal support is simply equal to the weight G of the body. We have

$$(1) \qquad\qquad F_{max} = \mu_0 N.$$

μ_0 is the *coefficient of static friction*; it depends on the nature and the state of the surfaces of the two materials in contact. If the two materials are the same, μ_0 is especially great (interpenetration).

By means of

$$(2) \qquad\qquad \mu_0 = \tan \phi$$

one can introduce an angle ϕ which can be thought of as the vertex angle of a " cone of friction." As long as the resultant of the two forces F and N

falls inside this cone, no motion takes place, cf. Fig. 20. Motion occurs
when their resultant lies in the surface of
the cone or outside it.

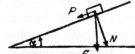

The significance of the angle of friction
is illustrated by experiments with the inclined
plane (Fig. 21) which go back to Galileo.
We write down without further explanation

FIG. 21. Equilibrium on
an inclined plane.

$$N = G \cos \alpha, \quad P = G \sin \alpha = -F.$$

From

$$F < F_{max} = \mu_0 N = N \tan \phi$$

we therefore obtain as the condition of rest

$$G \sin \alpha < \tan \phi \cos \alpha \cdot G$$

so that

$$\tan \alpha < \tan \phi$$

or

$$\alpha < \phi.$$

The body remains in a state of rest on the inclined plane as long as $\alpha < \phi$.
*The angle of friction ϕ is therefore that
inclination of a plane at which sliding
will set in.*

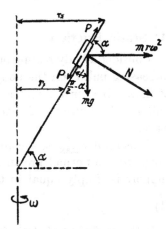

The following is a less trivial example. An oblique arm is attached to a
vertical axle at an angle $\frac{\pi}{2} - \alpha$. This arm
carries a movable sleeve or bead (cf.
Fig. 22). When the axle does not rotate the
bead is at rest or in motion depending on
whether $\alpha < \phi$ or $\alpha > \phi$. If now the axle
is set rotating, the centrifugal force
$mr\omega^2$ is added vectorially to the force of
gravity mg. The normal force N resulting
from these two and the pull P along the
guiding rod are, from the diagram,

FIG. 22. Movable sleeve or bead
on an oblique rotating rod.
Equilibrium under friction.

$$N = m(g \cos \alpha + r\omega^2 \sin \alpha), \quad P = \pm m(g \sin \alpha - r\omega^2 \cos \alpha).$$

The double sign in front of P means that we count the pull positive downward
as well as upward, so that we can take into consideration a downward as
well as an upward sliding of the bead.

From (1) and (2) the bead is in equilibrium if

$$\pm(g \sin \alpha - r\omega^2 \cos \alpha) < \tan \phi (g \cos \alpha + r\omega^2 \sin \alpha).$$

We now replace the $<$ sign by an $=$ sign, thereby obtaining the condition for "just sliding," i.e., the limit of equilibrium. By trigonometric transformation we carry out a separate calculation for the two cases $\pm$.

$+$ sign, downward sliding: $g \sin (\alpha + \phi) = r_2 \omega^2 \cos (\alpha + \phi)$,

$-$ sign, upward sliding: $g \sin (\alpha - \phi) = r_1 \omega^2 \cos (\alpha - \phi)$,

or, collected together,

$$\left.\begin{matrix} r_1 \\ r_2 \end{matrix}\right\} = \frac{g}{\omega^2} \tan (\alpha \mp \phi).$$

The force of friction hence results in a finite interval

$$r_1 < r < r_2$$

of r in which the bead is in equilibrium.

For $\alpha > \phi$ (the bead slides down as $\omega \rightarrow 0$) both r are positive; the smaller ω, the greater the interval between them. With $\alpha < \phi$ (the bead is in equilibrium under static friction for $\omega \rightarrow 0$) $r_1 = 0$ (even negative according to the equation) and only r_2 is positive; with increasing ω, r_2 approaches zero as well.

(2) Sliding Friction

Here the law of friction

(4) $$F = \mu N$$

applies.

The *coefficient of sliding friction* μ is roughly independent[7] of the velocity, and, like μ_0, a constant depending on the nature of the materials and conditions of the surfaces. It is universally true that

(5) $$\mu < \mu_0.$$

If the path along which the body slides is rectilinear, N equals the force of gravity (or its component perpendicular to the path); if the path is curved, we must, according to Eq. (11.15), add the effect of the centrifugal force.

We illustrate Eq. (5) by means of an extremely primitive experiment which is, however, very surprising in its result. Let us put a smooth cane

---

[7] Experience in railroad operation (sliding friction between wheel and brakeshoe) indicates that for high velocities v the factor μ decreases monotonically with increasing v.

or walking stick over the forefingers of the right and left hand, held some distance apart. From Fig. 11a the distribution of forces is

$$A = \frac{b}{a+b} G, \quad B = \frac{a}{a+b} G.$$

We now let the two fingers approach each other. Sliding occurs alternately on the right and left fingers, until the fingers meet. Where on the stick do they meet ?

Let $A > B$ initially. The sliding therefore begins at B. B remains in motion not only until $a = b$, but slides to the point $b_1 < a$ where the sliding friction of B equals the static friction of A. In general we have

$$F_{B,sl} = \mu a \frac{G}{a+b}, \quad F_{A,sl} = \mu_0 b \frac{G}{a+b}.$$

Putting these two expressions equal for $b = b_1$, we obtain

$$\mu a = \mu_0 b_1, \quad \frac{a}{b_1} = \frac{\mu_0}{\mu} > 1.$$

At this instant the stick must begin to move over A. At once the friction $F_{A,sl}$ falls to $F_{A,sl} < F_{A,st}$, so that in b_1 the friction $F_{B,sl}$ exceeds that in A; i.e., B comes to rest, and $F_{B,sl}$ changes to $F_{B,st}$.

This process is now repeated at each turning point. A and B thereby approach in geometric progression (since the quotient $\frac{\mu_0}{\mu}$ occurs each time) the center of mass of the stick for which $a = b = 0$. In the final state the stick balances in equilibrium over the juxtaposed fingers.

We now return to static friction, which plays a decisive role in pure rolling motion. Paradoxical as it may sound, it is the static friction which drives a railroad train forward. (The same is true of an automobile; a pedestrian on slippery ground likewise propels himself only by means of static friction.) The steam pressure is an internal force, and as such could never set the mass center of the locomotive in motion. To do this an external force is needed. This external force is the reaction between rail and wheel, i.e., just the static friction.

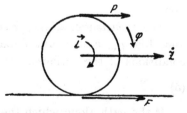

FIG. 23. Reaction between wheel and rail in a locomotive. For the case of pure rolling the static friction provides the driving force of the train.

Consider one of the driven wheels of the locomotive (Fig. 23). By means of the connecting rod the engine transmits a torque L to the wheel; its primary action would be to impart a rotational acceleration to the wheel. This is incompatible with the condition of pure rolling, Eq. (11.10),

(6) $$\dot{z} = r\omega.$$

Let M be the mass of the train per actuated wheel, R the resistance to motion (air resistance, frictional losses in the axle bearings, etc.), I the moment of inertia of the wheel, and F the force of static friction. The equations of motion become

(7)
$$M\ddot{z} = F - R;$$
$$I\ddot{\phi} = L - Fr.$$

The static friction F cannot be determined *a priori*; it can, however, be obtained from the foregoing equations as follows. Let us at first eliminate F from the equations

(8)
$$M\ddot{z} = F - R;$$
$$M_{\text{red}}\ddot{z} = P - F.$$

equivalent to (7). P is the peripheral force corresponding to the torque L, and M_{red}, as in (11.8), is the reduced mass corresponding to the moment of inertia I, i.e.,

$$L = Pr, \quad I = M_{\text{red}}r^2.$$

From (8) one obtains

(9) $$(M + M_{\text{red}})\ddot{z} = P - R$$

and, by virtue of the first Eq. (8),

(10) $$F = R + \frac{M}{M + M_{\text{red}}}(P - R) = \frac{MP + M_{\text{red}}R}{M + M_{\text{red}}}.$$

D'Alembert's principle could have furnished equation (9) directly. The first Eq. (8) contains the quantitative proof of our assertion that the static friction F is the driving force in the operation of a train. For in the case of uniform motion it gives

$$R = F.$$

As the second Eq. (8) shows, the peripheral force P resulting from the steam pressure has merely the function of calling into play the static friction at the rails.

Another evidence of this is the fact that as trains have become faster or the freight per train greater, locomotives have become constantly heavier. This circumstance points directly to Coulomb's law of friction, Eq. (1), which states that the limit of static friction available is proportional to the normal pressure N. The well-known fact that static friction fails and sliding occurs when the rails are too smooth (due to ice, or, for instance, to lubrication from run-over migrating caterpillars) points to the other factor μ_0 in Eq. (1), which, as emphasized, depends on the state of the surface of the rails. When the rails are too smooth, the factor μ_0 must be artificially increased; the sander serves this end.

CHAPTER III

OSCILLATION PROBLEMS

The investigations that are to follow will teach us nothing new about the principles of mechanics. So great, however, is the significance of oscillation processes for physics and engineering that their separate systematic treatment is deemed essential.

§ 15. The Simple Pendulum

The oscillating body is a particle of mass m which is attached to a fixed point O by means of a weightless rigid rod of length l; l is called the length of the pendulum. We may neglect friction at the point of suspension and air resistance, so that the only force acting is that of gravity, with a component $-mg \sin \phi$ in the direction of increasing ϕ (cf. Fig. 24). The general equation (11.14) for the guided motion along an arbitrary path gives us, with $v = l\dot{\phi}$ (circular path), the exact equation

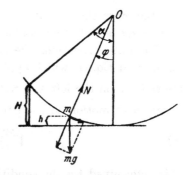

(1) $$ml\frac{d^2\phi}{dt^2} = -mg \sin \phi.$$

For sufficiently small oscillations, $\phi \ll 1$, we can put $\sin \phi = \phi$. With the abbreviation

(2) $$\frac{g}{l} = \omega^2$$

Fig. 24. Simple pendulum. Component of gravity along the direction of motion.

we then obtain the linear pendulum equation

(3) $$\frac{d^2\phi}{dt^2} + \omega^2\phi = 0.$$

This is the differential equation of " harmonic oscillations " as treated in § 3 (4). Apart from the designation for the dependent variable it is identical with Eq. (3.23). The circular frequency ω defined in (3.22) is now given by Eq. (2) above. We therefore have

(4) $$\omega = \frac{2\pi}{T} = \left(\frac{g}{l}\right)^{\frac{1}{2}}, \quad T = 2\pi\left(\frac{l}{g}\right)^{\frac{1}{2}}.$$

Notice that T is independent of the mass m, which dropped out already in (1). Thus different masses have the same period if the pendulum length l is the same. T is the full period, covering a complete swing to and fro. Sometimes one half of this time is designated as the period of oscillation. Thus one speaks of a " seconds pendulum " for which $\frac{1}{2} T$ equals one second. Its length is calculated from (4) to be

$$l = \frac{g}{\pi^2} \cong 1 \text{ meter.}$$

To the extent to which Eq. (3) is valid the period of oscillation is independent also of the amplitude of swing; i.e., small pendulum oscillations are *isochronous*.

The general solution of (3) has the form

$$\phi = a \sin \omega t + b \cos \omega t.$$

If we specify that $\phi = 0$ at $t = 0$ and $\phi = \alpha$ at $t = \frac{T}{4}$, we must put $b = 0$ and $a = \alpha$, so that

(5) $\phi = \alpha \sin \omega t.$

α is therefore the amplitude of ϕ, i.e., the maximum displacement of the particle measured in units of angle (radians).

For finite deflections the isochronism is destroyed because of the non-linearity of Eq. (1) which applies in that case. In order to integrate (1) we multiply it on the left and right by $\frac{d\phi}{dt}$; this amounts to passing from the equation of motion to the equation of energy. An integration yields

(6) $\left(\dfrac{d\phi}{dt}\right)^2 = 2\omega^2 \cos\phi + C.$

C is determined by the condition that $\frac{d\phi}{dt} = 0$ for $\phi = \alpha$, i.e.,

$$C = -2\omega^2 \cos \alpha.$$

Alternately we can proceed directly from the equation of energy. With the meaning of H indicated in Fig. 24 we obtain

(6a) $\dfrac{m}{2} l^2 \left(\dfrac{d\phi}{dt}\right)^2 + mgh = mgH$

$$\text{where} \begin{cases} h = l\,(1 - \cos\phi) \\ H = l\,(1 - \cos\alpha), \end{cases}$$

which is evidently identical to (6).

Consider now the equality

$$\cos\phi - \cos\alpha = 2\left(\sin^2\frac{\alpha}{2} - \sin^2\frac{\phi}{2}\right);$$

we substitute it in (6) to obtain

(7)
$$\frac{d\left(\frac{\phi}{2}\right)}{\left(\sin^2\frac{\alpha}{2} - \sin^2\frac{\phi}{2}\right)^{\frac{1}{2}}} = \omega\,dt$$

or

(8)
$$\int_0^{\frac{\phi}{2}} \frac{d\left(\frac{\phi}{2}\right)}{\left(\sin^2\frac{\alpha}{2} - \sin^2\frac{\phi}{2}\right)^{\frac{1}{2}}} = \omega t.$$

We have thus arrived at an *elliptic integral of the first kind.* In order to explain this name we shall have to speak in passing of the " rectification of the ellipse," i.e., the measurement of the length of an arc of an ellipse. Let us use the parametric form of the equation of an ellipse,

$$x = a \sin v$$

$$y = b \cos v$$

from which we calculate

$$ds^2 = dx^2 + dy^2 = (a^2 \cos^2 v + b^2 \sin^2 v)\,dv^2,$$

$$ds = \left[a^2 - (a^2 - b^2) \sin^2 v\right]^{\frac{1}{2}}dv.$$

We now put

$$k^2 = + \frac{a^2 - b^2}{a^2}\,(<1 \text{ for } a > b),$$

and obtain for the length of the arc of the ellipse between the endpoint $v = 0$ of the minor axis and an arbitrary point v of the ellipse

(9)
$$s = a\int_0^v (1 - k^2\sin^2 v)^{\frac{1}{2}}dv.$$

This is an " elliptic integral of the second kind."

The elliptic integral of the first kind is the simpler of the two from the viewpoint of function theory. In the " Legendre standard form " it is

$$\int_0^v \frac{dv}{(1 - k^2\sin^2 v)^{\frac{1}{2}}}.$$

We shall put our integral (8) in this form by means of the transformation

$$\sin\frac{\phi}{2} = \sin\frac{\alpha}{2} \cdot \sin v.$$

(10)
$$\left(\sin^2\frac{\alpha}{2}-\sin^2\frac{\phi}{2}\right)^{\frac{1}{2}}=\sin\frac{\alpha}{2}\,\cos\,v,$$

$$\frac{d\frac{\phi}{2}}{\left(\sin^2\frac{\alpha}{2}-\sin^2\frac{\phi}{2}\right)^{\frac{1}{2}}}=\frac{dv}{\cos\frac{\phi}{2}}=\frac{dv}{(1-k^2\sin^2 v)^{\frac{1}{2}}},$$

where the " modulus " k stands for

(11)
$$k=\sin\tfrac{1}{2}\alpha.$$

If we wish to calculate the period T, we must put in Eq. (8)

$$t=\frac{T}{4}\ \text{and}\ \phi=\alpha,$$

so that, according to (10), $v=\frac{\pi}{2}$. This yields the so-called " complete integral of the first kind," which is designated by the letter K,

(12)
$$K=\int_0^{\frac{\pi}{2}}\frac{dv}{(1-k^2\sin^2 v)^{\frac{1}{2}}}.$$

ω being defined by (2), we then obtain from (8) the period

(13)
$$T=4K\left(\frac{l}{g}\right)^{\frac{1}{2}}.$$

From (12) we can read off directly that
$K=\frac{\pi}{2}$ as $k\longrightarrow 0$, i.e., according to (11), for sufficiently small amplitudes α;
$K=\infty$ as $k\longrightarrow 1$, i.e., according to (11), for $\alpha=\pi$, 180° swing to upright position.

In the first case we obtain our former expression (4), as would be expected. In the latter case the deviation from this expression reaches an extreme.

In general a binomial expansion and term-by-term integration of (12) leads to

$$K=\frac{\pi}{2}\left(1+\frac{k^2}{4}+\frac{9k^4}{64}+\cdots\right).$$

The corresponding expression for T is

(14)
$$T=2\pi\left(\frac{l}{g}\right)^{\frac{1}{2}}\left(1+\frac{1}{4}\sin^2\frac{\alpha}{2}+\frac{9}{64}\sin^4\frac{\alpha}{2}+\cdots\right),$$

which gives the deviation from isochronism for finite deflections in quantitative fashion.

Astronomical clocks have simply-constructed pendulums with $\alpha\leqslant 1\tfrac{1}{2}°$. For them the first correction term in the parenthesis of (14) amounts to approximately 1 part in 20,000.

§ 16. *The Compound Pendulum*

This problem is essentially that of rotation of a rigid body about a fixed axis, treated already in § 11, subsec. 1, from which it differs only in that the external forces are now specified to be gravitational. Let s be the distance of the center of gravity G from the fixed axis O [we use the term " center of gravity " deliberately here, though, from (3.12), it coincides with the center of mass]; moreover, let ϕ be the angle which the line OG makes with the vertical. The total moment L of the gravitational forces acting on the individual elements of mass dm is evidently

(1) $$L = -mgs \sin \phi,$$

where m is the total mass; from (11.4) the equation of motion is then

(2) $$I\ddot{\phi} = -mgs \sin \phi.$$

A comparison with the equation of motion (15.1) of the simple pendulum shows that the length l of the equivalent simple pendulum, i.e., the simple pendulum having the same period of oscillation as our compound one, is

(3) $$l = \frac{I}{ms}.$$

Let us replace I by the so-called *radius of gyration a*, defined by

(4) $$I = ma^2.$$

The radius of gyration is therefore that distance from the point of suspension O of the pendulum at which we must concentrate the total mass m in order to obtain the moment of inertia I of the actual mass distribution. Note: in (11.8) we introduced a " reduced mass " for the distance r at which the initially unknown mass M_{red} was to be placed; here, *per contra*, the mass m is given and we are looking for the distance a at which this mass is to be located.

Comparison of (3) and (4) shows that a is the geometric mean of s and l,

(5) $$a^2 = ls.$$

Let us now lay off the equivalent pendulum length l from O along the center line OG of the pendulum. The point P thus obtained is called the *center of oscillation* (Huygens). Fig. 25 shows the relative positions of O, G and P and allows us to form a picture of the relation between s, a and l.

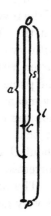

FIG. 25. Point of suspension O, center of gravity G, and center of oscillation P of a compound pendulum. The radius of gyration a is the geometric mean of the equivalent pendulum length l and distance from center of gravity s.

We now claim that the roles of O and P are interchangeable. So far O has been our point of suspension, P the center of oscillation. We shall now take P as the point of suspension and show that O becomes the center of oscillation. This is the idea underlying the *reversible pendulum*.

The scheme below tabulates the symbols so far used and completes the list for purposes of what is to follow.

Point of Suspension	Center of Oscillation	Equivalent Pend. Length	Moment of Inertia	Radius of Gyration	Distance of Mass Center
O	P	l	I	a	s
P	O'	l_P	I_P	a_P	$l-s$

Our assertion is that
$$l_P = l, \text{ i.e., } O' = O.$$

Proof: let us calculate l_P from equations (3) and (4) rewritten in terms of the corresponding new symbols. We have

(6) $$l_P = \frac{I_P}{m(l-s)} = \frac{a_P^2}{l-s}.$$

Now according to Eq. (10) of the supplement to this section

(6a) $$a_P^2 = l(l-s)$$

so that indeed the last member of (6) equals l.

The pendulum is used in the determination of the gravitational acceleration g at different points on or below the surface of the earth. Since in practice no simple pendulum is available and since in a compound pendulum the moment of inertia I cannot be calculated accurately (not only because of the complicated shape of the bob, but also because of possible internal inhomogeneities), one is forced to resort to the experimental method of the reversible pendulum for the determination of the equivalent pendulum length. We have to imagine that the pendulum of Fig. 25 is provided with two knife-edges for its points of support, one at O and one at P, the latter with its edge facing up, and both with their triangular cross-sections in the plane of the drawing. The knife-edge at P can be moved up and down by means of a micrometer screw. Given a sufficiently long period of observation the number of oscillations can be counted with very great accuracy, so that the equality or inequality of the periodic times for oscillations about O and P can be determined exceedingly precisely, and, if necessary, corrected by means of the micrometer screw.

The principle of the reversible pendulum is a first illustration of a type of very general reciprocity relation which recurs in all branches of physics.

Another example of such a relation is the interchangeability of source point and field point (" Aufpunkt ") in acoustics and electrodynamics.

We have in mind the rule of parallel axes, which states that the moment of inertia of a body of mass m about an axis through an arbitrary point O is equal to the sum of its moment of inertia about the parallel axis through the center of mass G and ms^2, where s is the distance between G and the axis through O.

If y is the direction of the axis in question and x the direction from O to G, the distance r from the axis through O of some element of mass dm must be

$$r^2 = x^2 + z^2.$$

Here x is measured from O.   If, instead, x is measured from G, and if, as in Fig. 25, $OG = s$, we have

$$r^2 = (x+s)^2 + z^2 = x^2 + z^2 + 2xs + s^2.$$

If we sum over all dm, it follows that

$$(7) \qquad I = I_G + 2s \int x\, dm + ms^2.$$

The middle term vanishes [cf., for instance, Eq. (13.3b)] provided the plane $x = 0$ passes through the center of mass.   If this is the case,

$$(8) \qquad I = I_G + ms^2,$$

as asserted above.

Accordingly we have from **Fig. 25** that

$$(8a) \qquad I_P = I_G + m(l-s)^2.$$

But from (8) and (8a)

$$I_P \quad I = ml^2 - 2mls$$

which, in view of (4), can be written

$$(9) \qquad a_P^2 - a^2 = l^2 - 2ls$$

or, by virtue of (5),

$$(10) \qquad a_P^2 = l^2 - ls = l(l-s).$$

This is the relation that was used in (6a).

§ 17.   The Cycloidal Pendulum

This pendulum was invented by Christian Huygens[1], the most ingenious watchmaker of all time.   Its purpose is to eliminate the lack of isochronism of the ordinary simple pendulum.   This is achieved by making the mass point move on a cycloidal instead of a circular arc.   Later on we shall see how this motion can be realized in practice.

The parametric representation of a common cycloid is

(1)
$$x = a(\phi - \sin \phi),$$
$$y = a(1 - \cos \phi).$$

The parameter ϕ is the angle through which a wheel of radius a rolling on the horizontal x-axis has turned from its initial position.   The common cycloid is generated by a point on the periphery of the wheel (Fig. 26).

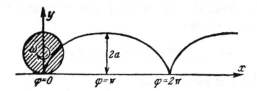

FIG. 26.   Generation of common cycloid by point on the periphery of a rolling wheel.   Definition of angle of rotation ϕ.

For our pendulum we need a cycloid that has its cusps at the top rather than on the bottom (cf. Fig. 27 on p. 96);   this is generated by having our wheel roll on the underside of the x-axis.   The x of such a curve is that given in (1) while its y is obtained by subtracting the y given in (1) from $2a$,

(2)
$$x = a(\phi - \sin \phi),$$
$$y = a(1 + \cos \phi).$$

The component of gravity mg along the tangent of the trajectory (in our case the cycloid) is

$$F_s = - mg \cos (y, s) = - mg \frac{dy}{ds}.$$

The general relation (11.14) therefore yields

(3)
$$m\dot{v} = - mg \frac{dy}{ds},$$

where, just as in the case of the circular pendulum, the mass m cancels on

[1] *Horologium Oscillatorium*, Paris (1673).   Collected Works, Vol. 18, The Hague (1934).

the left and right. Differentiation of (2) gives

$$dx = a(1 - \cos \phi)d\phi, \quad dy = - a \sin \phi \, d\phi.$$

$$ds^2 = a^2(2 - 2 \cos \phi)d\phi^2, \quad ds = 2 a \sin \frac{\phi}{2}d\phi.$$

Thus in our case

(4)
$$v = \frac{ds}{dt} = 2 a \sin \frac{\phi}{2} \frac{d\phi}{dt} = - 4 a \frac{d}{dt} \cos \frac{\phi}{2}$$

and

(5)
$$\frac{dy}{ds} = - \frac{1}{2} \frac{\sin \phi}{\sin \phi/2} = - \cos \frac{\phi}{2}.$$

If we replace (4) and (5) in (3), we obtain

(6)
$$\frac{d^2}{dt^2} \cos \frac{\phi}{2} = - \frac{g}{4a} \cos \frac{\phi}{2}.$$

This equation differs from Eq. (15.3) of the simple pendulum only in that the dependent variable is now called $\cos \frac{\phi}{2}$ rather than ϕ. This is of course of no consequence for the integration of (6). The earlier Eq. (15.4) therefore holds unchanged, viz.

(7)
$$T = 2\pi \left(\frac{l}{g}\right)^{\frac{1}{2}} \text{ with } l = 4a,$$

the latter because in (6) $4a$ took the place of our former l.

Eq. (15.3) described only the small displacements of a simple pendulum and was obtained from the exact relation (15.1) by an approximation; our present equation (6) and Eq. (7) resulting from an integration thereof are, on the other hand, exact for oscillations of arbitrary amplitude. The cycloidal pendulum is then rigorously isochronous; its periodic time is completely independent of the amplitude of oscillation.[2]

As regards the method used, we notice that in (6) the motion of our particle was represented not by its Cartesian coordinates or by some parameter bearing an immediate relation to the cycloidal curve, but by one half the angle of rotation ϕ of the wheel generating the cycloid. We see that

---

[2] The cycloid can also be called *tautochrone* (oscillations on a cycloid are "isochronous to each other"); it is also called *brachistochrone* (because it answers the question, "on what curve must a mass acted on by constant gravitational force slide in order to traverse the distance between two given endpoints in the least possible time?" It turns out that the mass takes less time on a cycloid than on a straight line or any other curve joining the same points). The brachistochrone problem is all the more notable because it was for it that the first principles of the Calculus of Variations were developed.

this parameter, although only indirectly connected with the cycloid, provides the simplest method of approach to the problem. Its introduction gives us a foretaste of the general Lagrange method of Chapter VI, which enables us to introduce arbitrary parameters as dependent variables in the equations of motion.

Just as remarkable as Huygens' discovery of the isochronism of the cycloidal pendulum is the way in which he actually achieved the frictionless motion of the bob on the cycloid. He availed himself of the rule that the evolute of a cycloid is another cycloid equal to the generating one. If, therefore, we tie a string of length $l=4a$ to the point O of Fig. 27 in which the two upper cycloid arcs form a cusp, and if this string be pulled taut so that it rests against the right part of the cycloid (or the left part if deflected to the left), the endpoint P of the string describes the lower cycloidal arc. The guiding of the bob along the lower cycloid effected in this manner is almost as frictionless as the guiding of the simple pendulum along a circular arc.

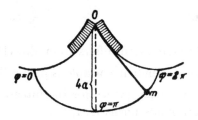

Actually Huygens' idea has been abandoned in the practice of pendulum clock construction; according to investigations of Bessel among others it is sufficient to install a spring — usually a short elastic lamina — at the upper end of the pendulum. If the length of the lamina and the mass of the bob are suitably chosen, a sufficient degree of isochronism is achieved.

Fig. 27. Huygens' isochronous cycloidal pendulum.

§ 18. The Spherical Pendulum

We require the pendulum to be suspended in such a fashion that the mass point m is able to move freely on the surface of a sphere of radius l (the length of the pendulum). It is then subject to the condition of constraint

$$(1) \qquad F=\frac{1}{2}(x^2+y^2+z^2-l^2)=0,$$

where the factor $\frac{1}{2}$ has been added for convenience's sake.

Here r, the number of conditions of constraint, equals 1, and $X_1=X_2=0$, $X_3=-mg$, so that the Lagrange equations of the first kind (12.9) take the form

$$(2) \qquad \begin{aligned} m\ddot{x} &= \lambda x, \\ m\ddot{y} &= \lambda y, \\ m\ddot{z} &= -mg+\lambda z. \end{aligned}$$

In view of Eqs. (13.13) and (13.13a), elimination of λ from the first two equations (2) yields the constancy of angular momentum about the z-axis, or, what amounts to the same thing, the conservation of the areal velocity

(3) $$x\frac{dy}{dt} - y\frac{dx}{dt} = 2\frac{dS}{dt} = C \quad (S = \text{area swept out}).$$

If, on the other hand, we multiply the Lagrange equations (2) by $\dot{x}$, $\dot{y}$, $\dot{z}$, we obtain the equation of energy, for condition (1) is independent of t (cf. p. 68). Addition yields

(4) $$m(\dot{x}\ddot{x} + \dot{y}\ddot{y} + \dot{z}\ddot{z}) = -mg\dot{z} + \lambda(x\dot{x} + y\dot{y} + z\dot{z}).$$

But from (1)

$$\frac{dF}{dt} = x\dot{x} + y\dot{y} + z\dot{z} = 0.$$

On the other hand we evidently have

$$\dot{x}\ddot{x} + \dot{y}\ddot{y} + \dot{z}\ddot{z} = \frac{1}{2}\frac{d}{dt}(\dot{x}^2 + \dot{y}^2 + \dot{z}^2) = \frac{1}{2}\frac{dv^2}{dt}.$$

Integration of (4) with respect to t then gives

(5) $$\frac{m}{2}v^2 = -mgz + \text{const.},$$

which we shall write in the form

(5a) $$T + V = E \quad \text{with} \quad V = mgz.$$

Let us finally multiply the Lagrange equations by x, y, z, respectively. With the aid of (1) this allows us to calculate λ,

$$\lambda l^2 - mgz = m(x\ddot{x} + y\ddot{y} + z\ddot{z})$$

or

(6) $$\lambda l = mg\frac{z}{l} + m\left(\frac{x}{l}\ddot{x} + \frac{y}{l}\ddot{y} + \frac{z}{l}\ddot{z}\right).$$

Now the normal to the surface of the sphere at the point x, y, z has direction cosines $\frac{x}{l}$, $\frac{y}{l}$, $\frac{z}{l}$, so that apart from sign the second term on the right is the inertial force $F_n{}^*$ normal to the spherical surface; similarly the first term on the right is, apart from sign, the component F_n of gravity in the same direction. According to d'Alembert the sum of these two must be equilibrated by the reaction R_n of the surface of the sphere, or, physically speaking, by the tension in the pendulum suspension. The meaning of Eq. (6) can hence be summed concisely by the equation

(7) $$\lambda l = -(F_n + F_n{}^*) = R_n.$$

We notice that within a factor l, λ is the constraint which is exerted on the motion by virtue of condition (1), this constraint acting in a direction normal to the motion. Corresponding statements hold in more general cases where several conditions of constraint and therefore several Lagrange multipliers are present.

In order to carry out a second integration of (5) we shall pass to spherical coordinates given by

$$x = l \cos \phi \sin \theta$$
$$y = l \sin \phi \sin \theta$$
$$z = l \cos \theta.$$

We form

$$\dot{x} = l \, \dot{\theta} \cos \phi \cos \theta - l \, \dot{\phi} \sin \phi \sin \theta,$$
$$\dot{y} = l \, \dot{\theta} \sin \phi \cos \theta + l \, \dot{\phi} \cos \phi \sin \theta,$$
$$\dot{z} = - l \, \dot{\theta} \sin \theta.$$

The equation of conservation of angular momentum (3) becomes

(8) $$2\frac{dS}{dt} = x\dot{y} - y\dot{x} = l^2 \sin^2 \theta \cdot \dot{\phi} = C$$

and the equation of energy (5a),

(9) $$\frac{ml^2}{2}(\dot{\theta}^2 + \sin^2 \theta \, \dot{\phi}^2) + mgl \cos \theta = E.$$

A further change of variables

$$u = \cos \theta, \; \dot{\theta} = - \frac{1}{(1-u^2)^{\frac{1}{2}}} \frac{du}{dt}$$

transforms (8) into

(10) $$\dot{\phi} = \frac{C}{l^2(1-u^2)}$$

and (9) into

(11) $$\left(\frac{du}{dt}\right)^2 = U(u) = \frac{2}{ml^2}(E - mglu)(1 - u^2) - \frac{C^2}{l^4}.$$

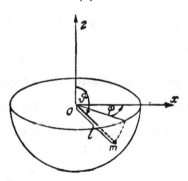

FIG. 28. Spherical pendulum treated as mass point m moving under gravity on the surface of a sphere of radius l.

This relation between t and u allows us to find t as a function of u,

(12) $$t = \int \frac{du}{U^{\frac{1}{2}}}.$$

Eq. (10) can now likewise be written in integrated form, for from (10) and (11)

$$\frac{d\phi}{du} = \dot{\phi} \cdot \frac{dt}{du} = \frac{C}{l^2(1-u^2)} \frac{1}{U^{\frac{1}{2}}},$$

so that one obtains

(13)
$$\phi = \frac{C}{l^2}\int \frac{du}{1-u^2}\cdot\frac{1}{U^{\frac{1}{2}}}.$$

U is a function of third degree in $u = \cos\theta$. $U^{\frac{1}{2}}$ is real only for $U > 0$. If then the constants of the equation correspond to a real physical problem, there must be two values $u = u_2 < u = u_1$ in the interval

$$-1 < u < +1$$

between which *U* is positive (cf. Fig. 29).

$u_1 = \cos\theta_1$ and $u_2 = \cos\theta_2$ are the two latitudes between which the mass point oscillates back and forth. If the integration of (12) or (13) reaches one of these limits of *u*, not only the direction of integration but also $U^{\frac{1}{2}}$ must change sign, in order that the integrals remain real and positive. Between two successive turning points one quarter of the full period of oscillation elapses, i.e.,

(14)
$$\frac{T}{4} = \int_{u_2}^{u_1}\frac{du}{U^{\frac{1}{2}}}.$$

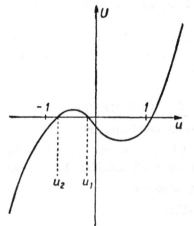

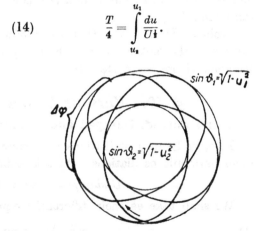

FIG. 29. Curve of third degree $U(u)$ and its intersections $u = u_1$ and $u = u_2$ with the abscissa. $u_2 < u_1 < 0$ means that the trajectory is located in the lower hemisphere.

FIG. 30. " Bird's-eye " view of the path of the spherical pendulum. Angle of precession $\Delta\phi$. A passage from θ_1 over θ_2 back to θ_1 corresponds to a half-period, $\Delta\phi$ therefore to a full cycle.

Note that now the oscillation is no longer periodic in space as in the case of the pendulum moving in a plane, but is modified by a slow *precession*. The angle of precession $\Delta\phi$ by which the mass advances (or recedes) in a full period *T* is calculated from (13) to be

(15)
$$2\pi + \Delta\phi = \frac{4C}{l^2}\int_{u_2}^{u_1}\frac{du}{(1-u^2)U^{\frac{1}{2}}}.$$

This precession is illustrated in Fig. 30, which is taken from A. G. Webster, " Dynamics of Particles," Leipzig, Teubner (1912), p. 51.

The integral (12) is an elliptic integral of the first kind, just like the integral (15.8) for the simple pendulum. This is the generic name applied to all integrals whose integrand contains the square root of a polynomial of the third or fourth degree in the variable of integration in the denominator. That Eq. (15.8) falls in this class can be seen by introducing the transformation $u=\sin\frac{\phi}{2}$, so that u becomes our variable of integration; if, moreover, we put $a=\sin\frac{\alpha}{2}$, (15.8) goes over into

$$\int \frac{du}{[(a^2-u^2)\,(1-u^2)]^{\frac{1}{2}}}.$$

In particular, expression (14) for T is, just like (15.12), a complete integral of the first kind. On the other hand, integral (13), which has the two factors $(1\pm u)$ in addition to $U^{\frac{1}{2}}$ in the denominator, is an " elliptic integral of the third kind," and (15) is a " complete elliptic integral of the third kind."

Problem III.1 shows that for infinitesimal oscillations the equation expressing the motion of the spherical pendulum becomes elementary and the angle of precession $\Delta\phi \to 0$.

§ 19. Various Types of Oscillations
Free and Forced, Damped and Undamped Oscillations

Free, undamped oscillations were treated in § 3, subsec. 4; we called them harmonic oscillations. At this point we shall consider, first of all,

Undamped, Forced Oscillations

We shall take as their differential equation

(1) $$m\ddot{x}+kx=c\sin\omega t,$$

where $\omega=\frac{2\pi}{T}$ is the circular frequency of the driving force.

We have here made the differential equation linear in the dependent variable x, which is permissible, at any rate, for small oscillations (cf. simple pendulum). The same remark applies to the remaining examples in this and the following section.

The restoring force is $-kx$ as in (3.19); c of Eq. (1) is the amplitude of the driving force causing our particle to oscillate.

By virtue of the addition of the right member, (1) is an inhomogeneous linear differential equation. The left side, when set equal to zero, gives the associated homogeneous differential equation, as previously mentioned in connection with Eq. (3.23).

A particular solution of the inhomogeneous differential equation is given by

$$x = C \sin \omega t,$$

where C must satisfy the equation

$$C(k - m\omega^2) = c.$$

If, with (3.20) as model, we put

(2)
$$\omega_0 = \left(\frac{k}{m}\right)^{\frac{1}{2}},$$

we obtain

(3)
$$C = \frac{c/m}{\omega_0^2 - \omega^2}.$$

The general solution of (1) is formed from this particular solution and the general solution of the associated homogeneous equation:

(4)
$$x = C \sin \omega t + A \cos \omega_0 t + B \sin \omega_0 t.$$

The amplitude C of the first term grows with increasing ω to become infinite for $\omega = \omega_0$; thereupon it jumps to negative infinity, and decreases slowly in absolute value toward 0 as $\omega \to \infty$.

Actually, when C becomes negative the amplitude does not change sign, for amplitudes are positive by definition. We therefore continue to define the amplitude by $|C|$ and put the change of sign that takes place into the sine factor, where it appears as a phase change of $\delta = \pm\pi$.

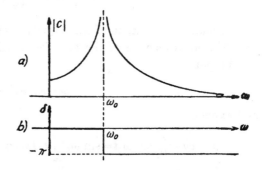

The foregoing is illustrated in Figs. 31a, b, where $|C|$ and δ have been plotted as functions of ω.

Fig. 31. Amplitude and phase of undamped forced oscillations.

In Fig. 31b we cannot *a priori* decide whether the phase leads or lags for $\omega > \omega_0$, i.e., whether we are to take $\delta = +\pi$ or $\delta = -\pi$. We shall, however, anticipate and consider undamped vibrations as a limiting case of damped vibrations (see below); this leads us to decide in favor of $-\pi$, so that the first term of (4) can be written in detail

(4a)
$$x = \frac{c/m}{\omega^2 - \omega_0^2} \sin(\omega t - \pi) \qquad (\omega > \omega_0).$$

The fact that the amplitude becomes infinite for $\omega = \omega_0$ illustrates the phenomenon of *resonance* between free and forced oscillations, a phenomenon that plays an important role in all of physics. The denominator of (3) and (4a) whose vanishing causes this infinite amplitude is called a " resonance denominator." It is intuitively clear that the closer the proper frequency of the oscillating system is to that of the driving force, the better the system will follow this force.

Incidentally we must keep in mind that we are guilty of gross extrapolation when we deduce infinite amplitudes at resonance, for in almost all cases our linear differential equation holds only for infinitesimal oscillations.

So far we have directed all our attention to the first term of the right member of Eq. (4). The other two terms are determined by the initial conditions. Let us take

$$x=0, \ \dot{x}=0 \ \text{at} \ t=0,$$

so that, from (4),

$$A=0, \ \omega \, C + \omega_0 \, B = 0, \quad \text{hence} \quad B = -\frac{\omega}{\omega_0} C.$$

It follows that

(5) $$x = C \left(\sin \omega t - \frac{\omega}{\omega_0} \sin \omega_0 t \right).$$

Let us make the content of this equation clearer by considering the special case of near resonance of the two frequencies ω and ω_0.

We put

$$\omega = \omega_0 + \Delta \omega$$

and expand

$$\sin \omega t - \frac{\omega}{\omega_0} \sin \omega_0 t = \sin \omega_0 t + t \, \Delta \omega \cos \omega_0 t - \sin \omega_0 t - \frac{\Delta \omega}{\omega_0} \sin \omega_0 t.$$

Eq. (5) then yields

$$x = C \, \Delta \omega \left(t \cos \omega_0 t - \frac{1}{\omega_0} \sin \omega_0 t \right).$$

and, by virtue of (3), in the limit $\Delta \omega = 0$,

(6) $$x = \frac{c}{2m \omega_0^2} (\sin \omega_0 t - \omega_0 t \cos \omega_0 t).$$

This type of oscillation, illustrated in Fig. 32, is no longer periodic as was that of free oscillations; indeed t appears in (6) as a secular term (i.e., no longer solely in the argument of a trigonometric function). For

$t \longrightarrow \infty$ the amplitude approaches the value $C = \infty$ as indicated in Fig. 31 for the case $\omega = \omega_0$.

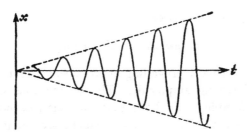

FIG. 32. Resonance of free and forced oscillations. Secular increase of amplitude.

Free, Damped Oscillations

These have the differential equation

$$(7) \qquad m\ddot{x} + kx = -w\dot{x} .$$

The frictional term on the right has been put proportional to the velocity, an assumption which finds its justification in the hydrodynamics of slow, laminar (= non-turbulent) flow (e.g., air friction).

Eq. (7) is a homogeneous linear differential equation. As before we put

$$(7\mathrm{a}) \qquad \frac{k}{m} = \omega_0^2, \ \omega_0 = \text{undamped proper frequency.}$$

Let us also make the convenient change of symbols

$$(7\mathrm{b}) \qquad \frac{w}{m} = 2\rho, \ \rho > 0.$$

Eq. (7) then takes the form

$$(8) \qquad \ddot{x} + 2\rho\dot{x} + \omega_0^2 x = 0.$$

The method described under Eq. (3.23) now proves its full worth. As there, we substitute

$$(8\mathrm{a}) \qquad x = Ce^{\lambda t}$$

in (8) and thus obtain the *characteristic equation* in λ,

$$\lambda^2 + 2\rho\lambda + \omega_0^2 = 0$$

with the two roots

$$\lambda = -\rho \pm (-\omega_0^2 + \rho^2)^{\frac{1}{2}} = \begin{cases} \lambda_1 \\ \lambda_2 \end{cases} .$$

Expression (8a) must therefore be generalized to

(8b) $$x = C_1 e^{\lambda_1 t} + C_2 e^{\lambda_2 t}.$$

We now distinguish two cases:

$$1. \ \rho < \omega_0, \qquad 2. \ \rho > \omega_0.$$

The first case is that usually prevailing in practice. The motion is a periodic oscillation with decaying amplitude. The second case is that of strong or " aperiodic " damping. In both cases we shall specialize the motion by imposing the condition, $x=0$ at $t=0$, which, according to (8b), leads to $C_2 = -C_1$.

$$1. \ \rho < \omega_0. \quad \lambda = -\rho \pm i(\omega_0^2 - \rho^2)^{\frac{1}{2}},$$
$$x = 2C_1' e^{-\rho t} \sin (\omega_0^2 - \rho^2)^{\frac{1}{2}} t.$$

For small ρ the periodic time

$$T = \frac{2\pi}{(\omega_0^2 - \rho^2)^{\frac{1}{2}}}$$

differs little from that of the undamped oscillation. $e^{-\rho t}$ is the damping factor, ρT the *logarithmic decrement*.

$$2. \ \rho > \omega_0. \quad \lambda_1 \text{ and } \lambda_2 \text{ are real and we obtain}$$

$$x = 2C_1 e^{-\rho t} \sinh(\rho^2 - \omega_0^2)^{\frac{1}{2}} t$$

where sinh is the hyperbolic sine.

We shall finally deal with a type of oscillation including all those so far considered, namely that of

Damped, Forced Oscillations

We may write their differential equation in the form

$$m\ddot{x} + w\dot{x} + kx = c \sin \omega t$$

or, with the abbreviations defined in (7a, b),

(9) $$\ddot{x} + 2\rho\dot{x} + \omega_0^2 x = \frac{c}{2mi}(e^{i\omega t} - e^{-i\omega t}).$$

To the general integral (8b) of the homogeneous equation we must now add a particular solution which we shall write in the form

$$x = |C| \sin (\omega t + \delta) = \frac{|C|}{2i}(e^{i(\omega t + \delta)} - e^{-i(\omega t + \delta)}).$$

Let us introduce this in (9). A comparison of the factors of $e^{\pm i\omega t}$ left and right yields

$$|C|\,(-\omega^2+2\,i\rho\omega+\omega_0^2)e^{i\delta}=\frac{c}{m},$$

$$|C|\,(-\omega^2-2\,i\rho\omega+\omega_0^2)e^{-i\delta}=\frac{c}{m}.$$

Multiplication and division of these two relations yields

$$|C|^2=\left(\frac{c}{m}\right)^2\frac{1}{(\omega_0^2-\omega^2)^2+4\rho^2\omega^2}$$

$$e^{2i\delta}=\frac{\omega_0^2-\omega^2-2i\rho\omega}{\omega_0^2-\omega^2+2i\rho\omega},$$

respectively. Accordingly

(10) $$|C|=\frac{c}{m}\frac{1}{[(\omega_0^2-\omega^2)^2+4\rho^2\omega^2]^{\frac{1}{2}}},$$

(11) $$\tan\delta=\frac{1}{i}\frac{e^{2i\delta}-1}{e^{2i\delta}+1}=-\frac{2\rho\omega}{\omega_0^2-\omega^2}.$$

Compare the plot of these two functions of ω in Fig. 33 with Figs. 31a, b.

Fig. 33 shows that our formerly infinite resonance maximum has been depressed to a finite value as a result of the damping (note, by the way, that the maximum value no longer occurs at the exact point $\omega=\omega_0$, but rather at a somewhat smaller ω; cf. problem III.2).

Fig. 33 also demonstrates that with increasing ω, δ goes from the value 0 at $\omega=0$ to negative values; for $\omega=\omega_0$ it exactly equals $-\frac{1}{2}\pi$, and it approaches $-\pi$ as $\omega\longrightarrow\infty$. Thus we have justified the arbitrary choice between $\pm\pi$, made earlier (in Fig. 31), when we were dealing with the undamped case. As a matter of fact we see now that the phase of the oscillation always *lags* behind that of the driving

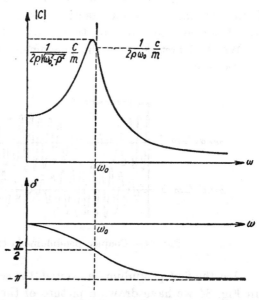

Fɪɢ. 33. Amplitude and phase of damped forced oscillations.

force. For further examples of forced vibrations see problems III.3 and III.4.

§ 20. Sympathetic Oscillations

The types of oscillation so far considered have concerned one mass point. We shall now deal with types of oscillation involving two masses capable of oscillation, these two masses being weakly coupled to each other. Sympathetic oscillations have for many years been important in electric measurements. There one speaks of a primary and a secondary circuit, the latter usually being " inductively " coupled to the former. The primary circuit is made to oscillate (" is excited "), whereupon the secondary circuit does likewise, and especially strongly so if resonance prevails. Indeed the " doubly tuned coupling stage " widely used in radio consists of a primary circuit and a secondary one tuned to the former. Here we shall of course restrict ourselves to coupled *mechanical* oscillations, which have often been used as models for electrical ones.

A particularly instructive example of sympathetic oscillations is furnished by the so-called " coupled pendulums." In the case of resonance these are two equally long and equally heavy pendulums. We may picture them most simply as oscillating in the same plane; their coupling may be effected by means of a helical spring as indicated in Fig. 35. If the spring offers but slight resistance to the relative motion of the two pendulums we speak of weak coupling; in the case of greater spring tension we speak of strong coupling. We assume that the coupling of our pendulums is *weak*. If the pendulums are not exactly equal in length or in weight, we shall say that they are " out of tune," or " detuned."

We shall first describe the phenomena which are observed in the case of resonance.

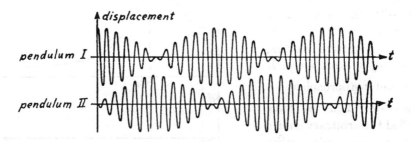

FIG. 34.   Coupled pendulums in the case of resonance.

Let the first pendulum be excited, the second one being initially at rest. In Fig. 34 we have drawn a picture of the resulting oscillations.

The oscillations of each pendulum are *modulated*. The energy alternates between one pendulum and the other. When one pendulum oscillates with maximum amplitude, the other one is at rest.

If, instead (cf. Fig. 35) both pendulums are set in motion simultaneously and with equal strength, either in the same direction (Fig. 35, left), or in opposite directions (Fig. 35, right), no energy is exchanged. These two oscillatory modes are called the *normal modes of oscillation* of our coupled system of two degrees of freedom. We have the general rule that *an oscillatory system of n degrees of freedom has n normal modes of oscillation.*

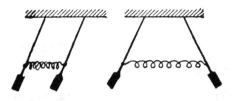

Fig. 35. The two normal modes of oscillation of coupled pendulums in resonance.

If, on the other hand, the pendulums are detuned, an energy exchange still takes place to be sure, but this exchange is of such a nature that the initially excited pendulum has a minimum amplitude different from zero. Only the pendulum initially at rest again reaches the state of rest in the course of the motion. Thus the " sympathy " of the two pendulums is upset by imperfect tuning.

We shall now sketch the theory for *complete resonance*, making the simplest possible assumptions: we neglect all damping, and approximate the circular trajectories of the bobs by the tangents at their lowest points, which is permissible for sufficiently small displacements. Let x_1 be the amplitude of oscillation of pendulum I, x_2 that of pendulum II; call k the " coupling coefficient," i.e., the spring tension caused by an elongation of unit length, divided by the mass of one of the pendulums. The simultaneous differential equations of the problem are

(1)
$$\ddot{x}_1 + \omega_0^2 x_1 = - k (x_1 - x_2)$$
$$\ddot{x}_2 + \omega_0^2 x_2 = - k (x_2 - x_1).$$

If we introduce in (1)

(2) $$z_1 = x_1 - x_2, \; z_2 = x_1 + x_2,$$

subtraction and addition yield the two equations for the normal modes,

(3)
$$\ddot{z}_1 + \omega_0^2 z_1 = - 2k z_1 \text{ or } \ddot{z}_1 + (\omega_0^2 + 2k) z_1 = 0,$$
$$\ddot{z}_2 + \omega_0^2 z_2 = 0$$

respectively, with the corresponding frequencies

(4)
$$\text{for } z_1: \; \omega = (\omega_0^2 + 2k)^{\frac{1}{2}} \approx \omega_0 + \frac{k}{\omega_0};$$
$$\text{for } z_2: \; \omega' = \omega_0.$$

The general solutions of Eqs. (3) are

(5)
$$z_1 = a_1 \cos \omega t + b_1 \sin \omega t;$$
$$z_2 = a_2 \cos \omega' t + b_2 \sin \omega' t.$$

At the moment of excitation $t=0$ let

(6)
$$x_2 = \dot{x}_2 = 0, \quad \dot{x}_1 = 0, \quad x_1 = C,$$

giving

(7)
$$\dot{z}_1 = \dot{z}_2 = 0, \quad z_1 = z_2 = C.$$

It follows that

(8)
$$b_1 = b_2 = 0, \quad a_1 = a_2 = C,$$

so that
$$z_1 = C \cos \omega t, \quad z_2 = C \cos \omega' t.$$

Finally

(9)
$$x_1 = \frac{z_2 + z_1}{2} = C \cos \frac{\omega' - \omega}{2} t \cdot \cos \frac{\omega' + \omega}{2} t$$

$$x_2 = \frac{z_2 - z_1}{2} = - C \sin \frac{\omega' - \omega}{2} t \cdot \sin \frac{\omega' + \omega}{2} t.$$

According to (4) $\frac{\omega - \omega'}{2} \cong \frac{k}{2\omega_0} \ll 1$ in the case of weak coupling. The first

factors of the right members of (9) therefore vary slowly with time; it is this circumstance which determines the *beats* in the oscillation illustrated in Fig. 34.

The theory is not quite so simple if the two pendulums are out of tune, i.e., if $l_1 \neq l_2$ or/and $m_1 \neq m_2$. Letting c be the tension of the spring due to unit elongation, we now put

$$\omega_1^2 = \frac{g}{l_1}, \quad \omega_2^2 = \frac{g}{l_2}, \quad k_1 = \frac{c}{m_1}, \quad k_2 = \frac{c}{m_2}$$

and instead of (1) we obtain the initial equations

(10)
$$\ddot{x}_1 + \omega_1^2 x_1 = - k_1 (x_1 - x_2)$$
$$\ddot{x}_2 + \omega_2^2 x_2 = - k_2 (x_2 - x_1).$$

Here again there are two normal modes, which can be obtained by an extension of the method set forth in (3.24). [In Eq. (1) we were able to use a more convenient method especially suited to that case; this method is not applicable in general.] We substitute

(11)
$$x_1 = A e^{i\lambda t}, \quad x_2 = B e^{i\lambda t}$$

and obtain from (10) the two characteristic equations

(12)
$$A\,(\omega_1^2-\lambda^2+k_1)=k_1B$$
$$B\,(\omega_2^2-\lambda^2+k_2)=k_2A.$$

The so-called *secular equation*[3] obtained from (12) is quadratic in λ^2, since

(13)
$$\frac{B}{A}=\frac{\omega_1{}^2-\lambda^2+k_1}{k_1}=\frac{k_2}{\omega_2{}^2-\lambda^2+k_2}$$

so that

(14)
$$\{\lambda^2-(\omega_1^2+k_1)\}\{\lambda^2-(\omega_2^2+k_2)\}=k_1\,k_2.$$

For small k_1, k_2 (14) has the two approximate roots

(15)
$$\lambda^2=\begin{cases}\omega_1^2+k_1+\dfrac{k_1k_2}{\omega_1{}^2-\omega_2{}^2}\\[2mm]\omega_2^2+k_2+\dfrac{k_1k_2}{\omega_2{}^2-\omega_1{}^2}.\end{cases}$$

We designate these two roots of the secular equation by ω^2 and ω'^2; furthermore we generalize the tentative solution (11) in the same manner as was done in (3.24b), using the principle of superposition of solutions of linear differential equations. Written in real form the general solution is then

(16)
$$x_1=a\cos\omega t+b\sin\omega t+a'\cos\omega't+b'\sin\omega't$$
$$x_2=\gamma a\cos\omega t+\gamma b\sin\omega t+\gamma'a'\cos\omega't+\gamma'b'\sin\omega't.$$

Here γ and γ' are the specific values of B/A which arise from (13) for $\lambda^2=\omega^2$ and $\lambda^2=\omega'^2$ respectively.

Let us once again take as the condition of excitation at $t=0$

$$x_2=0,\ \dot{x}_2=0,\ \dot{x}_1=0,\ x_1=C.$$

This yields

(17)
$$\gamma a+\gamma'a'=0,\ \gamma\omega b+\gamma'\omega'b'=0,$$
$$\omega b+\omega'b'=0,\ a+a'=C,$$

from which

$$b=b'=0$$

and

$$a=\frac{\gamma'}{\gamma'-\gamma}C,\ a'=\frac{\gamma}{\gamma-\gamma'}C.$$

---

[3] The word originated in the perturbation theory of celestial mechanics.

If we substitute these values in (16), we have

(18)
$$x_1 = \frac{C}{\gamma' - \gamma}(\gamma' \cos \omega t - \gamma \cos \omega't)$$

$$x_2 = \frac{C}{\gamma' - \gamma}\gamma\gamma' (\cos \omega t - \cos \omega't).$$

In the equation for x_2 we can perform the trigonometric transformation used in (9) to obtain

(19)
$$x_2 = \frac{2\gamma\gamma'}{\gamma - \gamma'}C \sin\frac{\omega' - \omega}{2}t \cdot \sin\frac{\omega' + \omega}{2}t$$

We see that the second pendulum still comes to rest at the times

$$\frac{\omega' - \omega}{2}t = n\pi;$$

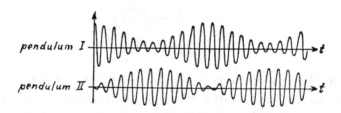

pendulum I

pendulum II

FIG. 36.   Oscillograph of two slightly detuned coupled pendulums.

not so the first pendulum, which [cf. the first Eq. (18) and Fig. 36] retains a finite amplitude when that of x_2 is at a maximum. Imperfect tuning results in an incomplete transfer of energy.

If we desire to apply the foregoing theory to electrical phenomena, we must extend it to include damping of the pendulums; damping has its electrical analogue in the Ohmic resistance (our acceleration term corresponds to the self-induction, our restoring force to capacitive effects); moreover the analysis of electrical oscillations in coupled circuits demands that we introduce " acceleration and velocity coupling " in addition to the " position coupling " [k multiplied by $\pm(x_2 - x_1)$] which was the only type of coupling taken into account in our mechanical problem.

In problem III.5 we shall investigate the motion of an experimentally convenient arrangement, in which the pendulums are suspended bifilarly from a flexible wire and oscillate not in the plane of their positions of rest, but perpendicularly to it.

An interesting arrangement, in which both coupled pendulums are, so to speak, realized in the same body, is that of an oscillating helical spring.[4]

---

[4] For details the reader is referred to the *Wüllner-Festschrift*, Teubner (1905): Lissajous Figures and Resonance Effects of Oscillating Helical Springs; Their Use in the Determination of the Poisson Ratio.

Such a spring (cf. Fig. 37) is capable not only of an oscillation (y) along its axis but also of a rotary oscillation (x) about this axis. For finite displacements the coupling between these two motions is produced by the spring itself. For if the spring is pulled vertically downward, a lateral force is experienced; the spring seeks to withdraw along the wire-direction in order to uncoil itself. If, on the other hand, the spring is coiled up, it will seek to shorten itself along the y-axis. In other words, if one excites an oscillation in the y-direction, an x-oscillation is induced, and conversely. (Note: as far as the elastic stress on the material is concerned, the y-oscillation is one of torsion, the x-oscillation one of deflection. For details about this consult Vol. II this series.)

By means of the adjustable mass Z, one can bring the vertical and horizontal oscillations into accurate or approximate resonance. If then one of the two vibrations is excited, an exchange of amplitudes of the type of Fig. 34 or Fig. 36 takes place.

Fig. 37. Torsional and deflection oscillations of a helical spring.

§ 21. The Double Pendulum

As at the beginning of the previous section, we shall first describe the empirical phenomena involved.

From a heavy pendulum (a chandelier, for instance) we suspend a light pendulum of about the same period of oscillation. Let us impart a sharp impulse to the heavy bob; the light bob will be set in vigorous motion, which suddenly subsides and stays at zero for a short time. At this instant one perceives that the heavy bob, which had previously remained practically at rest, now starts oscillating with noticeable amplitude. This oscillation soon ceases, however, whereupon in its turn the light pendulum again begins to move with considerable vigor, and so forth.

As mentioned, we demand that the masses of the two bobs, M and m, be very unequal, but that the equivalent lengths L, l be approximately the same. We let

$$\frac{m}{M} = \mu \ll 1.$$

We shall treat the displacements, X of the heavy pendulum, x of the light one, as small quantities, so that once again we can approximate arcs of circles by their tangents. Consequently we must also keep the angles ϕ and ψ (cf. Fig. 38, where ψ belongs to the relative

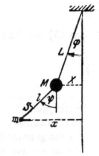

Fig. 38. Schematic arrangement of a double pendulum.

displacement $x - X$) small. We can therefore put

(1)
$$\sin \phi = \phi = \frac{X}{L}, \quad \sin \psi = \psi = \frac{x-X}{l} \quad \text{and} \quad \sin (\psi - \phi) = \psi - \phi = \frac{x-X}{l} - \frac{X}{L}.$$

$$\cos \phi = \cos \psi = \cos (\phi - \psi) = 1.$$

The upper pendulum is acted upon not only by the force of gravity, but also by the lower pendulum; the string tension[5] $S \approx mg \cos \psi$ contributes a component tangential to the motion of M of amount $- mg \cos \psi \sin (\phi - \psi)$. Thus we arrive at the equations of motion

(2)
$$M\ddot{X} = - M\frac{g}{L}X + mg\left(\frac{x-X}{l} - \frac{X}{L}\right)$$

$$m\ddot{x} = - m\frac{g}{l}(x - X)$$

or, in more convenient form,

(3)
$$\ddot{X} + \left(\frac{g}{L} + \mu\frac{g}{l} + \mu\frac{g}{L}\right)X = \mu\frac{g}{l}x,$$

$$\ddot{x} + \frac{g}{l}x = \frac{g}{l}X.$$

From now on we shall put $L = l$ and introduce the abbreviation

(4)
$$\omega_0^2 = \frac{g}{l}.$$

Our Eqs. (3) then become

(5)
$$\ddot{X} + \omega_0^2(1 + 2\mu)X = \mu\,\omega_0^2\,x,$$

$$\ddot{x} + \omega_0^2 x = \omega_0^2 X.$$

These equations of motion state that the upper pendulum is μ times more weakly coupled to the lower one than vice versa.

To integrate (5) we use a substitution similar to (20.11),

(6)
$$x = Ae^{i\lambda t}; \quad X = Be^{i\lambda t}.$$

From (5) we have as a result

(7)
$$A(\omega_0^2 - \lambda^2) = B\omega_0^2$$

$$B[\omega_0^2(1 + 2\mu) - \lambda^2] = A\mu\,\omega_0^2.$$

---

[5] In the present elementary treatment we have to introduce this tension S as a descriptive auxiliary quantity; later, when we analyse the same problem by means of the general Lagrange method, this procedure will become superfluous. In order to determine S we reason as follows: the tension in the suspension of the light bob is in equilibrium with gravity and inertial force (centrifugal force); the latter is a small quantity of second order and can therefore be neglected. We then have $S = mg \cos \psi$ as stated above.

If we put the two values of B/A obtained from these two equations equal, we arrive at the quadratic equation in λ^2,

(8) $$(\lambda^2 - \omega_0^2)^2 + 2\mu\,\omega_0^2(\omega_0^2 - \lambda^2) = \mu\omega_0^4.$$

Let its two roots be called $\lambda^2 = \omega^2$ and $\lambda^2 = \omega'^2$. Omission of higher powers of μ easily yields their approximate values

(9) $$\left.\begin{array}{c}\omega\\\omega'\end{array}\right\} = \omega_0(1 \pm \tfrac{1}{2}\mu^{\frac{1}{2}}).$$

Written in real form, the general solution of (5) is then

(10)
$$x = a\cos\omega t + b\sin\omega t + a'\cos\omega't + b'\sin\omega't,$$
$$X = \gamma a\cos\omega t + \gamma b\sin\omega t + \gamma'a'\cos\omega't + \gamma'b'\sin\omega't.$$

As in § 20, γ and γ' are here the values of B/A which result from (7) for $\lambda^2 = \omega^2$ and $\lambda^2 = \omega'^2$ respectively, viz.,

(11) $$\gamma = -\mu^{\frac{1}{2}}, \quad \gamma' = +\mu^{\frac{1}{2}} \quad \text{and hence} \quad \gamma' - \gamma = 2\mu^{\frac{1}{2}}.$$

Let the excitation of the system at $t=0$ be given by

(12) $$x = 0,\ \dot{x} = 0,\ X = 0,\ \dot{X} = C.$$

It follows that

$$\left.\begin{array}{c}a + a' = 0\\\gamma a + \gamma'a' = 0\end{array}\right\} a = a' = 0.$$

$$\left.\begin{array}{c}\omega b + \omega'b' = 0\\\gamma\omega b + \gamma'\omega'b' = C\end{array}\right\} b = \frac{C}{\omega(\gamma - \gamma')};\ b' = \frac{C}{\omega'(\gamma' - \gamma)}.$$

Thus we obtain the final solutions

(13)
$$x = \frac{C}{\gamma - \gamma'}\left(\frac{\sin\omega t}{\omega} - \frac{\sin\omega't}{\omega'}\right)$$
$$X = \frac{C}{\gamma - \gamma'}\left(\frac{\gamma}{\omega}\sin\omega t - \frac{\gamma'}{\omega'}\sin\omega't\right).$$

Let us pass from these to the velocities $\dot{x}$ and $\dot{X}$, taking (11) into account. We end up with

(14)
$$\dot{x} = \frac{C}{2\mu^{\frac{1}{2}}}(\cos\omega't - \cos\omega t),$$
$$\dot{X} = \frac{C}{2}(\cos\omega't + \cos\omega t).$$

Given the same phase, the velocity of the heavy upper bob is hence $\mu^{\frac{1}{2}}$ times smaller than that of the light lower one ; notice also that (14)

satisfies our initial conditions (12). The same can be said about the displacements themselves. Like the velocities, these are subject to beats because of the closeness of the values of ω and ω'. This modulation can be shown explicitly by writing the equations (13) and (14) in a form resembling Eq. (20.9).

We conclude the chapter with a problem which also pertains to the class of coupled oscillations and leads to oscillations very similar to the ones treated above. We shall, however, avail ourselves of a simpler mathematical method resembling that of the forced[6] undamped oscillations of § 19, so that we have to cope with the integration of only one differential equation rather than with that of a system of two simultaneous ones.

Let us suspend our pocket watch from a smooth nail, in such a way that the watch hangs completely free and friction is reduced to a minimum. By means of gentle contact with our fingers or a piece of cloth we bring the watch into a state of complete rest. When released, the timepiece at once begins to move, performing increasing oscillations about the vertical rest position. These oscillations reach a maximum, then gradually decrease once more to zero, after which the process repeats itself.

In these oscillations of the watch we are evidently confronted with a motion reacting against the rhythm of the balance wheel, i.e., a manifestation of the principle of conservation of angular momentum. The fluctuation of the oscillation amplitude, on the other hand, is caused by interference between the free pendulum oscillations of the watch in the gravitational field and the forced oscillations excited by the balance wheel.

We shall follow § 13, subsec. 2 in our notation. Accordingly we let M be the angular momentum of the total motion of the system. We decompose it into that of the pendulum motion (p) and that of the balance wheel oscillations (b),

$$(15) \qquad\qquad M = M_p + M_b.$$

M_p is calculated about the point of suspension O (nail), M_b about the center B of the balance wheel. The latter is permitted because a *pure* angular momentum (i.e., one caused by a motion in which the center of mass of the system remains fixed) can, just like a force couple (cf. p. 128), be shifted

---

[6] We can say quite generally that the excitation of forced oscillations in a system by means of an external force is equivalent to coupling with a second system on which the first one does not react. In the case about to be described it is certainly true that the reaction of the pendulum oscillations on the balance wheel is vanishingly small.

at will in its plane[7]; indeed, due to the symmetry of the balance wheel about B, the inertial action of the balance consists of a pure moment of momentum. Let ω be the circular frequency of the balance wheel; it is determined by the stiffness of the balance spring. Let ω_0 be the undisturbed, i.e., proper circular frequency of the pendulum oscillations. According to (11.6) and (16.4) we put

$$(16) \qquad M_p = I\dot{\phi}, \qquad I = m_p a^2;$$

m_p is the total mass of the watch, a its radius of gyration measured from O. We postulate a sinusoidal balance wheel oscillation which we shall therefore describe by $\phi_b = \alpha \sin \omega t$, B being the vertex of angle ϕ_b. The angular momentum of the balance wheel is then

$$(17) \qquad M_b = m_b \omega b^2 \alpha \cos \omega t,$$

where m_b is the mass of the balance wheel, b its radius of gyration measured from B.

As in the case of the compound pendulum [Eq. (16.1)] the moment of the external force is

$$(18) \qquad L = -m_p g s \phi,$$

where we have, as usual, made the approximation for small ϕ. Here s is the distance of the center of gravity of the watch from O, and ϕ the angle formed at O by the vertical and a line through the center of gravity. We now apply (13.9), use therein the values given by (15), (16), (17), and (18), and obtain the equation of motion

$$(19) \qquad \ddot{\phi} + \frac{gs}{a^2}\phi = \frac{m_b}{m_p}\left(\frac{b}{a}\right)^2 \alpha \omega^2 \sin \omega t$$

for our system.

This equation represents the type of oscillation which was treated in §19 as undamped forced oscillation. Again we put

$$\frac{gs}{a^2} = \omega_0^2$$

---

[7] This is a direct consequence of the fact that the angular momentum of a system about a given axis can be decomposed into the sum of the angular momentum of the system about a parallel axis through its mass center and the angular momentum of the mass center (containing the total mass of the system) about the given axis. In our case the latter term vanishes since the angular momentum of the mass center of the balance wheel due to the oscillation of the watch as a whole was included in M_{pend}.

where ω_0 is, we recall, the proper frequency of the pendulum motion; let us moreover abbreviate

$$c = \frac{m_b}{m_p} \left(\frac{b}{a}\right)^2 \alpha \omega^2 \ll 1.$$

Equation (19) becomes

(20) $$\ddot{\phi} + \omega_0^2 \phi = c \sin \omega t.$$

The solution satisfying the initial conditions $\phi = 0$, $\dot{\phi} = 0$ at $t = 0$ is

(21) $$\phi = \frac{c}{\omega_0^2 - \omega^2} \left(\sin \omega t - \frac{\omega}{\omega_0} \sin \omega_0 t\right).$$

The constant c is so small (factor m_b/m_p) that the oscillation is of visible magnitude only when the relation $\omega_0 = \omega$ is approximated, i.e., when approximate resonance exists between the external pendulum oscillations and the internal oscillations of the balance wheel. Surprisingly it turns out that this resonance is more or less well realized in pocket watches of not too small a size (ladies' watches are unsuitable for our purposes).

Eq. (21) further shows that amplitude modulation goes hand in hand with the approach to resonance $\omega_0 \longrightarrow \omega$. The period T of the beats is determined by the requirement

(22) $$\omega T = \omega_0 T \pm 2\pi,$$

and has therefore the value

(22a) $$T = \frac{2\pi}{|\omega - \omega_0|}.$$

It can be determined very accurately by counting the number of pendulum oscillations between two nodes of the beats, and furnishes therefore a convenient and precise measure of the degree of resonance. We can refer back to Fig. 32 which, as pointed out, represents the same differential equation as (20); we must, however, bear in mind that in the diagram we postulated complete resonance, i.e., $T = \infty$.

If one leaves the watch to itself for some time, one observes that the beats have ceased. The reason for this is evidently friction (at the point of suspension and in the air), which we have so far neglected. This friction damps the contribution of the free pendulum oscillations to the motion of the watch, leaving only the forced oscillations due to the motion of the balance wheel, the latter contribution (cf. Fig. 33, for instance) being somewhat reduced in amplitude due to friction. We can reason as follows: initially the forced oscillation is present in its full amount, and the free pendulum oscillation is excited to such a degree that at $t = 0$ it just cancels the forced one — in agreement with the initial conditions $\dot{\phi} = \phi = 0$.

Indeed the initially motionless state of the watch can be interpreted as being caused by an impulse exactly cancelling the balance wheel oscillation. The effect of this impulse is gradually used up by friction, so that only the forced oscillation due to the balance wheel remains.

The example of the watch appeared in the literature for the first time in the " Elektrotechnische Zeitschrift " of the year 1904, in connection with the phenomenon of " hunting " of synchronous machinery, then timely and surprising. Two synchronous alternators feeding the same power line and connected in parallel show undesirable fluctuations in their motions and their currents when resonance occurs. They provide a greatly magnified picture of the beats of our watch and of the coupling and resonance phenomena occurring in the coupled oscillations that we have just analyzed.

CHAPTER IV

THE RIGID BODY

§ 22. Kinematics of Rigid Bodies

At the beginning of § 7 we saw that a rigid body is endowed with six degrees of freedom; these we shall subdivide into three of translation and three of rotation.

Let us consider the body in two different positions, the " initial position " and the " final position." We pick out an arbitrary point of the body as " point of reference " O, and describe a sphere of reference (say of unit radius) about it. On this sphere we mark two points A and B. Once we have guided the three points OAB from their initial positions to their final ones, all other points of the rigid body have similarly reached their destinations.

First we take the point O from its initial position O_1 to its final position O_2. Let this be achieved by means of a parallel displacement or *translation* in which each point of the body is subjected to the same rectilinear displacement $O_1 \rightarrow O_2$. We have thus described the three degrees of freedom of translation.

The sphere K_1 described about O_1 is now in coincidence with the corresponding sphere K_2 described about O_2. In general this is not true of the position of the points A, B, which we designate by A_1, B_1 on K_1 and A_2, B_2 on K_2. We shall show that there is one definite *rotation* about the point $O_1 = O_2$ which will take points A_1, B_1 over into A_2, B_2. Axis and angle of this rotation define the three degrees of freedom of rotation to be added to those of translation.

In order to construct the axis of rotation, i.e., the point Ω at which the axis cuts the unit sphere, we connect A_1 to A_2 and B_1 to B_2 by means of arcs of great circles. At the centers A' and B' of these arcs we erect their perpendicular bisectors whose intersection is the point Ω in question. The angle of rotation, which we shall also call Ω, is

(1) $$\Omega = \sphericalangle A_1 \Omega A_2 = \sphericalangle B_1 \Omega B_2.$$

The equality of these two angles results from the congruence of the shaded spherical triangles $A_1 \Omega B_1$ and $A_2 \Omega B_2$ of Fig. 39, whose three corresponding sides are equal to each other. It follows that the two angles designated by γ in Fig. 39 are equal. If we subtract one or the other of these angles from the total angle $A_1 \Omega B_2$ we obtain the right or middle member of Eq. (1). This equation evidently states that the same rotation Ω not only takes point A_1 over into A_2, but also point B_1 into B_2.

So far the magnitude and direction of the translation are still arbitrary[1] within wide limits, for we have free choice over the reference point O. The magnitude and axis of the rotation, on the other hand, are *independent* of the choice of the reference point. For let us substitute for O a new reference point O'. The difference between the translations associated with O' and O for a given total displacement of the rigid body is again a translation. This latter translation, however, does not affect the positions of the points A, B on the spheres K_1 and K_2. It follows that the construction of Fig. 39 carries over unchanged to the present case and

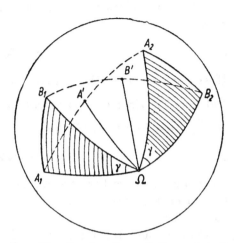

Fig. 39. Construction of the point Ω determining the axis of rotation for a rigid body revolving about a fixed point O. This diagram also suggests how the resultant of two finite rotations can be found.

yields not only the same angle of rotation Ω as previously, but also an axis of rotation passing through the point of reference O' and parallel to our former axis.

Of much greater importance than finite displacements of the rigid body are its infinitesimal displacements which succeed each other continuously to result in a finite motion. We shall therefore assume that now the magnitude $O_1 O_2$ of the translation and the angle Ω of rotation are arbitrarily small. Let us divide them by the correspondingly small interval of time Δt. We then obtain the velocity **u** of translation and the angular velocity ω of rotation,

$$(2) \qquad \mathbf{u} = \frac{\overrightarrow{O_1 O_2}}{\Delta t}, \quad \omega = \frac{\Omega}{\Delta t}.$$

As before, the angular velocity is independent of the choice of reference point O, whereas **u** depends on this choice. The heavy type indicates that

---

[1] In the supplement to § 23 we shall see that we can, in particular, make the direction of the translation parallel to the axis of rotation. We then speak of a " screw displacement."

ω is to be regarded as a vector which expresses not only the magnitude, but also the axial direction of rotation of the angular velocity.

We can easily show that ω does indeed possess vector character. In Fig. 15 and Eq. (13.4), while discussing virtual rotations, we derived the relation

(3) $$\delta s = \delta \phi \times r.$$

If we now pass from the virtual rotation $\delta \phi$ to the angular velocity $\omega = \dfrac{d\phi}{dt}$ and from the virtual displacement δs caused by the rotation to the velocity $w = \dfrac{ds}{dt}$, we obtain from (3) that

(4) $$w = \omega \times r.$$

As in Fig. 15, r is here the radius vector from the point of reference O located on the axis of rotation to the point P whose velocity w is to be determined.

Consider now the total effect of two successive infinitesimal rotations $\omega_1 dt$ and $\omega_2 dt$ on the motion of the point P of the rigid body, reference point O being common to both axes ω_1 and ω_2. We have

(4a) $$w_1 = \omega_1 \times r, \quad w_2 = \omega_2 \times r, \quad w_1 + w_2 = (\omega_1 + \omega_2) \times r.$$

In the last of these equations the left member is the velocity w_r resulting from w_1 and w_2. A comparison with (4) shows that

(5) $$\omega_r = \omega_1 + \omega_2$$

is likewise the resultant angular velocity, equivalent to the two rotations $\omega_1 dt$ and $\omega_2 dt$ in its effect on the rigid body. We conclude that *angular velocities add like vectors*. As in the case of vectors their order in addition is immaterial, i.e., *their addition is commutative*, for

(6) $$\omega_1 + \omega_2 = \omega_2 + \omega_1.$$

Neither of these two laws is valid for finite rotations. Their composition does not follow the simple rules of vector algebra, but those of the algebra of *quaternions* invented by Hamilton. Moreover the effect of two finite rotations depends on their order; two such rotations do not commute.

At this point it is convenient to discuss the difference between polar and axial vectors.

Examples of *polar vectors* are velocity, acceleration, force, radius vector, etc. They can be represented by directed segments provided with an arrow head. In a rotation of the system of coordinates their rectangular components transform like the coordinates themselves, i.e., according to the scheme of orthogonal transformations with determinant $+1$. In an inversion of the coordinate system through the origin, in which x, y, z are replaced by $-x$, $-y$, $-z$ respectively so that the transformation has determinant -1, the components of polar vectors change sign.

Angular velocity, angular acceleration, torque and angular momentum are examples of *axial vectors*. In accordance with their nature they are represented by an axis on which sense and magnitude of rotation are indicated (e.g., by a curved arrow and a number). If, instead, we represent them by means of an arrow of corresponding magnitude laid off on the axis, we must make some arbitrary agreement about the direction of this arrow, such as the rule of the right-handed screw. In a pure rotation of the coordinate system the rectangular components of axial vectors transform like the components of their associated arrows, i.e., orthogonally ; in an inversion of the coordinates through the origin, however, these rectangular components do not change sign. In such a transformation the rule of the right-handed screw must be replaced by that of the left-handed screw, in agreement with the fact that an inversion through the origin takes a right-handed coordinate system over into a left-handed one.

The vector product of two polar vectors is an axial vector (e.g., the moment of a force). The vector product of an axial and a polar vector is a polar vector [e.g., the velocity **w** in Eq. (4)]. The reader may easily convince himself of this by checking the behavior of these products under inversion of coordinates.[2]

After this digression we return to the kinematics of the rigid body. The motion of each one of its points is composed of the velocity **u** of Eq. (2) connected with translation and the velocity **w** in Eq. (4) connected with rotation. The velocity **v** of an arbitrary point of the rigid body is hence given by

(7) $$\mathbf{v} = \mathbf{u} + \boldsymbol{\omega} \times \mathbf{r}.$$

The choice of the reference point O is completely up to us; for it we have

(7a) $$\mathbf{v} = \mathbf{u}.$$

For many purposes it is advantageous to put O at the mass center G. This becomes evident if, for instance, we wish to calculate the kinetic

---

[2] From now on we shall simply talk about the torque **L**, the angular velocity $\boldsymbol{\omega}$, where the reader should bear in mind that we mean by this the axial vectors representing the torque and the angular velocity respectively. When, on the other hand, we speak of the plane of the torque and the plane of the angular velocity, we mean, of course, the planes perpendicular to the axial vectors **L** and $\boldsymbol{\omega}$ respectively.

energy of the body,

(8) $$T = \int \frac{dm}{2}\, \mathbf{v}^2.$$

To this end we form, with the help of (7),

(8a) $$v^2 = u^2 + (\boldsymbol{\omega} \times \mathbf{r})^2 + 2\mathbf{u} \cdot (\boldsymbol{\omega} \times \mathbf{r})$$

and accordingly break up T into three parts,

(9) $$T = T_{\text{transl}} + T_{\text{rot}} + T_m,$$

where T_m is a "mixed" energy which is determined by the translation and the rotation combined.

Since $\mathbf{u}$ has the same value for all points dm, we evidently have

(10) $$T_{\text{transl}} = \frac{u^2}{2} \int dm = \frac{m}{2} u^2.$$

In order to calculate T_m we perform the transformation

(11) $$T_m = \int \mathbf{u} \cdot \boldsymbol{\omega} \times \mathbf{r}\, dm = \mathbf{u} \cdot \boldsymbol{\omega} \times \int \mathbf{r}\, dm = m\, \mathbf{u} \cdot \boldsymbol{\omega} \times \mathbf{R},$$

where $\mathbf{R}$ is the directed segment from O to the mass center G,

(11a) $$\mathbf{R} = \frac{1}{m} \int \mathbf{r}\, dm$$

as in Eq. (13.3b).   If now we let O coincide with G, we have $\mathbf{R} = 0$ and, from (11),

(11b) $$T_m = 0.$$

The kinetic energy T then becomes simply the sum of T_{transl} and T_{rot}. Notice, in passing, that if the body rotates about a fixed point and if one chooses this fixed point as reference point O, not only T_m, but also T_{transl} vanishes (in both cases because $\mathbf{u} = 0$), so that

(11c) $$T = T_{\text{rot}}.$$

We shall now focus our attention on the rotational contribution to the kinetic energy.   If we square the components of $\boldsymbol{\omega} \times \mathbf{r}$, we obtain from the middle term of the right member of (8a)

(12) $$2T_{\text{rot}} = \omega_x^2 \int (y^2 + z^2)\, dm + \omega_y^2 \int (z^2 + x^2)\, dm + \omega_z^2 \int (x^2 + y^2)\, dm$$
$$- 2\omega_y \omega_z \int yz\, dm - 2\omega_z \omega_x \int zx\, dm - 2\omega_x \omega_y \int xy\, dm.$$

With the notation

(12a)
$$I_{xx} = \int (y^2 + z^2)\, dm \, \ldots$$

$$I_{xy} = \int xy\, dm \, \ldots$$

this yields

(12b) $2T_{\text{rot}} = I_{xx}\omega_x^2 + I_{yy}\omega_y^2 + I_{zz}\omega_z^2 - 2I_{yz}\omega_y\omega_z - 2I_{zx}\omega_z\omega_x - 2I_{xy}\omega_x\omega_y.$

According to the definition introduced in (11.3), I_{xx} is the moment of inertia of the mass distribution about the x-axis; a corresponding statement holds for I_{yy} and I_{zz}. We shall call I_{xy}, I_{yz}, I_{zx} the *products of inertia* (the name " centrifugal moments " is sometimes used synonymously). We can also abbreviate I_{xx}, $\ldots$ without ambiguity to I_x, $\ldots$

In accordance with (11.5) we put the left member of (12) equal to $I\omega^2$ and with the abbreviations

(13) $\dfrac{\omega_x}{\omega} = \alpha, \quad \dfrac{\omega_y}{\omega} = \beta, \quad \dfrac{\omega_z}{\omega} = \gamma$

obtain

(13a) $I = I_{xx}\alpha^2 + I_{yy}\beta^2 + I_{zz}\gamma^2 - 2I_{yz}\beta\gamma - 2I_{zx}\gamma\alpha - 2I_{xy}\alpha\beta.$

α, β, γ are the direction cosines of the vector $\boldsymbol{\omega}$ whose axis is arbitrarily located in the rigid body. It follows from (13a) that the moment of inertia about any axis is completely determined once the six magnitudes I_{ik} are given.

A sextet of magnitudes of the type of our I_{ik} is called a *tensor*, or, more precisely, a *symmetrical tensor*. The name originated in the theory of elasticity where stress and strain tensors play a central role. In general a tensor is very aptly written as a square scheme, which in our case would be

(13b) $I_{ik} = \begin{pmatrix} I_{xx} & -I_{xy} & -I_{xz} \\ -I_{yx} & I_{yy} & -I_{yz} \\ -I_{zx} & -I_{zy} & I_{zz} \end{pmatrix}$

where $I_{xy} = I_{yx}$, $\ldots$

From an elementary viewpoint the mathematics of tensors is less concrete and easily intelligible than that of vectors. Whereas a vector is represented by a line segment, we must resort to a surface of second degree for the geometrical representation of a tensor. In our case this " tensor surface " is obtained as follows: we put

(14) $\alpha = \dfrac{\xi}{\rho}, \quad \beta = \dfrac{\eta}{\rho}, \quad \gamma = \dfrac{\zeta}{\rho},$

where ξ, η, ζ are interpreted as Cartesian coordinates, hence $\rho = (\xi^2 + \eta^2 + \zeta^2)^{\frac{1}{2}}$

as radius vector from the point O. We now set ρ equal to $I^{-\frac{1}{2}}$, so that along every axis through O we lay off, not I, but rather the reciprocal of $I^{\frac{1}{2}}$ (else we should not obtain a surface of *second* degree). In this manner we obtain from (13a)

(15) $$1 = I_{xx}\xi^2 + I_{yy}\eta^2 + I_{zz}\zeta^2 - 2I_{yz}\eta\zeta - 2I_{zx}\zeta\xi - 2I_{xy}\xi\eta.$$

Apart from possible degeneracies this is the equation of an ellipsoid, since for a finite mass distribution I is, in general, greater than zero. The surface represented by (15) is called the *momental ellipsoid*.

If one transforms the coordinates so that they coincide with the principal axes of the ellipsoid, one obtains an equation of the form

(15a) $$1 = I_1\xi_1^2 + I_2\xi_2^2 + I_3\xi_3^2,$$

where I_1, I_2, I_3 are the three *principal moments of inertia*. The products of inertia vanish for the principal axes, which can be regarded as a definition of the latter. The tensor scheme (13b) reduces to diagonal form. When the tensor is described in a system of coordinates different from that of the principal axes one must mentally add the three direction parameters of the principal axes; thus we are again lead to the six magnitudes characterizing a symmetrical tensor.

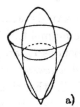 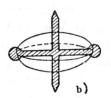 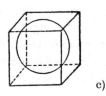

FIG. 40a—c.   (a) Momental ellipsoid of the toy top.   (b) Momental ellipsoid of the flywheel top.   (c) An example of a spherical top.

Every plane of symmetry of the mass distribution is of course also a plane of symmetry of the momental ellipsoid. A mass distribution with rotational symmetry has a momental ellipsoid of revolution, i.e., in addition to the principal axis along the " axis of figure " it possesses infinitely many other " equatorial " principal axes. As examples we may mention two types of tops; one is of the conical type used as toy, the other has the shape of a flywheel and is usually employed for demonstration purposes (Figs. 40a and b). In the first type the moment of inertia about the axis of the body is a minimum, so that the corresponding principal axis is longer than the equatorial ones (by virtue of the relation $\rho = I^{-\frac{1}{2}}$); we have a *prolate spheroid*. In the second case the moment of inertia about the axis of figure is a maximum, hence the corresponding principal axis is, for the same reason, smaller than the equatorial ones; the result is an *oblate spheroid*.

Incidentally a momental ellipsoid becomes one of revolution not only for mass distributions with rotational symmetry, but also whenever more than two planes of symmetry pass through an axis, as for example in the case of a square or hexagonal prism.

Similarly the ellipsoid degenerates into a sphere not only in the case of a spherically symmetrical distribution, but · also in cases such as that of a cubical distribution, for instance, because here there exist more planes of symmetry than are compatible with the ellipsoidal shape of the tensor surface. In such a case we speak of a "spherical top." In a spherical top (cf. Fig. 40c) any axis is a principal axis.

§ 23. *Statics of Rigid Bodies*

This subject forms the theoretical basis for the whole field of structural mechanics dealing with such topics as the construction of bridges, trusses, arches, etc., and for this reason it is treated with the greatest detail in the texts of mechanical engineering, both analytically and graphically. Here we shall restrict ourselves to the general features of the subject.

(1) The Conditions of Equilibrium

These, like all questions of equilibrium, are governed by the principle of virtual work. Since this principle can be regarded as the special case of d'Alembert's principle in which the inertial forces vanish, our present analysis can be directly modeled after that of the principles of linear and angular momentum of § 13. Indeed the virtual displacements (translation and rotation) used there are evidently compatible with the internal connections of the rigid body and correspond to the two component parts of the general motion of a rigid body considered in the preceding section.

By deleting the inertial forces in Eqs. (13.3) and (13.9), we obtain the general conditions of equilibrium of a rigid body,

(1) $$\sum \mathbf{F}_k = 0, \quad \sum \mathbf{L}_k = 0.$$

The $\mathbf{F}_k$ are external forces acting at arbitrary points P_k of the rigid body. The first Eq. (1) asks us to lay off the force vectors end to end in arbitrary order and with no regard to their points of application, and to examine the resulting *force polygon*. According to Eq. (1) *for equilibrium the polygon of forces must be closed*.

The $\mathbf{L}_k$ are the moments of the $\mathbf{F}_k$ about a reference point O whose choice is arbitrary but which must be the same for all the $\mathbf{F}_k$. The second Eq. (1) asks us to replace these $\mathbf{L}_k$ by their (axial) vector representations (cf. p. 37) and to examine the polygon of torques arising when all these vectors are added vectorially. According to the second Eq. (1) *the torque polygon must also be closed for equilibrium.*

In analogy to Eqs. (13.12) and (13.13) we can pass from the two vector equations (1) to the following six component equations:

(2)
$$\sum X_k = \sum Y_k = \sum Z_k = 0$$
$$\sum(y_k Z_k - z_k Y_k) = \sum(z_k X_k - x_k Z_k) = \sum(x_k Y_k - y_k X_k) = 0.$$

These represent the projections of the vector equations (1) on the coordinate axes; the x_k, y_k, z_k are the coordinates of the points of application, measured from O as origin.

(2) Equipollence; the Reduction of Force Systems

If the external forces (or torques) are not in equilibrium, we can ask whether there exists a single force (or single torque) of such properties that under its action alone the rigid body moves in the same way as it would under the action of the given system of forces (or torques).

Posing this question is, among other things, useful (even though in general not sufficient) for the determination of the forces which are exerted on a rigid body by its supports if the rigid body is acted on by a system of forces which themselves are not sufficient to bring about a state of equilibrium.

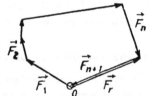

FIG. 41. Construction of the resultant force for an "open" polygon of forces.

We obtain the answer by drawing the closing segment in the now "open" polygon $F_1, F_2, \ldots F_n$, once in the direction in which the polygon is traced (F_{n+1}) and once (cf. Fig. 41) in the opposite direction (F_r, resultant force). Nothing is changed thereby. We have now a closed force polygon $F_1, \ldots F_{n+1}$, and a single force F_r, which, taken together, are *equipollent* to the "open" polygon of forces $F_1, \ldots F_n$. The forces $F_1, \ldots F_{n+1}$ are, however, in equilibrium and can therefore be left out, so that the single force F_r is equipollent to the given system of forces $F_1, \ldots F_n$. Mathematically,

(3)
$$F_r = \sum_{k=1}^{n} F_k.$$

The same process of reasoning can be carried out with an "open" torque polygon. One thereby obtains a resultant force moment L_r which is equipollent to the given system of moments $L_1, L_2, \ldots L_n$, i.e.,

$$(4) \qquad\qquad L_r = \sum_{k=1}^{n} L_k.$$

Let us mention in passing that there is nothing to stop us from making the single force F_r act at the same point O which serves as reference point in the calculation of moments L_k. This choice is indicated in Fig. 41.

(3) Change of Reference Point

Eq. (3) immediately shows that F_r is independent of the choice of reference point O. If F_r' is the resultant single force associated with a different reference point O', we therefore have

$$(5) \qquad\qquad F_r' = F_r.$$

From Eq. (4), on the other hand, we have with corresponding meaning of L_r'

$$(6) \qquad\qquad L_r' = \sum_{k=1}^{n} L_k' \quad \text{with} \quad L_k' = r_k' \times F_k,$$

where r_k' is the radius vector from O' to the point of application P_k of F_k. Let a be the vector distance from O' to O. Then

$$(6a) \qquad r_k' = a + r_k, \qquad L_k' = a \times F_k + r_k \times F_k = a \times F_k + L_k.$$

Therefore

$$(6b) \qquad L_r' = \sum_{k=1}^{n} a \times F_k + \sum_{k=1}^{n} L_k = a \times \sum_{k=1}^{n} F_k + L_r.$$

But in view of (3)

$$a \times \sum_{k=1}^{n} F_k = a \times F_r.$$

Thus we have

$$(7) \qquad\qquad L_r' = L_r + a \times F_r.$$

(4) Comparison of Kinematics and Statics

As remarked in connection with Eq. (22.2), in kinematics ω is independent of the choice of reference point, whereas u depends on that choice. We write

$$(8) \qquad\qquad \omega' = \omega$$

and, from (22.7), with $\mathbf{v}=\mathbf{u}'$ and $\mathbf{r}=\mathbf{a}$,

(9) $$\mathbf{u}'=\mathbf{u}+\omega\times\mathbf{a}.$$

This equation has the same structure as the preceding Eq. (7) provided we disregard the sequence of the factors in the corresponding vector products. If we also take into consideration Eqs. (5) and (8), we arrive at a remarkable reciprocity between statics and kinematics which can be expressed by the scheme below:

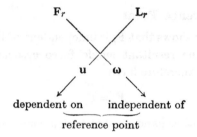

This crosswise reciprocity holds as well between the concepts of force couple and rotational couple which we shall now take up.

The *force couple* (or " couple," for short) is a basic element in elementary statics. As is well known, a couple consists of two parallel and opposite forces of equal magnitude, $\pm\mathbf{F}$, whose lines of action are a finite distance, say l, apart. If we carry out the reduction of such a couple in the sense of subsec. 2, we obtain

(10) $$\mathbf{F}_r=0, \quad \mathbf{L}_r=\overrightarrow{L}, \quad |\overrightarrow{L}|=|\mathbf{F}|\ l,$$

where one should think of the vector $\overrightarrow{L}$ as directed normal to the plane of the two forces. Whereas, however, the former $\mathbf{L}_r$ was, so to speak, attached to the reference point O, our present $\overrightarrow{L}$ is the same for all reference points and completely free to move in space; i.e., two given couples can be added vectorially to yield a third couple ; two couples of equal and opposite moments located in parallel planes cancel, etc.

Let us follow up the crosswise reciprocity indicated by our scheme by defining a rotational couple. By a *rotational couple* we understand two equal and opposite rotational velocities $\pm\omega$, whose axes are parallel to each other and a distance l apart. According to the rule of addition (22.5), the reduction of a rotational couple yields a resultant rotational velocity $\omega_r=0$. Our rotational couple generates then a pure translation perpendicular to the plane of the two axes of rotation. The magnitude of the velocity of translation is easily found to be $|\overrightarrow{u}|=\omega l$. The analogy to Eqs. (10) in the sense of our reciprocity scheme is therefore complete. Whereas our former **u** depended on the choice of the reference point O, the $\overrightarrow{u}$ equivalent to a rotational couple is independent of O, and can be translated parallel to itself in space in any manner whatever. From this it follows that two arbitrarily located rotational couples add vectorially just like their velocities of translation $\overrightarrow{u}$; two rotational couples of equal and opposite moment $\pm \omega l$ located in parallel planes cancel, etc.

SUPPLEMENT: WRENCHES AND SCREW DISPLACEMENTS

From (7) we see that $\mathbf{L}_r$ depends on the reference point. We are therefore tempted to choose this point in such a way that $\mathbf{L}_r$ and $\mathbf{F}_r$ become parallel. We then obtain an especially simple picture of our system of forces called a *wrench*, i.e., a single force and a moment acting about this force or, equivalently, a couple located in a plane perpendicular to the force. If our initial reference point is O, the position of O' required for a wrench is obtained as follows: in Eq. (7) we decompose $\mathbf{L}_r$ into $\mathbf{L}_p$ parallel to $\mathbf{F}_r$ and $\mathbf{L}_n$ perpendicular to it and determine **a** from the equation

(11) $$\mathbf{L}_n = -\mathbf{a} \times \mathbf{F}_r.$$

From (5) and (7) we then have for the reference point O',

$$\mathbf{F}'_r = \mathbf{F}_r, \qquad \mathbf{L}'_r = \mathbf{L}_p \parallel \mathbf{F}_r$$

as demanded by the definition of a wrench. Eq. (11) states that for this purpose the reference point O must be displaced a certain distance

$$a = -\frac{|\mathbf{L}_n|}{|\mathbf{F}_r|}$$

normal to $\mathbf{F}_r$ and $\mathbf{L}_n$.

A line of reasoning exactly reciprocal in the sense of the preceding discussion leads to the *screw displacement*. With Eq. (9) as starting point we decompose **u** into $\mathbf{u}_p$ parallel to ω and $\mathbf{u}_n$ perpendicular to it. The displacement **a** of the reference point required for a screw is determined by the equation

(12) $$\mathbf{u}_n = -\omega \times \mathbf{a}.$$

From (8) and (9) we then obtain for the reference point O',

(13) $$\omega' = \omega, \qquad \mathbf{u}' = \mathbf{u}_p \parallel \omega,$$

which in fact represents a screw displacement. Eq. (12) states that the reference point O must here be displaced by a certain distance normal to ω and $\mathbf{u}_n$.

Attractive as the concept of the wrench and the screw displacement may be, it is of no great practical value in the treatment of specific problems involving rotation. For this reason mention of them has been relegated to a supplement.

§ 24. Linear and Angular Momentum of a Rigid Body. Their Connection with Linear and Angular Velocity

Let us imagine that a momentum of translation (linear momentum, impulsive force) and a momentum of rotation (moment of momentum, impulsive torque) have been imparted to a rigid body. Let the first one of these be designated by the letter $\mathbf{p}$, the latter by $\mathbf{M}$.

$\mathbf{p}$ is calculated as a sum over all linear momenta $d\mathbf{p} = \mathbf{v}\,dm$, i.e.,

$$(1) \qquad \mathbf{p} = \int d\mathbf{p} = \int \mathbf{v}\,dm.$$

With the help of Eq. (22.7) we get

$$\mathbf{p} = \mathbf{u}\int dm + \omega \mathbf{X} \int \mathbf{r}\,dm$$

or, with introduction of the radius vector $\mathbf{R}$ from O to the center of mass, cf. (22.11a),

$$(2) \qquad \mathbf{p} = m\mathbf{u} + m\omega \mathbf{X} \mathbf{R}.$$

In particular, if we choose $O = G$, we have $\mathbf{R} = 0$ and

$$(3) \qquad \mathbf{p} = m\mathbf{u}.$$

The angular momentum $\mathbf{M}$ of the rigid body, on the other hand, is composed of the moments of all the elements of linear momentum taken about the common reference point O. We therefore have

$$(4) \qquad \mathbf{M} = \int \mathbf{r}\mathbf{X}\,d\mathbf{p} = \int dm\,(\mathbf{r}\mathbf{X}\mathbf{v}),$$

from which, because of (22.7) and (22.11a),

$$(5) \qquad \mathbf{M} = \int dm\,(\mathbf{r}\mathbf{X}\mathbf{u}) + \int dm\,\mathbf{r}\mathbf{X}(\omega\mathbf{X}\mathbf{r}) = m\mathbf{R}\mathbf{X}\mathbf{u} + \int dm\,\mathbf{r}\mathbf{X}(\omega\mathbf{X}\mathbf{r}).$$

The first term on the right vanishes for $O = G$ as well as for $\mathbf{u} = 0$, so that in both these cases

$$(6) \qquad \mathbf{M} = \int dm\,\mathbf{r}\mathbf{X}(\omega\mathbf{X}\mathbf{r}).$$

In order to evaluate this integral we remind the reader of the vector rule for the triple cross-product, valid for any three vectors $\mathbf{A}$, $\mathbf{B}$, $\mathbf{C}$,

$$(7) \qquad \mathbf{A}\mathbf{X}(\mathbf{B}\mathbf{X}\mathbf{C}) = \mathbf{B}(\mathbf{A}\cdot\mathbf{C}) - \mathbf{C}(\mathbf{A}\cdot\mathbf{B}).$$

It follows that

$$\mathbf{r} \times (\boldsymbol{\omega} \times \mathbf{r}) = \boldsymbol{\omega}\, r^2 - \mathbf{r}(\boldsymbol{\omega} \cdot \mathbf{r})$$

and therefore, taking the x-component as example,

(8)
$$M_x = \int [\mathbf{r} \times (\boldsymbol{\omega} \times \mathbf{r})]_x\, dm$$
$$= \omega_x \int (x^2 + y^2 + z^2)\, dm - \omega_x \int x^2\, dm - \omega_y \int x\, y\, dm - \omega_z \int x\, z\, dm.$$

By introducing the moments and products of inertia from (22.12a), we can then write (6) in the form

(9)
$$\begin{aligned}
M_x &= I_{xx}\omega_x - I_{xy}\omega_y - I_{xz}\omega_z \\
M_y &= -I_{yx}\omega_x + I_{yy}\omega_y - I_{yz}\omega_z \\
M_z &= -I_{zx}\omega_x - I_{zy}\omega_y + I_{zz}\omega_z.
\end{aligned}$$

We have thus arrived at a linear relation between the dynamic vector $\mathbf{M}$ and the kinematic vector $\boldsymbol{\omega}$; this relation is achieved by means of the tensor I of Eq. (22.13b). We therefore say that $\mathbf{M}$ is a "linear vector function" of $\boldsymbol{\omega}$. Such linear vector functions play an important role in all aspects of the tensor calculus, especially in the theory of elasticity (cf. Vol. II, this series).

Eqs. (9) can be put into instructive form if we make use of expression (22.12b) for the kinetic energy of rotation. For then we simply have

(10)
$$M_i = \frac{\partial T_{\text{rot}}}{\partial \omega_i}, \quad i = x, y, z.$$

Notice, moreover, that this expression is valid not only for the case $O = G$ or $\mathbf{u} = 0$ presupposed in (9), but also for $\mathbf{u} \neq 0$ and arbitrary position of O. For in the more general case one need only complete expression (22.12b) for T_{rot} by adding expression (22.11) for T_m, so that the term

$$\frac{\partial T_m}{\partial \omega_i} = m(\mathbf{R} \times \mathbf{u})_i$$

will be added on the right of Eq. (10). But this is the same term which appears on the right side of Eq. (5) for $\mathbf{M}$ whenever O and G do not coincide. Since, finally, the total kinetic energy T differs from $T_{\text{rot}} + T_m$ only by the term T_{transl} independent of $\boldsymbol{\omega}$ [cf. (22.9) and (22.10)], we can generalize (10) in the form

(10a)
$$M_i = \frac{\partial T}{\partial \omega_i}, \quad i = x, y, z$$

valid for arbitrary position of O.

What has been said of the angular momentum $\mathbf{M}$ is also valid for the linear momentum **p**. Here we consider at once the general case $O \neq G$ and from Eqs. (22.9), (22.10) and (22.11) form

$$\frac{\partial T}{\partial u_i} = m u_i + m(\omega \times \mathbf{R})_i$$

which is in agreement with Eq. (2) for **p**. The equation complementary to (10a) is therefore

(11) $$p_i = \frac{\partial T}{\partial u_i}, \quad i = x, y, z.$$

Eqs. (10a) and (11) are special cases of a much more general relationship connecting momentum and velocity coordinates of an arbitrary mechanical system. The proof of this must be postponed to Chapter VI, § 36. Here we shall only concern ourselves with the geometrical meaning of Eq. (10), which leads us to the celebrated geometrical construction of Poinsot. The *Poinsot method* tells us how to find the position of the axis of angular momentum **M** with reference to a given axis of rotation. The same can be said of this method as of the foregoing equations, namely, that it is not restricted to the case of the rigid body, but is applicable whenever one deals with a symmetric tensor ; one represents this tensor by means of a tensor surface of second degree, and asks for the linear vector function given by means of this tensor.

The Poinsot construction runs as follows : from the center O of the momental ellipsoid we lay off the angular velocity vector ω and construct the tangent plane to the ellipsoid at the point where ω intersects it. The perpendicular from O to this tangent plane gives the direction of **M**. As proof we need merely recall that for an arbitrary surface $f(\xi, \eta, \zeta) = \text{const.}$, the direction cosines of the normal to the tangent plane are proportional to

(12) $$\frac{\partial f}{\partial \xi}, \frac{\partial f}{\partial \eta}, \frac{\partial f}{\partial \zeta}.$$

In our case $f(\xi, \eta, \zeta) = \text{const.}$ is the equation (22.15) of the momental ellipsoid and its derivatives with respect to ξ, η, ζ are indeed proportional to the components of **M** of Eq. (9).

We may also interpret the Poinsot construction as the direct geometrical expression of our Eq. (10), for the momental ellipsoid is essentially identical to the surface $T_{\text{rot}} = \text{const.}$

Our Figs. 42a, b represent the case of the symmetrical momental ellipsoid, where ω, **M** and the axis of symmetry (" axis of figure ") **f** are coplanar ; so that the tangent plane can be represented as the tangent to the cross-section of the ellipse in this plane. In the prolate ellipsoid of revolution (i.e., spheroid), Fig. 42b, **M** and **f**, the axis, lie on opposite sides of ω ; in the oblate spheroid of Fig. 42a, **M** lies between **f** and ω. The case of the ellipsoid with three unequal axes presents a more difficult graphical problem.

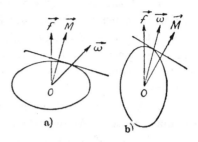

In conclusion we emphasize that the relations discussed in this section are basically nothing but expressions of the Newtonian definition, "the quantity of motion is the measure of the same, arising from the velocity and the quantity of matter conjunctly," extended to the rigid body. The reason why our present relations are so much more involved than the one between momentum and velocity of a single particle is that in particle

Fig. 42. Poinsot construction giving the relative position of angular velocity ω and angular momentum **M** for the two cases where the momental ellipsoid degenerates into a) an oblate spheroid, b) a prolate spheroid.

mechanics the "quantity of matter," i.e., the mass, is a *scalar*, whereas in the case of the rigid body the moment of inertia that takes its place is a *tensor*.

§ 25. *Dynamics of a Rigid Body. Survey of its Forms of Motion*

Let us first consider the rigid body moving freely in space. As reference point we choose its center of mass, and reduce all the forces acting on the body to forces acting on this point, in agreement with the prescription of § 23. We need then only deal with a single resultant force **F** and a resultant torque **L**. The equations of motion are the equations of momentum and of moment of momentum of § 13 ; they read

$$(1) \qquad\qquad \dot{\mathbf{p}} = \mathbf{F},$$

$$(2) \qquad\qquad \dot{\mathbf{M}} = \mathbf{L}.$$

Since the rigid body possesses but six degrees of freedom, these two vector equations suffice for the complete description of its state of motion.

Eqs. (1) and (2) can be treated separately whenever **F** is independent of the angular velocity and **L** independent of the translational velocity. In ballistics, for instance, this is not the case. If it is the case, (1) becomes a problem of pure particle mechanics, (2) a problem of rotation about a fixed point or, as we shall say for brevity, a " problem of the spinning top."

At this point we shall be interested principally in the latter. With the choice of reference point made above, we can disregard the force of gravity since it has no moment about the mass center. If, furthermore, we neglect air resistance, friction and the like, we are confronted with the problem of the *spinning top under no forces*. Thus the gyroscope in a Cardan suspension (cf. Fig. 47) is a top under no forces provided we can neglect the mass of the gimbals in comparison with that of the flywheel, which is approximately valid in the usual constructions. Otherwise we would be confronted with a considerably more involved mathematical problem.

We shall also deal with rotation about a fixed point other than the mass center. As remarked on p. 122, it is then advisable to take this fixed point as reference point O and introduce the gravitational moment $\mathbf{L}$ acting about it. In that case we speak of a *heavy top*. Subsecs. 4 and 5 are devoted to its discussion.

We shall postpone the complete analytical treatment of the top under no forces until the following section, where we shall become acquainted with the tool provided by *Euler's equations*. The complete treatment of the heavy top — to the extent to which it can be carried through at all — must be postponed even further, namely to § 35. There we shall have at our command the yet more powerful method of the *generalized Lagrange equations*.

For the top under no forces Eq. (2) yields $\dot{\mathbf{M}} = 0$. This can be integrated at once to yield

$$(3) \qquad\qquad \mathbf{M} = \text{const.}$$

The angular momentum of a top under no forces is constant in magnitude and spatial direction. This statement completely parallels Galileo's law of inertia, but in general does not lead to an expression for velocity and position in space which is as simple as in the other case.

(1) The Spherical Top Under No Forces

Only in the case of a spherical momental ellipsoid do we have $\mathbf{M} = I\boldsymbol{\omega}$, from which $\mathbf{M} = \text{const.}$ leads to $\boldsymbol{\omega} = \text{constant}$. The axis of rotation is in permanent coincidence with the fixed axis of angular momentum. Each point of the body, no matter what the external shape of the same (cf. Fig. 40c, for instance), describes a circle about this axis with constant velocity.

(2) The Symmetrical Top Under No Forces

Here a simple rotational motion occurs only if the direction of $\mathbf{M}$ coincides with one of the principal axes, that is, either with the axis of the body or an equatorial axis. The *general* form of motion of the symmetrical top under no forces is the so-called *regular precession*.

We explain this form of motion with the aid of Fig. 43. We have drawn the axis of the angular momentum, which is fixed in space, vertically upward; let M be the point at which it intersects a unit sphere described about the center of the momental ellipsoid. Call R and F the points of intersection of this sphere and the axes of rotation and of symmetry at an arbitrary instant. Since by the Poinsot method these three axes lie in a meridian plane through F, the three points M, R and F are located on a great circle passing through the fixed point M; in the case of a momental oblate spheroid, which we shall postulate for definiteness, M is situated between F and R. At any instant the motion consists of a rotation about OR. In this process F advances normally to the arc of the great circle just mentioned. The angular distance between F and M is not changed thereby; thus we can draw the instantaneous path of F as a short arc of a circle of latitude about M (arrow at the left in Fig. 43). Now R too must change its position — it must move to the great circle defined by M and the new position of F. In this motion the angular distance between M and R is conserved, since it is determined by the Poinsot construction. Thus R, too, advances on the arc of a circle of latitude about M (arrow at right of Fig. 43). The relative position of points F, M, and R is now the same as initially, so that our process of reasoning can be repeated. It follows that *axes of symmetry and rotation each describe a circular cone about the spatially fixed angular momentum, each cone being traced with constant angular velocity;*

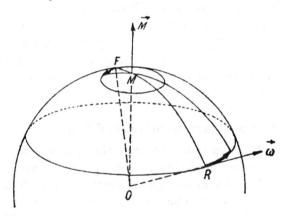

FIG. 43. Regular precession of the symmetrical top under no forces.

the latter because the angular velocity is completely determined by the magnitude of **M** and its position with respect to the momental ellipsoid. Thus the character of the regular precession has been fully described.

The same applies of course to a momental prolate spheroid, with the only difference (cf. Fig. 42b) that R would now be located between F and M.

(3) The Unsymmetrical Top Under No Forces

The form of motion of the symmetrical top just derived could have been described with more brevity, but less clarity of detail, as follows:

through the terminus of the angular momentum vector **M** we pass the "invariable plane" ϵ (cf. p. 73) normal to **M**. About the origin of **M** we construct the ellipsoid of twice the kinetic energy ("Poinsot ellipsoid") which is similar to the momental ellipsoid. The Poinsot ellipsoid is tangent to ϵ [3], and the point of tangency is the terminus of the angular velocity vector ω. The instantaneous motion of the top consists of a rotation of this ellipsoid about ω. In this process the ellipsoid rolls without slipping on the plane ϵ [4]. If the Poinsot ellipsoid is one of revolution, the curve of the point of tangency becomes a circle about **M**; the cones described by ω ("space cone") and the axis of figure therefore become circular cones. Thus we have again the regular precession of the top.

The same construction now leads at once to the Poinsot picture of the force-free motion of a general ("unsymmetrical") top of three distinct principal moments of inertia. Again we let the Poinsot ellipsoid roll on the invariable plane ϵ (cf. footnote 3 below). Now the curve of contact is no longer a circle, but a transcendental curve which in general does not close on itself. Similarly the cones which describe the motion of the axis of rotation and of the body "axis" in space are now transcendental cones. The analysis of the unsymmetrical top, even when under no forces, leads to elliptic integrals [cf. § 26, (3)], while that of the symmetrical top under no forces requires only elementary functions. Of course even for the unsymmetrical top a pure rotation about one of the three principal axes is a steady rotation whose representation is elementary.

(4) The Heavy Symmetrical Top

Here we shall not treat the spherical top separately, since its motion is hardly simpler than that of the symmetrical top.

For the heavy symmetrical top the fixed point O (point of support in the socket) no longer coincides with the center of mass G (located on the axis of symmetry); call s the distance OG. The magnitude of the gravitational torque is then

(4) $$|\mathbf{L}| = mgs \sin \theta,$$

where θ is the angle between the vertical and the axis of figure. **L** is normal to the vertical and to the axis of symmetry or, in other words, it lies along the line of intersection of the horizontal plane with the equatorial plane of the momental ellipsoid. This line of intersection is called the *line of nodes*,

---

[3] This follows from the Poinsot construction of p. 132 and from Eq. (26.17a) soon to be encountered.

[4] Rolling without sliding is equivalent in meaning to the equality of the rate of change of the angular velocity vector ω as observed from space and from the body. In this connection refer to Eq. (26.8a), where this equality is proved.

a term borrowed from astronomy. For a more precise definition of signs refer to p. 141.

Our general Eq. (2) can no longer be integrated immediately as in the case of a top under no forces ; rather, the angular momentum is subject to continuous change given by the law

(5) $$dM = L\,dt.$$

Thus the infinitesimal vector $L\,dt$ adds to the vector M at any given instant t to give the angular momentum at $t+dt$. The terminus of M advances in the direction of the instantaneous line of nodes, i.e., normal to the vertical and the axis of symmetry. From this it follows that the projections of M on the vertical as well as on this axis must be constant. Let us call the two constants

(6) $$M' = M_{\text{vert}} \text{ and } M'' = M_{\text{fig}}.$$

The two quantities M' and M'', which can be prescribed arbitrarily, are two constants of integration of the equations of motion.

A third constant is that of total energy E. Corresponding to Eq. (6.18) we have the gravitational potential energy

(6a) $$V = mgs\cos\theta$$

so that

(7) $$T + mgs\cos\theta = E.$$

In order to pass to an analytical description of the motion we must express T and the projections of M mentioned in (6) in terms of suitable position parameters of the top (the Eulerian angles); this will be carried out in § 35. The calculation of the motion is there shown to lead to elliptic integrals.

The *regular precession* is now no longer the general form of motion as in the case of the top under no forces, but only results for specially chosen values of M', M'' and E. The precessional motion usually observed with a heavy top excited in the customary way appears to be — but is not — regular; it can be called *pseudo-regular precession*. Finally a *pure rotation* about the vertically oriented axis of figure is also a possible (stable or unstable) form of motion, no matter what the magnitude of ω.

So far we have considered only the equation of angular momentum (2). We must throw a quick glance at the equation of linear momentum (1). Its right member consists of the force F acting at the fixed point O, which is composed of the force of gravity $m\mathbf{g}$ acting vertically downward, and the reaction of the support, $\mathbf{F}_{sup}$. The change in momentum in the left member is

$$\dot{\mathbf{p}} = m\frac{d}{dt}(\boldsymbol{\omega} \times \mathbf{R}) = m\dot{\mathbf{V}}$$

from Eq. (24.2) with $\mathbf{u} = 0$, where $\mathbf{V}$ is the velocity of the center of mass. Eq. (1) then makes the simple statement that

$$\mathbf{F}_{sup} = m(\dot{\mathbf{V}} - \mathbf{g}).$$

In other words, the law of linear momentum demands that at any given instant the support furnish a force equal to the mass of the top $\times$ the acceleration of the mass center diminished by the gravitational acceleration.

(5) The Heavy Unsymmetrical Top

In spite of the efforts of many great mathematicians, all attempts to integrate the differential equations of this problem in the most general form have failed so far. Among the integrals of angular momentum (6) the first remains in effect, to be sure, because even here the gravitational torque acts about a horizontal axis so that the terminus of the vector $\mathbf{M}$ remains in a horizontal plane fixed in space. The second integral (6) is, however, invalidated, because it is based on the symmetry of the momental ellipsoid. Of course the energy integral (7) is valid also for a general momental ellipsoid.

The soluble special cases of the problem postulate either a particular mass distribution or a particular form of motion.

The best-known case is that of Kowalewski. The momental ellipsoid is here assumed to be symmetrical; the center of mass no longer lies on the axis of the body, but in the equatorial plane defined as the plane perpendicular to the axis and passing through the fixed point; in addition it is required that the moment of inertia about the axis of the body be one half the equatorial one. In that case the form of motion need not be restricted.

The case of Staude concerns the question as to which axes can serve as axes of steady rotation when directed vertically. It turns out that these axes lie in the body on a cone of second degree which, in addition to the three principal axes, contains also the axis through the center of mass. To each axis belongs a (to within a sign) definite angular velocity. Neither mass distribution nor position of the center of mass need be specialized in this problem.

The case of Hesse, finally, is concerned with the analogue to the simple motion of a pendulum (spherical pendulum or, in particular, ordinary pendulum). For such motion the mass center must lie on a certain axis in the momental ellipsoid, and the initial excitation must be of proper form, just as in the case of the symmetrical top, whose mass center describes a pure pendulum motion only if the initial angular momentum has no component along the axis of symmetry.

§ 26. Euler's Equations.
Quantitative Treatment of the Top Under No Forces

(1) Euler's Equations of Motion

We distinguish between a reference system x, y, z fixed in space and a second one, X, Y, Z fixed in the body. In the (x, y, z)-system the angular momentum for motion under no forces has an invariable position: $\mathbf{M}=$ constant [Eq. (25.3)]; seen from the body, the position of $\mathbf{M}$ varies continuously. We want to study the law of this variation.

Let us therefore focus our attention on a point P fixed in the body, and a point Q fixed in space, the two points being momentarily in coincidence. Let $\mathbf{v}$ be the velocity of P in space, $\mathbf{V}$ that of Q in the body. According to the kinematic Eq. (22.4), $\mathbf{v} = \boldsymbol{\omega} \mathbf{X} \mathbf{r}$. As seen from the body, Q moves with equal but opposite velocity to that of P as seen from space, so that

$$\mathbf{V} = -\boldsymbol{\omega}\mathbf{X}\mathbf{r} = \mathbf{r}\mathbf{X}\boldsymbol{\omega}.$$

In tabular form we have

	Seen from Space	Seen from Body
P	$\mathbf{v} = \boldsymbol{\omega}\mathbf{X}\mathbf{r}$	$\mathbf{V} = 0$
Q	$\mathbf{v} = 0$	$\mathbf{V} = \mathbf{r}\mathbf{X}\boldsymbol{\omega}$

For point Q we choose the spatially fixed terminus of the vector $\mathbf{M}$ and hence write

$$\mathbf{r} = \mathbf{M}, \quad \mathbf{V} = \frac{d\mathbf{M}}{dt}.$$

Thus $\frac{d\mathbf{M}}{dt}$ means " change in the body " (we called the change in space $\dot{\mathbf{M}}$; it is equal to zero here).

From the second line of our table we then read off

$$(1) \qquad\qquad \frac{d\mathbf{M}}{dt} = \mathbf{M}\mathbf{X}\boldsymbol{\omega}.$$

This completes the derivation of *Euler's equations for a rotating body under no forces.*

We shall rewrite them in terms of their components in the (X, Y, Z)-system. We shall call ω_1, ω_2, ω_3 the components of ω, and M_1, M_2, M_3 those of $\mathbf{M}$. Eq. (1) yields

$$\frac{dM_1}{dt} = M_2\,\omega_3 - M_3\,\omega_2,$$

(2) $$\frac{dM_2}{dt} = M_3\,\omega_1 - M_1\,\omega_3,$$

$$\frac{dM_3}{dt} = M_1\,\omega_2 - M_2\,\omega_1.$$

The system of the X, Y, Z is so far completely arbitrary. If, now, we take the directions X, Y, Z along the principal moments of inertia of Eq. (22.15a) and call these I_1, I_2, I_3, we obtain, by virtue of the general relation (24.9),

(3) $$M_1 = I_1\,\omega_1,\;\; M_2 = I_2\,\omega_2,\;\; M_3 = I_3\,\omega_3\,;$$

and (2) takes the simple form

$$I_1\frac{d\omega_1}{dt} = (I_2 - I_3)\omega_2\,\omega_3,$$

(4) $$I_2\frac{d\omega_2}{dt} = (I_3 - I_1)\omega_3\,\omega_1,$$

$$I_3\frac{d\omega_3}{dt} = (I_1 - I_2)\omega_1\,\omega_2.$$

It is these remarkably symmetrical and elegant equations one usually thinks of when one speaks of Euler's equations.

Let us now extend them to include the case that an external torque $\mathbf{L}$ is in effect. In that case the terminus of $\mathbf{M}$ is no longer fixed in space, but, according to (25.2), has the velocity $\mathbf{v} = \mathbf{L}$.

As seen from the body, our point Q now moves with a velocity composed of $\mathbf{v} = \mathbf{L}$ and $\mathbf{V} = \mathbf{r} \times \omega$. It follows that Eq. (1) must be changed to

(5) $$\frac{d\mathbf{M}}{dt} = \mathbf{M} \times \omega + \mathbf{L}$$

and the components of $\mathbf{L}$ with respect to X, Y, Z must be added to the right members of (2) and (4). This yields *Euler's equations of motion for a rigid body with a fixed point.*

We shall write these equations explicitly only for the case of the heavy symmetrical top, where $\mathbf{L}$ acts about the line of nodes and, from (25.4), has the magnitude

$$|\mathbf{L}| = mgs\sin\theta.$$

In order to dispel all ambiguities contained in the meaning of the words vertical, axis of symmetry, line of nodes, we agree that

> the positive side of the spatially fixed z-axis points up and defines the vertical;
>
> the positive side of the Z-axis passes through the mass center and defines the axis of symmetry; it makes an angle θ with the vertical;
>
> the line of nodes is the semi-infinite line normal to the positive z- and Z-axes and in the direction of advance of a right-handed screw as θ increases.

We further specify that the distance s is to be a positive quantity. Call ϕ the angle which the line of nodes makes with the positive X-axis. The components of $\mathbf{L}$ with respect to X, Y, Z are then given by

(5a) $$mgs \sin \theta \cos \phi, \quad -mgs \sin \theta \sin \phi, \quad 0$$

respectively, and with $I_1 = I_2$ equations (4) go over into

$$I_1 \frac{d\omega_1}{dt} = (I_1 - I_3)\omega_2 \omega_3 + mgs \sin \theta \cos \phi$$

(6) $$I_1 \frac{d\omega_2}{dt} = (I_3 - I_1)\omega_3 \omega_1 - mgs \sin \theta \sin \phi$$

$$I_3 \frac{d\omega_3}{dt} = 0.$$

The last equation shows that for the heavy symmetrical top (and therefore *a fortiori* for one under no forces) we have

(7) $$I_3 \omega_3 = M_3 = \text{const.,}$$

which we already knew. We see at the same time that Euler's equations are not suited for a further integration for the heavy top, since as yet we are ignorant of the relation between the ω_1, ω_2 and the θ, ϕ.

As far as the ω_1, ω_2, ω_3 are concerned, we wish to emphasize very strongly that they are not velocities in the ordinary sense, i.e., not derivatives with respect to time of spatial measurements of some sort. Indeed, in view of the expression defined on p. 50, we can aptly designate them as " non-holonomic velocity components."

We shall finally write (5) in a somewhat different form. Since $\mathbf{v}$ is the velocity as seen from space, we can generalize our expression by substituting $\mathbf{v} = \dot{\mathbf{M}}$ for $\mathbf{v} = \mathbf{L}$. We thus obtain

(8) $$\mathbf{M} = \frac{d\mathbf{M}}{dt} + \boldsymbol{\omega} \times \mathbf{M},$$

an equation which, by the analysis of p. 139, is valid for all (axial or polar) vectors. If, specifically, we apply it to the angular velocity vector ω, it simply yields

(8a)
$$\dot{\omega} = \frac{d\omega}{dt}.$$

For the angular velocity vector ω and only for this vector the spatial change is equal to the change as judged from the body. It is this rule to which we referred in the footnote on p. 136.

(2) Regular Precession of the Symmetrical Top Under No Forces and Euler's Theory of Polar Fluctuations

We need not say any more about the spherical top. Its general motion is a pure rotation about an axis fixed in the body. This follows at once from Eqs. (4) if we put $I_1 = I_2 = I_3$. As we know from § 25, subsec. 1, this axis is at the same time fixed in space and coincides with the angular momentum direction.

Let us now turn to the symmetrical top, $I_1 = I_2 \neq I_3$. The third Eq. (4) yields

$$\omega_3 = \text{const.}$$

as we already know from Eq. (7). The first two equations are

(9)
$$I_1 \frac{d\omega_1}{dt} = (I_1 - I_3)\omega_2\omega_3$$

$$I_1 \frac{d\omega_2}{dt} = (I_3 - I_1)\omega_1\omega_3$$

It is convenient to consolidate them into one by introducing a complex variable. Multiply the second equation by i and add to the first to obtain

(10)
$$I_1 \frac{ds}{dt} = i(I_3 - I_1)s\omega_3, \quad s = \omega_1 + i\omega_2.$$

Let us abbreviate this by putting

(11)
$$\alpha = \frac{I_3 - I_1}{I_1}\omega_3,$$

so that an integration of (10) gives

(12)
$$s = s_0 e^{i\alpha t}, \quad s_0 = \text{constant of integration.}$$

s is the projection of the angular velocity vector ω on the equatorial plane of the top, if we use this plane as the complex plane of s. Eq. (12) states that this projection describes a circle of radius s_0 with the constant angular velocity α. At the same time the total angular velocity vector

ω describes a circular cone about the axis of figure. The vertex angle β of the cone is given by

(12a)
$$\tan \beta = \frac{(\omega_1{}^2 + \omega_2{}^2)^{\frac{1}{2}}}{\omega_3} = \frac{|s_0|}{\omega_3}.$$

This is the picture of the regular precession which is seen by an observer located on the top. (To an observer fixed in space the axis of the top rotates of course about the instantaneous axis of rotation which, as we saw earlier describes in its turn a circular cone about the spatially fixed angular momentum vector **M**.) Since it is our intention to apply the foregoing to the earth, the viewpoint of the observer located on the top rather than that of the one fixed in space will be useful, as it corresponds to the viewpoint of a human being located on the earth.

The earth is a top whose momental ellipsoid is an oblate spheroid. We call the *geometric North Pole* the point at which the axis of symmetry pierces the surface of the earth ; it is, in general, distinct from the *celestial North Pole* which is the point at which the angular velocity vector cuts through the earth's surface. According to the Euler theory reproduced above, the celestial North Pole describes a circle about the geometric North Pole, a phenomenon called *Eulerian motion*. Inasmuch as it is the path of the rotational pole, this circle is also referred to as the *polhode*.

A suitable measure of the flattening of the earth is the so-called ellipticity

(13)
$$\frac{I_3 - I_1}{I_1} \sim \frac{1}{300}.$$

The angular velocity of the earth is determined by the length of the day ; we have

(14)
$$\omega_3 \sim \omega = \frac{2\pi}{\text{day}}$$

from which, according to (11),

(15)
$$\alpha - \frac{I_3 - I_1}{I_1}\,\omega_3 - \frac{2\pi}{300}\,\text{day}^{-1}.$$

Thus Euler's period for the precession amounts to

(16)
$$\frac{2\pi}{\alpha} = 300 \text{ days} = 10 \text{ months}.$$

We are accustomed to think of the axis of rotation of the earth as fixed in the globe and passing through the geometrical poles. This is not rigorously true. Every movement of mass on the earth along a longitude must change the position of the axis of rotation[5], and every movement of mass along a circle of latitude must change the angular velocity, that is, the length of the day; both changes are a result of the law of conservation of angular momentum. Let us imagine that this movement has ceased and that the celestial pole is deviated from the geometric one. In that case the axis of rotation would, by virtue of the Eulerian motion, commence a circular motion about the geometric pole.

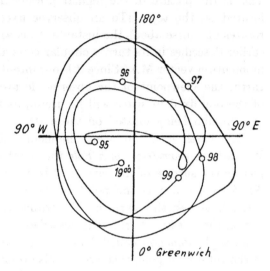

Let us now compare our theoretical results with the observations of polar fluctuations, which have been gathered by international cooperation. In Fig. 44 we have sketched the polhode obtained between the years 1895 and 1900.

The average deviation of the celestial pole, i.e., the mean radius of Euler's circle, amounts to about $\frac{1}{8}''$ of arc or 4 meters on the earth's sur-

FIG. 44. Polar fluctuations between the years 1895 and 1900. Confirmation of Chandler's period.

face, according to observations between these years. But instead of a period of 10 months we have, according to Fig. 44, $3\frac{1}{2}$ complete revolutions for the four years 1896–1900, which corresponds to a period of 14 months.

The fourteen-month period is called Chandler's period after its discoverer. Its explanation lies in the elastic deformations that the earth suffers as a result of the changed centrifugal effect caused by polar fluctuations. The modulus of elasticity of the earth compares in magnitude to that of steel.

The observed polhode, as drawn in Fig. 44, can now be explained as a superposition of 1) fluctuations occurring with Chandler's period, 2) annual fluctuations evidently of meteorological origin, and 3) deviations at irregular intervals which may point to isolated and unrelated mass transports. No trace remains of Euler's ten-month period which was derived by assuming the earth to be an ideal rigid body.

---

[5] The terrestrial mass transport most important for this effect seems to be the yearly migration of the air pressure maximum from the continent of Asia to the Pacific Ocean and back.

In agreement with usage in gyroscopic theory we have here described the motion of the earth's axis first investigated by Euler as a " precession under no forces." We have thus usurped a word having an entirely different meaning in astronomical usage. There, " precession " denotes a slow rotation of the earth's axis about the normal to the ecliptic which causes an advance of the equinoctial points of 50″ per year. This *precession of the equinoxes* has a period of $\frac{360°}{50″} = 26,000$ years. Instead of " precession of the equinoxes " we could also speak of an " advance of the line of nodes " (line of inter-section of the plane of the ecliptic with the equatorial plane of the earth); as mentioned earlier, our designation, " line of nodes," was borrowed from astronomy.

The precession of the equinoxes is not a free one, but rather a motion *forced* on the global top by the joint effect of the attractions of sun and moon.

We shall clarify this effect by means of Fig. 45, where we have, at least qualitatively, anticipated the theory of the heavy symmetrical top.

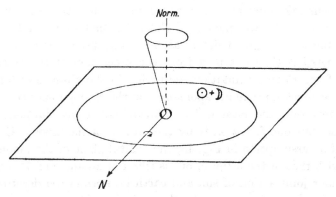

Fig. 45.   Precession of the earth's axis, called " precession of the equinoxes."

The diagram shows the plane of the ecliptic on which a circle is drawn. One should think of the circumference of this circle as being uniformly " smeared " with the masses of sun ⊙ and moon ☽ (actually we should draw two circles, one for the sun and one for the moon[6]; we have fused these two circles into one). The uniform mass distribution represents a time average over the instantaneous positions relative to the earth of sun and moon during their revolutions (in the sense of a Gaussian perturbation method). We justify the taking of this time average by the experimental fact that the periods of sun and moon are very small compared with the precession period mentioned above, so that this precession can in no way

---

[6] As a matter of fact the moon is so close to the earth that its effect is about twice as great as that of the sun.

depend on the instantaneous positions of sun and moon. At the center of the ☉ + ☽ circle we see a cross-section of the earth with its two protuberances at the equator. Only these latter have a part in the phenomenon in question; for the attraction of the ☉ + ☽ ring tends to pull the two protuberances into the plane of the ecliptic, an effect which is intuitively almost obvious. We therefore have a torque about the line of nodes N in the sense of the arrow drawn about N. Now this torque is of the same type as the gravitational torque acting on a top whose mass center lies below the fixed point of support. The result is therefore similar to that in the case of the top. Rather than yield to the torque the axis of figure " escapes " in a perpendicular direction and describes a cone of precession about the vertical, here the normal to the ecliptic.

To be sure, the *regular* precession is only a special form of motion of the heavy top (cf. p. 137); under the present circumstances one would therefore expect the more general *pseudo-regular* precession consisting of a regular precession on which small " nutations " are superposed. Now these small nutations are nothing but the conical oscillations of the axis of figure occurring under no forces, hence, in our case, the polar fluctuations that take place with the period of Euler (or that of Chandler, obtained from the former by global deformation). The pseudo-regular precession to be expected is thus obtained from the precession of the equinoxes by addition of the Eulerian nutations occurring in the absence of forces.

Here we must once more apologize for the ambiguous use of a term. In astronomy one understands by nutation not a *free* fluctuation of the earth's axis, but one *forced* on it by the motion of the moon. Contrary to our preceding assumption in Fig. 45, the orbital plane of the moon does not coincide with that of the ecliptic, but is tilted at an angle of 5° with respect to it. Under joint action of sun and earth its normal too describes a cone of precession about the normal to the ecliptic. This precession is tantamount to a *recession of the lunar nodes* (intersection of moon's orbit with ecliptic) which, however, occurs at a much livelier rate than the advance of the line of nodes of the earth, viz., in $18\frac{2}{3}$ years. It is understandable that the earth's axis is in its turn implicated in this precession; the recession of the lunar nodes results in the *astronomical nutation of the earth's axis*, which takes place with the same period.

(3) Motion of an Unsymmetrical Top Under No Forces. Examination of its Permanent Rotations as to Stability

We turn to the integration of Eqs. (4) in the case $I_1 \neq I_2 \neq I_3$. Multiplication of these equations by ω_1, ω_2, ω_3 and addition yields

$$I_1\omega_1\frac{d\omega_1}{dt}+I_2\omega_2\frac{d\omega_2}{dt}+I_3\omega_3\frac{d\omega_3}{dt}=0$$

or, integrated,

(17) $$\frac{1}{2}(I_1\omega_1{}^2+I_2\omega_2{}^2+I_3\omega_3{}^2)=\text{const.}=E.$$

E is the energy constant, and the left member is the kinetic energy, in agreement with Eq. (22.12b) specialized to principal axes. Instead of (17) one can evidently also write

(17a) $$E_{\text{kin}}=\frac{1}{2}\mathbf{M}\cdot\boldsymbol{\omega}.$$

We can instead multiply Eqs. (4) by $I_1\omega_1$, $I_2\omega_2$, $I_3\omega_3$; addition once more yields zero on the right. The result of the integration can be written

(18) $$(I_1\omega_1)^2+(I_2\omega_2)^2+(I_3\omega_3)^2=\text{const.}=|\mathbf{M}|^2.$$

On the left we have the sum of the squares of the angular momentum components. This sum, as we know, remains invariant in the absence of forces, even if the components themselves vary in the course of the motion.

In (17) and (18) we have two linear homogeneous equations for $\omega_1{}^2$, $\omega_2{}^2$, $\omega_3{}^2$, from which we can, for instance, solve for $\omega_2{}^2$ and $\omega_3{}^2$ in terms of $\omega_1{}^2$:

(19) $$\omega_2{}^2=\beta_1-\beta_2\omega_1{}^2,\quad \beta_1=\frac{2EI_3-|\mathbf{M}|^2}{I_2(I_3-I_2)},\quad \beta_2=\frac{I_1(I_3-I_1)}{I_2(I_3-I_2)};$$

$$\omega_3{}^2=\gamma_1-\gamma_2\omega_1{}^2,\quad \gamma_1=\frac{2EI_2-|\mathbf{M}|^2}{I_3(I_2-I_3)},\quad \gamma_2=\frac{I_1(I_2-I_1)}{I_3(I_2-I_3)}.$$

If we replace these values of ω_2 and ω_3 in the first Eq. (4), we have

(20) $$\frac{d\omega_1}{[(\beta_1-\beta_2\omega_1{}^2)(\gamma_1-\gamma_2\omega_1{}^2)]^{\frac{1}{2}}}=\frac{I_2-I_3}{I_1}dt.$$

t is therefore an elliptic integral of the first kind in ω_1 (cf. p. 100); function theory allows us to state conversely that ω_1 is an elliptic function of the time. The same holds of course for ω_2 and ω_3.

We furthermore deduce from Eqs. (17) and (18) that the polhode cone or *body cone* is no longer a circular cone as in the case of the symmetrical top, but a cone of fourth degree.

We shall finally consider the rotations of the unsymmetrical top about one of its three principal axes which, as we know [cf. § 25, toward end of (3)], are steady rotations. Let us, for definiteness, put

$$A>B>C.$$

We shall show that the rotations about the axes of the *greatest* and *smallest* principal moment of inertia are stable, those about the axis of the *intermediate* principal moment are unstable. We choose Eqs. (17) and (18) as starting point. It will be convenient in connection with the diagrams below to rewrite these in terms of the angular momentum components M_1, M_2, M_3,

(21a) $$\frac{M_1^2}{I_1} + \frac{M_2^2}{I_2} + \frac{M_3^2}{I_3} = \text{const.},$$

(21b) $$M_1^2 + M_2^2 + M_3^2 = \text{const.} = |\mathbf{M}|^2.$$

Eq. (21b) describes a sphere of radius $|\mathbf{M}|$, (21a) an ellipsoid with three distinct axes (a " non-degenerate " ellipsoid).

Case 1. Rotation about the longest axis of the ellipsoid (21a). In a pure rotation the sphere is tangent to the ellipsoid from the *outside* at point A, Fig. 46a. A small jolt will in general alter both the sphere and the ellipsoid. The point of tangency A will change to a small curve of intersection which remains, however, in the neighborhood of A. A narrow body cone is the result; the original rotation proves to be stable.

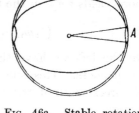

FIG. 46a. Stable rotation of unsymmetrical top about the longest axis of the momental ellipsoid.

The same is true in case 3, rotation about the shortest axis of the ellipsoid (21a). The sphere now lies inside the ellipsoid and is hence tangent to it from the *inside*. A small jolt will again cause the point of tangency to transform into a neighboring curve; again the original rotation is stable.

Case 2. Rotation about the intermediate axis. The sphere *intersects* the ellipsoid in a curve of the fourth degree; its singular point B (foremost point of Fig. 46b) represents the original rotation. If the top is given a small impulse, the curve of intersection splits into two branches. The axis of rotation wanders off along one of these branches and moves further and further from its initial position in the body. The rotation is unstable.

It is instructive to prove this analytically; one proceeds from the differential equations (4). One can show (problem IV.2) that the lateral components generated by a small perturbation of the original rotation satisfy two simultaneous differential equations of first order. These have solutions of trigonometric character in cases 1 and 3, exponential character in case 2 (method of infinitesimal oscillations as stability criterion).

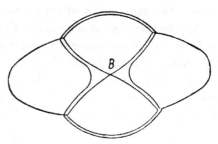

FIG. 46b. Unstable rotation of unsymmetrical top about the intermediate axis of the momental ellipsoid.

Let us perform the following experiment with a (full) matchbox: we hold the box between thumb and forefinger at opposite ends of its shortest edge and flip it into the air, thus imparting to it considerable angular momentum about this shortest edge. We notice that if the box originally shows its label, it will continue to do so throughout the motion. The same phenomenon occurs, though less clearly, if we hold the box at opposite ends of its longest edge, and flip it as before. If, instead, we hold it at opposite ends of the intermediate edge, with the striking surface showing, and repeat the procedure, we shall not see this surface throughout the motion, but rather a distinct change of colors.

Another striking example of instability of a state of motion is the following: occasionally one finds smooth-worn, flat pebbles in nature which, if spun about their vertical axis on a flat support, show stability of motion only for one sense of rotation; if made to spin in the opposite sense, they will start to wobble more and more violently, and finally end up by spinning in the stable direction opposite to their original angular momentum. The same can often be observed with small pocketknives (penknives) set to stand edgewise with blade folded in, when one gives them a gentle impulse.

We can perform a geometrically well-defined, instructive experiment in this connection. Let us take the wooden model of a non-degenerate flat ellipsoid of principal axes a, b, c (a and b much larger than c) and equip it with a heavy metal strip which, in its original position, hugs the upper surface of the ellipsoid in its (ac)-section. The strip can be rotated about the short c-axis, but is clamped down during each experiment. In the position ac the strip does not disturb the symmetry of the mass distribution. Both senses of spin about c are therefore equally stable. Let us now turn the strip by a small angle from this position. The two principal axes of inertia a and b are then each displaced by a small angle γ; the symmetry of the lower surface facing the plane support is determined by the two principal radii of curvature in the planes ac and bc; thus the symmetry of this surface remains unchanged. The direction of spin in the sense of the acute angle γ is now geometrically " distinguishable " from that in the opposite sense. Indeed the former is stable, the latter unstable since it is accompanied by rolling motions which increase with time.

A more elegant, though less easily achievable, form of the experiment is the following (G. T. Walker demonstrated it to us at Trinity College, Cambridge, in 1899): the non-degenerate ellipsoid is made of brass sheet; a certain circular region about the point of support has been stamped out and can be moved with respect to the remaining ellipsoidal shell. By a small angular displacement of this circular plug the curvature relations of the lower surface near the point of support are altered with respect to the inertial distribution of the shell, which remains sensibly unchanged. This alteration is so slight that it passes unnoticed when the ellipsoid is examined. Nevertheless one sense of spin is again preferred to the other.

These experiments with the non-degenerate ellipsoid, while enlightening in themselves, also furnish an adequate substitute for the analytical theory of the phenomenon. Such a theory would have to investigate the rolling oscillations which might accompany the spin in one direction or the other when a small perturbation is superposed on this spin; it would show that the characteristic equation for the frequency of these oscillations has only real roots in the one case, some complex roots in the other. In the first case one would decide that the spin was stable, in the second, that it was unstable, i.e., subject to secular increase of the perturbation. Equations for this treatment are set up in the treatise of Routh (Advanced Part, Art. 241 and ff.) cited in § 42.

§ 27. Demonstration Experiments Illustrating the Theory of the Spinning Top; Practical Applications

We begin by describing the well-known device known as *Cardan's suspension*, which affords an unusually effective means of demonstrating the properties of tops and gyroscopes.

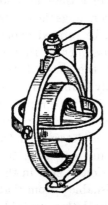

FIG. 47. Gyroscope in Cardan's suspension. Axis of rotation of outer ring = vertical, axis of rotation of inner ring = horizontal perpendicular to paper, axis of rotation of gyroscope = horizontal in plane of paper.

The suspension consists of an outer and an inner ring. The outer ring has a vertical axle borne by the outer frame or cage; the inner ring has an horizontal axle with bearings in the outer ring. The flywheel-shaped top revolves with its axis perpendicular to the axis of rotation of the inner ring. Fig. 47 shows the flywheel axle pointing normal to that of the outer ring, which causes the inner ring to lie in a horizontal plane. We shall designate this arrangement of the apparatus as its normal position.

On the axle of the flywheel provision is made for a means by which angular momentum can be imparted to the wheel while in its normal position, with the gimbals at rest. This angular momentum must be so great that all phenomena are essentially dominated by it and the effect of the mass of the gimbals becomes negligible.

In the following experiments a considerable angular momentum and the initial normal position are presupposed.

1. We exert a slight pressure downward on the *inner ring*. This ring does not give way; instead it is the *outer* ring that turns. Thus the axis of the flywheel moves backward or forward in an horizontal plane, depending on the position of the point at which the pressure is exerted. Instead of pressing on the inner ring we can load it unilaterally by means of a small weight. As long as the angular momentum remains sufficiently great, the top then describes a *regular precession* with horizontal axis.

2. We press on the *outer ring*. It remains motionless, whereas the *inner* ring turns upward or downward from its horizontal position depending on the sense in which pressure is exerted on the outer one. We can even deliver a vigorous blow to the outer ring without its yielding noticeably. All one perceives in that case is a rapid conical oscillation of the axis of the top about an axis close to that of the normal position.

3. If the pressure on the outer ring continues so that, with continual rotation of the inner ring, the axis of the top approaches the vertical, we notice that the resistance of the outer ring weakens more and more. One can then without effort set the outer ring spinning rapidly, but only in that sense which corresponds to the direction of the pressure originally exerted on the ring. If one attempts to rotate the outer ring in the opposite sense, the flywheel " rebels "; its axis suddenly tends in the opposite direction, thus causing the inner ring to flip through an angle of 180°. Now we can turn the outer ring without effort in this opposite direction, but another flipping of the top occurs if we return to the original sense of rotation.

4. This is the *tendency of the spins to align parallel to each other* which was emphasized by Foucault. The axis of the top is stable in the vertical position as long as its spin is *homologous* (= in the same sense) to that of the outer ring. If the spins are anti-parallel, this position is, on the contrary, unstable to a high degree and the axis comes to rest only when the opposite direction has been attained; in this latter direction homologous parallelism of the two spin axes prevails again. If we exert pressure on alternate sides of the outer ring in the proper rhythm, we can cause the top to revolve continually about the axis of the inner ring.

5. If we tie the inner ring to the outer one, so that the movability of the inner ring is destroyed, the resistance of the top to motion is destroyed as well. Seemingly without a will of its own, the top then obeys all pressures exerted on the outer ring, just as if it did not have any spin. Thus, typical gyroscopic effects occur only in the case of the top with three degrees of freedom, and are completely absent in that of two degrees. One can, however, restitute the missing degree of freedom by clamping the top to the rotating surface of the turning stool described on p. 74; this must be done in such a way that the axis of the outer ring, which has so far been

kept vertical, is tilted with respect to the axis of the stool (which remains vertical) at not too small an angle. Then the axis of the top with two degrees of freedom tends to align itself with the axis of the rotating support, just as a compass needle turns towards the North Pole, i.e., in the sense of the homologous parallelism described above. Thus the single ring containing the top will come to lie in a vertical plane, with one or the other of the axle pins of the top uppermost, depending on the sense in which the stool is rotated.

The explanation of all these phenomena is contained in the fundamental principle (25.5),

(1) $$dM = L\,dt.$$

1. If we press on the inner ring, L is horizontal and coincides with the axis of rotation of the inner ring. The angular momentum M is directed toward the left or right of Fig. 47 and is hence deflected laterally by L. If then we are allowed to assume that the axis of the top, originally in coincidence with the angular momentum, tends to remain in coincidence by following it, we have explained the lateral deflection of the axis of figure, that is, the rotation of the outer ring. That the assumption made here is actually valid for sufficiently rapid spin of the top will be justified in § 35 (cf. the discussion about the pseudo-regular precession in that section).

2. If we exert pressure on the outer ring, L is directed vertically. The angular momentum, originally directed horizontally to the right or left, is deflected upward or downward. Under the same assumption as in 1, we therefore obtain a rotation of the inner ring. If we impart a very strong blow to the outer ring, our assumption regarding the coincidence of angular momentum and axis of the top is only approximately satisfied; we then obtain the small conical oscillations mentioned earlier, which betray a small dislocation of the two axes.

3 and 4. By the same token we see that if the axis of angular momentum is almost vertical and if we rotate the outer ring in a sense homologous to that of the spin of the top, the axis of angular momentum becomes more nearly vertical. Gimbals and flywheel then rotate as a whole about the vertical. The resistance of the outer ring vanishes. If we rotate the outer ring in the non-homologous or anti-parallel sense, a small deviation of the axis of angular momentum from the vertical suffices to make the former recede further and further from the vertical; the almost-vertical position of the top proves to be unstable with respect to such a non-homologous rotation.

5. If we tie inner and outer ring together, the axis of angular momentum can no longer move in a vertical plane when a vertical torque L is imposed thereon by a rotation of the outer wheel. The torque is therefore transmitted to the whole system. This is possible because the horizontal change in direction that the vector M suffers can be compensated by the bearings of the outer ring, since inner and outer ring are now rigidly connected. Not so on the turning stool, where the angular momentum can follow the imposed L at least to some extent, which explains why the axis of the top tends to point in the direction of the axis of the stool.

We shall now discuss some practical applications. Let it be remarked in advance that details on many points of the discussion can be found in the older literature from which much of the following is borrowed.

(1) The Gyrostabilizer and Related Topics

Around the year 1870 Henry Bessemer, whose name is renowned in metallurgy, built a drawing room cabin destined for navigation on the English Channel. The cabin was suspended so that it could move about a fore-aft axis of the ship and was to be stabilized against the ship's roll by means of a flywheel. The axis of the flywheel was, however, rigidly fixed in the cabin, and therefore lacked the required third degree of freedom (cf. above under 5). As a result the construction was a failure soon to be abandoned.

It was O. Schlick, mentioned in connection with the mass balancing of piston engines (cf. p. 76), who successfully worked out the present problem. His method was applied to several steamers, including the " Silvana " of the Hamburg-America Line, and the Italian " Conte di Savoia " (considerable literature on the latter exists in American publications). In the " Silvana " the flywheel had a weight of 5,100 kg., a diameter of 1.6 m., and made 1,800 r.p.m. (a peripheral velocity of 150 m. per sec.). It was fixed in a cage which could, like a pendulum, swing about an axis in the port-starboard direction, so that the axis of symmetry of the flywheel oscillated in the vertical fore-aft plane of the ship. This cage corresponds to the inner ring of our demonstration top, the ship's hull itself to the outer one. The vertical of Fig. 47 is replaced by the long axis of the ship; instead of the former rotations about the vertical there is now the rolling of the vessel. The required three degrees of freedom then consist in the rolling of the ship, the oscillations of the cage, and the spin of the flywheel. When the vessel rolls, the axis of the flywheel, vertical in its normal position, alternately swings fore and aft in its cage, so that the energy contained in the rolling is converted to energy of motion and position of the cage. The rolling of the ship and the swinging of the cage are now coupled to each

other; if, in particular, their corresponding proper oscillations are in
resonance, conditions resembling those of coupled pendulums obtain.
To be sure, no damping of the ship's oscillation has so far been achieved.
But it is now possible to absorb the oscillation energy of the cage and thus
the energy of roll of the vessel by a braking device acting at the axle of the
cage, just as the velocity of a car is reduced by a brakeshoe tangent to the
wheel. Of course the braking action at the cage must not be so strong as
to prevent the deflection of the flywheel axis altogether; for then we should
again be confronted with the ineffective top of two degrees of freedom.
Graphs of the rolling motion, similar to seismograms in an earthquake,
show that there exists an optimum or " best compromise " value of braking
action; in the " Silvana " the amplitude of roll was reduced to $\frac{1}{10}$ to $\frac{1}{20}$ of
its original value almost as soon as the flywheel was put in action; the
amplitude of oscillation of the frame hovered around 30° to 40° under these
circumstances.

Nevertheless the gyrostabilizer has not been applied extensively. This
is partly due to the danger inherent in the construction — a rapidly rotating,
massive flywheel is an unpleasant passenger —, partly to the invention of
an even more successful competitor, the Frahm stabilization tank, a device
based on an entirely different principle.

A problem connected with the foregoing is that of stabilizing by gyroscopic action a turntable on board a ship. We do not know to what extent
this problem has been solved for practical use; for obvious reasons work
on it has been going on in all countries.

(2) The Gyrocompass

This is the finest and most nearly perfect gyroscopic device. Its conception goes back to Foucault. After Foucault had demonstrated the rotation
of the earth by means of his pendulum experiments (cf. Ch. V, § 31), he
made plans to achieve the same end by means of spinning tops. Of his
several attempts we mention only the gyrocompass which was to replace
the magnetic compass. The Foucault gyrocompass consists of a spinning
top of two degrees of freedom constrained to the horizontal plane, which
points, not to the magnetic North Pole, but to the actual celestial North
Pole, the axis of rotation of the earth. Actually we dealt with this
arrangement already in the fifth of our demonstration experiments, where
we put the top with fixed inner ring on the turning stool. The rotating
earth now takes the place of the turning platform of the stool. The only
difference between the two cases lies in the fact that we were able to impart
an arbitrarily large angular velocity to the rotating platform, resulting in a
very strong orientation effect on the top, whereas the angular velocity of

the earth is very small, so that the alignment of the Foucault gyroscope takes a considerable time. In the earlier arrangement we mentioned that the angle between the axes of rotation of the outer ring and the stool should not be too small. In the present case this angle is the complement of the geographic latitude, the " co-latitude " at the point of observation. At the two poles of the earth, where this angle is zero, the orientation power of the gyrocompass vanishes. In general it is proportional to the angular velocity of the earth, the angular momentum of the top and the sine of the co-latitude.

Foucault's experiments lead only to rough indications of the effect. Its full realization was achieved by Hermann Anschütz-Kaempfe, by means of successive improvements in construction. His original goal was to reach the North Pole by means of a submarine passing under the drift ice. Since the readings of a magnetic compass become very unreliable near the North Pole, failing altogether inside a submarine, he had the idea of making the top serve as his direction-finder. It is true that in the pursuit of this idea through several decades he did not reach the North Pole; but his experimentation lead to an ideal instrument which has become indispensable in navigation.

The Anschütz gyroscope, unlike that of Foucault, is not constrained to a horizontal plane, but is merely pulled back into this plane by its weight, like a pendulum. Originally it was arranged so as to swim in a bath of mercury. Later constructions made use of two or three tops whose effects strengthened and corrected each other. The angular momentum of the spinning tops is kept constant by electric drive. In the latest Anschütz construction the whole system is enclosed in a sphere which floats with almost no friction in a second sphere of only slightly larger radius. Since the gyroscope is taken along on trips during which it may not be touched for several months, provision must be made for a particularly ingenious automatic lubrication method.

Measures to eliminate the harmful effects of the ship's own motion are of special importance. When the ship travels in a curve or changes its speed, the gyrocompass, with its ability to oscillate about the horizontal plane, is sensitive to the corresponding inertial forces. These exert pressures on the axis of spin, causing it to deflect from its undisturbed position, with the result that erroneous readings are obtained. One can show that the motion of the vessel becomes harmless if the free oscillation of the compass needle about the meridian has the period

$$T = 2\pi \left(\frac{l}{g}\right)^{\frac{1}{2}} = (8\pi)^{\frac{1}{2}} 10^3 \text{ sec} = 84.4 \text{ min.}$$

which is the same as that of a pendulum of length equal to the earth's
radius

$$l = \frac{2}{\pi} \cdot 10^7 \, m \,.$$

(Law of Schuler, completed by Glitscher[7]).

A further beautiful application of the gyroscope concerns the
automatic steering mechanism of large steamers. If a ship is to retain
its course in spite of the motion of waves and ocean currents, the uninter-
rupted attention of the helmsman and the corresponding corrective action
of the steering mechanism are required. This corrective action is, however,
always too late by a certain amount of time, therefore causing losses in
mileage and time. The gyrocompass is, on the contrary, a sense organ
which "feels" much more accurately and swiftly than man, and takes
instantaneous countermeasures. As a result of these countermeasures
the line of travel becomes almost rigorously rectilinear (actually loxodromic,
i.e., a rhumb line), which results in a considerable saving in energy. For
this reason every passenger ship of good size is now equipped with such
an automatic steering mechanism.

(3) Gyroscopic Effects in Railroad Wheels and Bicycles

A set of rolling wheels of a railroad car is a spinning top whose angular
momentum can become considerable for fast trains. When the wheels
go around a curve, the angular momentum must, at any instant, be deflected
to a position determined by the normal to the curve. For this, according
to Eq. (1), a torque is required whose axis lies along the direction of travel.
Since such a torque (often called "gyroscopic couple") is not present, the
"gyroscopic effect" will result in a countertorque which presses the set
of wheels against the outer rail and pulls it off the inner one. This countertorque adds to the moment of the centrifugal force about the direction of
travel. The latter effect is compensated, as we know, by adequate banking
of the roadbed. Both moments have the form

$$m v \omega$$

where v is the velocity of travel, and ω the angular velocity of the train
in the curve; m is, in the present case, the mass of the set of wheels reduced
to the wheel periphery, whereas in the centrifugal effect m is the total
mass of the car carried by the wheels. Our gyroscopic couple and its equal
and opposite countertorque are therefore extremely small compared with
the moment of the centrifugal force; one could compensate for it by lifting
the outer rail a very slight additional amount.

---

[7] Cf. *Wissensch. Veröffentl. aus den Siemenswerken,* **19,** 57 (1940).

More serious effects may result from any vertical irregularities in the rails, such as, for instance, a " hump " on one of the rails (to this category also belong the increasing and decreasing elevation of one of the rails at the beginning and end of a banked curve). Such a hump causes a deviation of the angular momentum in a vertical direction, and hence a countertorque which seeks to twist the set of wheels out of the rail-bed by pressing, say, the front wheel of the set against the rail, and pushing the last wheel of the set away from the rail. The play allowed by the rails will thus cause the flanges of the wheels to bite now into one rail, now into the other. This has indeed been observed on test runs with fast electric trains. In order to control the condition and exact position of the rails at all times, the German Reichsbahn uses test cars equipped with gyroscopic instruments, the latter manufactured by the Anschütz company.

A bicycle is a doubly non-holonomic system ; for, like the wheel in problem II.1, it has five degrees of freedom in finite motion, but only three such degrees in infinitesimal motion (rotation of the rear wheel in its instantaneous plane, to which the rotation of the front wheel is coupled by the condition of pure rolling ; rotation about the handle bar axis; and common rotation of front and rear wheel about the line connecting their points of contact with the ground), as long as we do not consider the degrees of freedom of the cyclist himself. It is well-known that given sufficient velocity the stability of this system relies on the fact that either by means of rotations of the handle bar, or by means of unconsciously released motions of the body, the cyclist calls forth suitable centrifugal effects. That the gyroscopic effects of the wheels are very small compared with these can be seen from the construction of the wheel; if one wanted to strengthen the gyroscopic effects, one should provide the wheels with heavy rims and tires instead of making them as light as possible. It can nevertheless be shown[8] that these weak effects contribute their share to the stability of the system. This is the case because, just as in the automatic steering mechanism of ships, they react more quickly against a sinking of the center of gravity than do the centrifugal effects. In the small oscillations which one has to consider in testing the stability of the motion, the gyroscopic action lags the oscillations of the center of gravity by a quarter period, whereas the centrifugal action lags them by a half period.

---

[8] Cf. F. Klein and A. Sommerfeld, *Theorie des Kreisels*, Vol. IV, p. 880 and ff. In order to carry out the stability considerations we must of course exclude all participating action on the part of the cyclist. Not only must he be assumed to ride without hands, but also with motionless body; he should act only by means of his weight. This work also offers detailed material on other applications and on the mathematical foundations of the theory of the spinning top.

SUPPLEMENT: THE MECHANICS OF BILLIARDS

The beautiful game of billiards opens up a rich field for applications of the dynamics of rigid bodies. One of the illustrious names in the history of mechanics, that of Coriolis[9], is connected with it.

The following explanations have as their main object the clarification of some problems which we shall pose on the subject. In these problems not only the dynamics of the rolling and sliding ball, but also the theory of friction on the billiard cloth will come into its own.

(a) High and Low Shots

The experienced player almost always gives the ball a "side" or "English." For the time being we shall, however, consider only shots without English, in which the cue therefore hits the ball in its vertical median plane, and in a horizontal direction. We distinguish high and low shots.

We speak of a *high shot* if the point of impact between cue and ball lies above $\frac{7}{5}a$ ($a =$ radius of the cue ball), as measured from the plane of the table; of a *low shot* if the ball is hit at a height less than $\frac{7}{5}a$ (cf. problem IV.3 in connection with this and the following). Only if the ball is hit at exactly this height does pure rolling take place from the very start. By virtue of the moment of inertia of a sphere given on p. 65, the rotation transmitted to the ball is then of such magnitude that the peripheral velocity corresponding to it is just equal and opposite at the point of support to the forward motion of the ball, so that the condition (11.10) of pure rolling is fulfilled.

For high shots the peripheral velocity at the point of contact generated by the rotation is opposite to that of the center of mass of the ball and exceeds the latter. The friction at the cloth opposes the excess velocity (peripheral velocity – forward velocity), thus augmenting the original velocity of the mass center: for high shots, friction acts on the ball in the direction of the shot. The final velocity under pure rolling, which sets in once the friction has consumed the excess velocity, is greater than the initial one. Balls that are hit high run for a long time and in general betray the experienced player.

For low shots the peripheral velocity at the point of contact is opposite to that of the center of mass, but outweighed by it; for even lower shots it is directed forward. In both cases friction acts in a direction opposite to that of the original impact. The final velocity under pure rolling is smaller than the initial one.

---

[9] G. Coriolis, *Théorie mathématique des effets du jeu de billard*. Paris, 1835.

As for the *impulse Z* (dimensions dyne-sec), it is of course to be interpreted as the time integral of a very great force F in the direction of the cue over the very short time of duration τ,

$$Z = \int_0^\tau F\, dt.$$

The *impulsive torque* about the center of the ball is accordingly given by

$$Zl = \int_0^\tau Fl\, dt$$

where l is the distance of the center from the axis of the cue. The impulsive torque vector is directed perpendicularly to the plane passing through center and cue axis. For the shots without English so far considered, it is directed horizontally and is normal to the median plane mentioned above.

(b) Follow Shots and Draw Shots

If the ball, after being struck high, meets one of the other two balls in central impact, it transfers all its forward motion to the latter because of the equality of the two masses involved [cf. Eq. (3.27a)]; but it retains its rotational motion if we neglect the friction between the two balls during the short time of contact. The instant after the impact the center of the striking ball is therefore momentarily at rest, while its lowest point glides over the billiard cloth. The friction thus arising is constant in time and acts on the ball in the sense of the original forward motion, while its moment about the center simultaneously slows down the existing rotation. Thus the ball is accelerated from the state of rest, while its rotation decreases accordingly. The acceleration ceases as soon as the peripheral velocity at the cloth has become equal to the forward velocity of the center, whereupon pure rolling sets in. Once this stage is reached, the ball rolls on with constant final velocity (we shall neglect the very slow effect of the rolling friction). This is the theory of the *follow shot.*

The ball which is hit low similarly transfers its center of mass velocity to the struck ball and is momentarily at rest. We shall assume that the ball was hit very low, at any rate below the center, so that the peripheral velocity at the point of contact remaining after collision is directed forward. The friction now acts backward. The ball begins to move with constant backward acceleration, while at the same time its rotational velocity decreases, until pure rolling sets in. This is the theory of the *draw shot.*

Since sliding friction is independent of the velocity, the variation with time of the center of mass velocity v as well as of the peripheral velocity $u = a\omega$ is a linear one. The exercises so far considered can therefore be treated more conveniently by graphical than by mathematical methods.

To do the former we may construct a diagram in which we plot the instantaneous values of v and u as ordinates against the time (problem IV.3).

(c) Trajectories with " English " Under Horizontal Impact

If the ball is not hit in the vertical median plane, but to either side of it, we speak of " right English " and " left English." As long as the cue is advanced horizontally against the ball, the trajectory remains a straight line in the direction of the initial impact.

The plane of the impulsive torque is now inclined to the vertical median plane, in high shots either to the right for right English, or to the left for left English, this inclination being such that the normal to the plane of the impulsive torque (this normal is parallel to the axial vector torque) is contained in the vertical plane through the center of the ball normal to the median plane. We can decompose the torque into a vertical component and a horizontal one at right angles to the direction of the impact. The first component causes a spin about the vertical diameter of the ball and generates a small " boring friction " at the cloth which has, however, no effect on the path of the ball. The lateral component on the other hand acts in the same way that it did in the shots considered under 1 and 2, so that the phenomena there observed apply without change to shots with English. In particular the trajectory remains rectilinear.

The spin about the vertical diameter makes itself felt in the collision of the ball with a cushion or with a second ball. In the first case friction at the cushion occurs which deviates the ball to the left for right English and to the right for left English as seen by the player. The angle of reflection, which, for shots without English, is equal to the angle of incidence, is thereby altered; as a matter of fact the actual reflected path is generated from the equiangular reflected path by a rotation of the latter in the sense of the vertical spin imparted to the ball. This phenomenon is familiar to every billiards player. Together with the frictional force at the cushion there appears a frictional torque about the vertical which weakens the spin about the vertical diameter. The original English therefore gradually disappears after several impacts, a fact which is likewise known to every player. In a collision of ball against ball the effect of the English is similar, acting in the same sense as in a ball-cushion impact.

(d) Parabolic Path Due to Shot with Vertical Component

The plane of the impulsive torque is now not only inclined as under (c) but also tilted forward as seen by the player. The vector torque has therefore not only components along the vertical and lateral directions, but also a component in the direction of motion. Thus the point of contact has a component of sliding velocity perpendicular to the initial motion.

The friction, which is opposed to the resultant velocity of the point of contact, therefore makes an angle different from zero with the initial motion. If we convince ourselves (cf. problem IV.4) that this angle formed with the original motion remains constant during the motion, and if we remember that the magnitude of the friction likewise remains constant, we conclude that the path of the ball is a parabola in the horizontal plane, since it is under the influence of a single force of constant magnitude and direction (principle of J. A. Euler, son of the great Leonhard).

Shots of this type are very surprising to a player who does not have full knowledge of the laws of friction and the vectorial decomposition of angular momentum. They are especially useful when the two balls to be hit are at the two opposite ends of the short side of the table. In that case the vertical component of the impulse must be very strong, i.e., the cue must be guided at a small angle to the vertical.

CHAPTER V

RELATIVE MOTION

The interest in the subject matter of this chapter derives mainly from the fact that we make all our observations on the rotating earth, which is not an allowable frame of reference, either in the sense of classical mechanics or in the sense of the special theory of relativity. In general relativity, on the other hand, all systems of reference are permitted (cf. p. 15), so that a separate theory of relative motion becomes meaningless.

In this chapter we shall adopt the viewpoint that in every theoretically admitted reference system the mechanics of Newton holds rigorously. We shall then ask for the deviations from Newtonian mechanics that result from the motion of the reference system to which, for practical reasons, we are chained.

§ 28. Derivation of the Coriolis Force in a Special Case

Let a mass point move along a meridian of the terrestrial globe, of radius a, with the constant angular velocity μ, while at the same time the earth rotates about its axis with constant angular velocity ω. As usual, we call θ the colatitude, ϕ the (celestial !) longitude. Apart from arbitrary initial values the motion of our mass point is then given by

$$(1) \qquad \theta = \mu t, \quad \phi = \omega t.$$

From the Cartesian coordinates of the point,

$$(2) \qquad \begin{aligned} x &= a \sin \theta \cos \phi \\ y &= a \sin \theta \sin \phi \\ z &= a \cos \theta, \end{aligned}$$

we obtain by differentiation with respect to t,

$$(3) \qquad \begin{aligned} \dot{x} &= a\mu \cos \theta \cos \phi - a\omega \sin \theta \sin \phi \\ \dot{y} &= a\mu \cos \theta \sin \phi + a\omega \sin \theta \cos \phi \\ \dot{z} &= -a\mu \sin \theta. \end{aligned}$$

$$(4) \qquad \begin{aligned} \ddot{x} &= -a\mu^2 \sin \theta \cos \phi - a\omega^2 \sin \theta \cos \phi - 2\,a\mu\omega \cos \theta \sin \phi \\ \ddot{y} &= -a\mu^2 \sin \theta \sin \phi - a\omega^2 \sin \theta \sin \phi + 2\,a\mu\omega \cos \theta \cos \phi \\ \ddot{z} &= -a\mu^2 \cos \theta. \end{aligned}$$

In the triplet of equations (4) the first terms on the right represent the usual centripetal acceleration which is associated with the motion along the meridian if the latter is at rest in space. The second terms give the centripetal acceleration resulting from the motion of a fixed point of the meridian in a circle of latitude (due to the earth's rotation about its axis). The third terms, however, constitute something new, for they represent the kinematic interplay of both motions. If we multiply (4) by $-m$, we obtain the inertial force $\mathbf{F}^*$ of our mass point in the compound rotation. In vector form it is

(5) $$\mathbf{F}^* = \mathbf{C}_1 + \mathbf{C}_2 + \mathbf{F}_c.$$

The symbols $\mathbf{C}_1$ and $\mathbf{C}_2$ refer, as in (10.3), to " ordinary centrifugal forces." $\mathbf{C}_1$ is directed radially outward from the earth's center and has the magnitude

$$|\mathbf{C}_1| = ma\mu^2 = m\frac{v_1^2}{a}, \quad v_1 = a\mu.$$

$\mathbf{C}_2$ is directed outward normal to the earth's axis, and has the magnitude

$$|\mathbf{C}_2| = ma\,\omega^2 \sin\,\theta = m\frac{v_2^2}{a\sin\theta}, \quad v_2 = a\,\omega\sin\,\theta.$$

We can call the third constituent $\mathbf{F}_c$ the " composite centrifugal force " (force centrifuge composée) or *Coriolis force*. Its complete vector expression [cf. Eq. (29.4a)] is given by

(6) $$\mathbf{F}_c = 2m\,\mathbf{v}_{\text{rel}}\mathbf{X}\omega.$$

We have here written $\mathbf{v}_{\text{rel}}$ instead of the vector $\mathbf{v}_1$ corresponding to the preceding v_1; by this we wish to indicate that quite generally it is the velocity relative to the rotating reference system that gives rise to $\mathbf{F}_c$.

According to (6) the magnitude of $\mathbf{F}_c$ is

(6a) $$|\mathbf{F}_c| = 2m\,v_{\text{rel}}\,\omega\sin\,(\mathbf{v}_{\text{rel}},\,\omega),$$

so that, in our case,

(6b) $$|\mathbf{F}_c| = 2m\,v_{\text{rel}}\,\omega\cos\,\theta.$$

$\cos\,\theta$ is of course just the sine of the geographic latitude. As for direction, $\mathbf{F}_c$ is perpendicular to both $\mathbf{v}_{\text{rel}}$ and ω, or, equivalently, to $\mathbf{C}_1$ and $\mathbf{C}_2$. The sense of $\mathbf{F}_c$ is given by the direction of advance of a right-handed screw turning from $\mathbf{v}_{\text{rel}}$ to ω. This is illustrated in Fig. 48 for a particle moving from south to north. Two positions, one in the southern and one in the northern hemisphere, are shown. In the former, corresponding to the sense of the right-handed screw $\mathbf{v}_{\text{rel}} \to \omega$, $\mathbf{F}_c$ acts from east to west; in the latter, from west to east.

Instead of a single particle we can also consider a continuous sequence of such particles, hence a river flowing along the meridian. Fig. 48 then tells us that the inertial force of the water moving from south to north presses against the *right bank in the northern hemisphere, against the left bank in the southern hemisphere.* The change of sign in the pressure is evidently connected with the sine of the geographic latitude occurring in (6b). This rule is valid not only for south-north flow, but, as will be shown in the next section, for any direction of v_{rel}, and therefore, in particular, also for the north-south direction of flow. This is intuitively obvious in our example. The west-east velocity of the water deriving from the earth's rotation depends on its distance from the axis of rotation, hence on the geographic latitude. If the stream moves from south to north, the water in the northern hemisphere has an excess of west-east momentum imported from more southerly latitudes; this excess manifests itself as a pressure eastward, that is, against the right bank. But similar reasoning must hold in the case of north-south motion. In that case the water imports a deficiency of west-east motion from the northern latitudes.

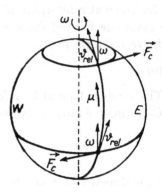

Fig. 48. Special derivation of Coriolis force: a mass point moves along a meridian of the rotating earth with constant velocity v_{rel} corresponding to the constant angular velocity μ as seen from the earth's center.

Let us mentally add the deficient amount in the sense of Fig. 41, once with + sign, once with − sign. The part added with − sign has an east-west direction, and therefore exerts a pressure westward, i.e., again on the right bank. The same process of reasoning shows that in the southern hemisphere the river exerts excess pressure on its left bank, for south-north as well as north-south motion of the water.

Geographers have proved by numerous examples that the pressure against the right bank in the northern hemisphere manifests itself in a stronger erosion of the right embankment (Baer law of river displacements); in addition the water stands slightly but measurably higher at the right shore of the river.

Of much greater significance are the effects of the Coriolis force on ocean currents (deviation of the Gulf Stream and tidal currents of the northern hemisphere to the right).

It is, however, in the atmosphere that its effects are most pronounced. The well-known law of Buys-Ballot states that the wind does not blow in the direction of the pressure gradient, but is deviated considerably, to the right in the northern, to the left in the southern hemisphere; it is only at the equator that it follows the pressure gradient exactly.

All these phenomena are immediate results of Newton's first law and in the last analysis derive from the fact that in mechanics the rotating earth is not an admissible reference frame.

In this section we have calculated the Coriolis force with the help of spherical polar coordinates. In problem V.1 we shall derive it in cylindrical coordinates.

§ 29. The General Differential Equations of Relative Motion

We replace the earth by an arbitrary rigid body B, which rotates with instantaneous angular velocity ω about a fixed point O. Let P be a particle which moves with arbitrarily varying velocity relative to B. Its velocity with respect to space is then composed of this relative velocity and the velocity in space of a point of the body instantaneously in coincidence with P. According to (22.4) the latter is given by $\omega \times r$. As in (22.4) we shall designate by w the velocity of P with respect to space; furthermore we shall call v (instead of v_{rel}) the relative velocity of P with respect to B. We then have

$$(1) \qquad w = v + \omega \times r.$$

Let us agree that temporal changes be designated by an overhead dot if observed from space, by $\frac{d}{dt}$ if observed from the body B. We can then write

$$(2a) \qquad w = \dot{r},$$

$$(2b) \qquad v = \frac{dr}{dt},$$

$$(2c) \qquad \dot{r} = \frac{dr}{dt} + \omega \times r.$$

The acceleration of our point P in space is given from (1) as

$$(3) \qquad \dot{w} = \dot{v} + \omega \times \dot{r} + \dot{\omega} \times r.$$

In the middle term of the right member we substitute the value of $\dot{r}$ from (2a) and (1) to obtain

$$(3a) \qquad \omega \times \dot{r} = \omega \times v + \omega \times (\omega \times r).$$

We shall transform the first term on the right of (3) by replacing the arbitrary vector $\mathbf{r}$ in (2c) by $\mathbf{v}$. This yields

(3b) $$\dot{\mathbf{v}} = \frac{d\mathbf{v}}{dt} + \boldsymbol{\omega} \times \mathbf{v}.$$

Substitution of (3a) and (3b) into (3) gives

(4) $$\dot{\mathbf{w}} = \frac{d\mathbf{v}}{dt} + 2\boldsymbol{\omega} \times \mathbf{v} + \boldsymbol{\omega} \times (\boldsymbol{\omega} \times \mathbf{r}) + \dot{\boldsymbol{\omega}} \times \mathbf{r}.$$

We notice that according to (26.8a) we can write either $\dot{\boldsymbol{\omega}}$ or $\frac{d\boldsymbol{\omega}}{dt}$ in the last term of Eq. (4).

From (4) we proceed to the inertial force acting on our particle by multiplying both sides by $-m$. At the left we then have the inertial force $\mathbf{F}^*$ in space; the first term on the right is the inertial force observed in the non-inertial reference system B, which we shall call $\mathbf{F}^*_{\text{rel}}$. The second term on the right gives the expression for the Coriolis force which we met in (28.6), viz.,

(4a) $$-2m\,\boldsymbol{\omega} \times \mathbf{v} = +2m\,\mathbf{v} \times \boldsymbol{\omega} = \mathbf{F}_c.$$

Our present treatment therefore supplements that of the preceding section by furnishing a general derivation of the Coriolis force. In the next to last term of Eq. (4) one easily recognizes (after multiplication by $-m$) the ordinary centrifugal force $\mathbf{C}$, which appears to act on our particle by virtue of the rotation of the reference system B, and which was designated by $\mathbf{C}_2$ in Eq. (28.5).

From (4) we therefore have, collecting all the terms,

(5) $$\mathbf{F}^* = \mathbf{F}^*_{\text{rel}} + \mathbf{C} + \mathbf{F}_c + m\,\mathbf{r} \times \dot{\boldsymbol{\omega}}.$$

Here we replace $\mathbf{F}^*_{\text{rel}}$ by its value from the definition

$$\mathbf{F}^*_{\text{rel}} = -m\,\frac{d\mathbf{v}}{dt}$$

and recall that due to the equilibrium of external and inertial forces in the system fixed in space we must have

$$\mathbf{F} + \mathbf{F}^* = 0.$$

Thus we obtain *the general differential equation of relative motion*

(6) $$m\frac{d\mathbf{v}}{dt} = \mathbf{F} + \mathbf{C} + \mathbf{F}_c + m\,\mathbf{r} \times \dot{\boldsymbol{\omega}}.$$

We see that in the system B there appear, in addition to the actual external force, the *fictitious forces* $\mathbf{C}$ and $\mathbf{F}_c$; from the standpoint of an observer moving with B, they act in the same manner as the external force **F**; actually, they result solely from the inertia of the particle m fixed in, or moving relative to, a non-Newtonian reference frame. The last term on the right of (6) is of similar origin; it stems from a possible acceleration or change in direction of the rotation. Applied to the earth it corresponds to the polar fluctuations and can certainly be neglected as vanishingly small. The differential equation (6) will be used in the three following sections and in problems V.1 and V.2.

§ 30. Free Fall on the Rotating Earth ; Nature of the Gyroscopic Terms

Whenever we try to measure the effect of gravity, it is not just the gravitational attraction itself, but the resultant of the earth's attraction **F** and the centrifugal force **C** that is observed. The flattening of the geoid, i.e., of the mean terrestrial surface, is itself determined by this resultant, and, in fact, in such a manner that the geoid is everywhere normal to it. If we put

(1)
$$\mathbf{F} + \mathbf{C} = - m\, \mathfrak{g},$$

the gravitational acceleration $\mathfrak{g}$ is a vector which has the magnitude g, but a direction along the normal to the geoid, rather than along the produced radius of the earth.

From (29.6) we obtain, in view of (1) and (28.6) and with neglect of the term in $\dot{\omega}$,

(2)
$$\frac{d\mathbf{v}}{dt} = - \mathfrak{g} + 2\mathbf{v}\times\boldsymbol{\omega}.$$

Let us now resolve this vector equation into coordinate equations by introducing an orthogonal system ξ, η, ζ, fixed in the earth, and defined as follows (cf. Fig. 49):

(3) ξ = north-south direction on the earth,

  η = west-east direction on the earth,

  ζ = point of observation → zenith = normal to geoid.

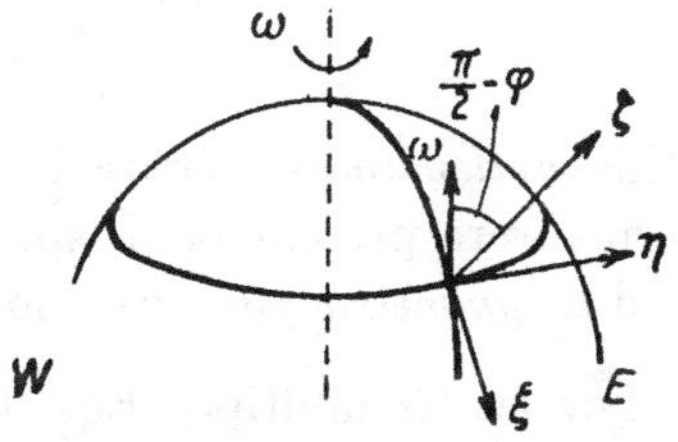

FIG. 49. Free fall on rotating earth. System of coordinates: ξ along a meridian, η along a circle of latitude, ζ along the normal to the geoid.

We then have, in component form,

(4)
$$\mathbf{v} = \left(\frac{d\xi}{dt}, \qquad \frac{d\eta}{dt}, \qquad \frac{d\zeta}{dt}\right);$$
$$\mathfrak{g} = (0, \qquad\qquad 0, \qquad\qquad g);$$
$$\boldsymbol{\omega} = (-\omega\cos\phi, \qquad 0, \qquad\qquad \omega\sin\phi);$$

ϕ being the geographic latitude as in Fig. 49.　It then follows from (2) that

$$
\begin{aligned}
\frac{d^2\xi}{dt^2} &= && 2\,\omega \sin\phi\,\frac{d\eta}{dt} \\[4pt]
(5) \qquad \frac{d^2\eta}{dt^2} &= -2\,\omega \sin\phi\,\frac{d\xi}{dt} && -2\,\omega \cos\phi\,\frac{d\zeta}{dt} \\[4pt]
\frac{d^2\zeta}{dt^2} + g &= && 2\,\omega \cos\phi\,\frac{d\eta}{dt}
\end{aligned}
$$

Before proceeding to integrate (5) we wish to examine the general character of these equations.　They are distinguished by the fact that the array of coefficients of the right-hand side is antisymmetric.　Let us introduce abbreviations

$$
(6) \qquad \alpha = 2\omega \sin\phi, \quad \beta = 0, \quad \gamma = -2\omega \cos\phi.
$$

The array then is clearly antisymmetric about the diagonal, as shown below:

(7)

	$\dfrac{d\xi}{dt}$	$\dfrac{d\eta}{dt}$	$\dfrac{d\zeta}{dt}$
$\dfrac{d^2\xi}{dt^2}$	0	α	β
$\dfrac{d^2\eta}{dt^2}$	$-\alpha$	0	γ
$g+\dfrac{d^2\zeta}{dt^2}$	$-\beta$	$-\gamma$	0

This *antisymmetric* character indicates *conservation of energy*.　If diagonal terms were present or if, speaking more generally, the array of coefficients had a *symmetric part*, we should have *dissipation of energy*.

For let us multiply Eqs. (5) row by row by $\frac{d\xi}{dt}$, $\frac{d\eta}{dt}$, $\frac{d\zeta}{dt}$, and add.　All the coefficients of α, β and γ on the right vanish, and we are left with

$$
\frac{1}{2}\frac{d}{dt}\left[\left(\frac{d\xi}{dt}\right)^2 + \left(\frac{d\eta}{dt}\right)^2 + \left(\frac{d\zeta}{dt}\right)^2\right] + g\frac{d\zeta}{dt} = 0,
$$

that is,

$$
(8) \qquad\qquad\qquad T + V = \text{const.}
$$

Here T and V are the kinetic and potential energy of relative motion (where we have put the mass $=1$).　This conservative character of our array of coefficients can be made evident even without calculation; for by virtue of the factor $\mathbf{v} \times \boldsymbol{\omega}$, $\mathbf{F}_c$ is perpendicular to the motion and therefore does no work, in analogy with magnetic forces in electrodynamics.

If, on the other hand, the array of coefficients had a symmetric contribution, we should have

$$\text{(9)} \qquad \frac{d}{dt}(T+V) < 0$$

where the $<$ sign results from the assumption that the signs of the coefficients satisfy the physically necessary conditions corresponding to a damping of the motion. It is seen that (9) results not in conservation, but rather, as asserted, in dissipation of energy. An example (only one-dimensional, it must be admitted) of the dissipative character of an even array of coefficients is furnished by Eqs. (8) and (9) in the treatment of damped oscillations of Chapter II, § 19.

With Lord Kelvin we call the terms of an antisymmetric array of coefficients *gyroscopic terms*. The name suggests that they indicate an internal gyration of the system (in our case the earth) which has not been taken into account explicitly in setting up the problem, but has instead been incorporated in the choice of coordinates (in our case the ξ, η, ζ). Such gyroscopic terms play an important role in general laws concerning the stability of equilibria and motions.

We shall now proceed with the integration of Eqs. (5). Let us postulate a free fall from height h without initial velocity. We therefore require at $t=0$:

$$\text{(10)} \qquad \xi = \eta = 0, \; \zeta = h$$

$$\frac{d\xi}{dt} = \frac{d\eta}{dt} = \frac{d\zeta}{dt} = 0.$$

From the first and third Eq. (5) we then have

$$\text{(11)} \qquad \frac{d\xi}{dt} = 2\omega\eta \sin\phi, \qquad \frac{d\zeta}{dt} + gt = 2\omega\eta \cos\phi.$$

Replacing these in the second Eq. (5), we obtain

$$\text{(12)} \qquad \frac{d^2\eta}{dt^2} + 4\,\omega^2\eta = C\,t, \qquad C = 2\omega g \cos\phi.$$

The integral of this equation is found by the general rule laid down in connection with Eq. (19.4), viz., " particular solution of the inhomogeneous equation + general solution of the homogeneous equation." In the present case this leads to

$$\eta = \frac{C}{4\,\omega^2}\,t + A\,\sin 2\omega t + B\,\cos 2\omega t.$$

Conditions (10) require that we put

$$B=0, \qquad 2\omega A = -\frac{C}{4\omega^2}, \quad \text{i.e.,}$$

(13) $$\eta = \frac{C}{4\omega^2}\left(t - \frac{\sin 2\omega t}{2\omega}\right) = \frac{g\cos\phi}{2\omega}\left(t - \frac{\sin 2\omega t}{2\omega}\right).$$

According to the meaning of η, cf. (3), this is the *eastward deflection*. ξ is the *southward deflection*. From (11) and (13) it satisfies

$$\frac{d\xi}{dt} = g\sin\phi\cos\phi\left(t - \frac{\sin 2\omega t}{2\omega}\right)$$

whose solution, with due regard for (10), is

(14) $$\xi = g\sin\phi\cos\phi\left(\frac{t^2}{2} - \frac{1-\cos 2\omega t}{4\omega^2}\right).$$

With the help of (13) and (10) we finally obtain from the second Eq. (11) the motion along the vertical,

(15) $$\zeta = h - \frac{gt^2}{2} + g\cos^2\phi\left(\frac{t^2}{2} - \frac{1-\cos 2\omega t}{4\omega^2}\right).$$

ωt is a very small number of order of magnitude (time of fall) $\div$ (1 day). We can therefore develop the solutions in powers of ωt. In lieu of (13), (14), and (15) we then obtain

$$\eta = \frac{gt^2}{3}\cos\phi\;\omega t, \qquad \xi = \frac{gt^2}{6}\sin\phi\cos\phi\;(\omega t)^2$$

$$\zeta = h - \frac{gt^2}{2}\left(1 - \frac{\cos^2\phi}{3}(\omega t)^2\right).$$

The eastward deflection is accordingly of the first, the southward deflection of the second order in ωt. The deviation from the law of freely falling bodies along the vertical caused by the earth's rotation is likewise only of the second order in ωt. The eastward deflection has been observed in several instances and found to be in agreement with theory; under favorable circumstances (deep mine shaft) it amounts to several centimeters.

Evidently these (observable or unobservable) deflections are due to the fact that the initial conditions (10) which lie at the very basis of both theory and experiment prescribe *rest with respect to the earth*. They hence imply a certain velocity in space, which is of the magnitude (earth's angular velocity) · (distance from axis of earth). This velocity is somewhat different from the velocity with which the earth's surface moves away under the falling body. It is then clear that the body does not hit the earth at the exact vertical projection of its initial position.

§ 31. Foucault's Pendulum

Once more Eqs. (30.5) are in force, but with the added condition that the mass point have the constant distance l from the point of suspension of the pendulum. We write this condition in a form similar to that used for the spherical pendulum (18.1), i.e.,

$$(1) \qquad F = \frac{m}{2}(\xi^2 + \eta^2 + \zeta^2 - l^2) = 0$$

and introduce the Lagrange multiplier associated with it. Eqs. (30.5) then read

$$\frac{d^2\xi}{dt^2} = \qquad\qquad 2\,\omega\sin\phi\,\frac{d\eta}{dt} \qquad\qquad +\lambda\xi$$

$$(2) \qquad \frac{d^2\eta}{dt^2} = -2\,\omega\sin\phi\,\frac{d\xi}{dt} \qquad\qquad -2\,\omega\cos\phi\,\frac{d\zeta}{dt} + \lambda\eta$$

$$\frac{d^2\zeta}{dt^2} + g = \qquad\qquad 2\,\omega\cos\phi\,\frac{d\eta}{dt} \qquad\qquad +\lambda\zeta.$$

We shall of course restrict ourselves to small oscillations. We therefore regard $\frac{\xi}{l}$ and $\frac{\eta}{l}$ as small quantities of the first order; from (1) it follows that $\frac{\zeta^2}{l^2} = 1$ up to quantities of the second order. More precisely, for points in the neighborhood of the rest position we can write

$$\zeta = -l\,(1 + \text{quantities of second order}),$$

since ζ is of course directed vertically upward. The third Eq. (12) then shows that up to quantities of the first order

$$(3) \qquad g = -\lambda l, \text{ hence } \lambda = -\frac{g}{l}.$$

Once more we write down the first two Eqs. (2), neglecting the term in $\frac{d\zeta}{dt}$ because small of second order, and using the abbreviation

$$(4) \qquad u = \omega\sin\phi,$$

to obtain

$$\frac{d^2\xi}{dt^2} - 2u\frac{d\eta}{dt} + \frac{g}{l}\xi = 0$$

$$(5)$$

$$\frac{d^2\eta}{dt^2} + 2u\frac{d\xi}{dt} + \frac{g}{l}\eta = 0.$$

It is convenient to consolidate them in complex form by multiplying the second Eq. (5) by i, adding it to the first one, and, as on p. 142, Eq. (26.10), introducing the new variable

(6) $$s = \xi + i\eta.$$

We obtain

(7) $$\frac{d^2 s}{dt^2} + 2iu \frac{ds}{dt} + \frac{g}{l} s = 0,$$

which is a homogeneous linear differential equation of second order with constant coefficients. Note that it is the gyroscopic character of the middle terms of Eqs. (5) which made step (5) → (7) possible.

Eq. (7) is solved by putting

$$s = A \, e^{i\alpha t}.$$

Substitution in (7) gives

$$\alpha^2 + 2u\,\alpha - \frac{g}{l} = 0,$$

a quadratic equation in α with the roots

(8) $$\alpha_1 = -u + \left(u^2 + \frac{g}{l}\right)^{\frac{1}{2}} \quad \text{and} \quad \alpha_2 = -u - \left(u^2 + \frac{g}{l}\right)^{\frac{1}{2}}.$$

It follows that the general solution of (7) is

(9) $$s = A_1 e^{i\alpha_1 t} + A_2 e^{i\alpha_2 t}.$$

The constants A_1 and A_2 are determined from the initial conditions. In agreement with the experimental arrangement we shall stipulate that these be

(10) $$\xi = a, \quad \eta = 0, \quad \frac{d\xi}{dt} = \frac{d\eta}{dt} = 0 \quad \text{at } t = 0.$$

We therefore imagine that the bob is pulled by an amount a out of its plumb line position along the positive ξ-axis, i.e. (cf. Fig. 50), southward along the meridian, and then released without impulse. From (10) the initial values of our complex variables are

(10a) $$s = a, \quad \frac{ds}{dt} = 0 \quad \text{at } t = 0.$$

Eq. (9) then gives

(11) $$A_1 + A_2 = a,$$

(11a) $$A_1 \alpha_1 + A_2 \alpha_2 = 0,$$

(11b) $$A_1 = \frac{a}{2}\left[1 + \frac{u}{\left(u^2 + \frac{g}{l}\right)^{\frac{1}{2}}}\right], \quad A_2 = \frac{a}{2}\left[1 - \frac{u}{\left(u^2 + \frac{g}{l}\right)^{\frac{1}{2}}}\right].$$

Next we calculate the expression for $\frac{ds}{dt}$; it is somewhat less involved than that for s itself. Recalling (11a) we have

$$\frac{ds}{dt}=i\,\alpha_1 A_1 e^{-iut}\left[e^{i\left(u^2+\frac{g}{l}\right)^{\frac{1}{2}}t}-e^{-i\left(u^2+\frac{g}{l}\right)^{\frac{1}{2}}t}\right],$$

from which, according to (8) and (11b),

(12) $$\frac{ds}{dt}=-a\frac{g}{l}\frac{1}{\left(u^2+\frac{g}{l}\right)^{\frac{1}{2}}}e^{-iut}\sin\left(u^2+\frac{g}{l}\right)^{\frac{1}{2}}t.$$

We arrive at the following conclusions: whenever the sine factor vanishes, we have

$$\frac{ds}{dt}=0\quad\text{and hence}\quad\frac{d\xi}{dt}=\frac{d\eta}{dt}=0.$$

This represents the occurrence of a turning point or cusp in the trajectory of the bob. According to our initial conditions (10), the first one of these occurs at $t=0$. If we put

(13) $$T=\frac{2\pi}{\left(u^2+\frac{g}{l}\right)^{\frac{1}{2}}},$$

succeeding cusps occur at

$$t=\frac{T}{2},\ t=T,\ t=\frac{3T}{2},\ \cdots$$

$t=T$ is the duration of a complete to-and-fro motion. Putting $u=0$ (that is, $\omega=0$) makes Eq. (13) agree with the period of oscillation of a simple pendulum without terrestrial rotation — as would be expected.

In order to see where the bob of our Foucault pendulum is located at $t=T$, we make use of (13) and (11) to obtain from (9)

$$s_{t=T}=A_1 e^{-iuT+2\pi i}+A_2 e^{-iuT-2\pi i}=(A_1+A_2)e^{-iuT}=ae^{-iuT}.$$

The bob therefore has the same distance a from its rest position as it had at the outset of the motion, but its azimuth no longer coincides with the southward meridian, as initially, but has acquired a lag with respect to this direction given by the angle

Fig. 50. Foucault's pendulum. Bird's-eye view of the trajectory of the bob. Initial displacement to the south; westward deflection in a complete oscillation.

$$uT=2\pi\frac{u}{\left(u^2+\frac{g}{l}\right)^{\frac{1}{2}}}\cong 2\pi\left(\frac{l}{g}\right)^{\frac{1}{2}}\omega\sin\phi.$$

The bob is thus deflected westward, cf. Fig. 50. We can explain this by saying that for zero rotation of the earth the pendulum bob would pursue a straight south-north-south course. In our case, however, the Coriolis force, through its "pressure on the right bank," displaces the trajectory by an angle $\frac{1}{2}uT$ eastward while the bob is swinging out, $\frac{1}{2}uT$ westward while it is moving back.

Foucault's experiments of 1851 and those of his countless successors yielded only qualitative results; a quantitative investigation of all sources of error was carried out by H. Kamerlingh Onnes, later leading authority in the field of low temperatures and discoverer of superconductivity, in his Groningen thesis of 1879.

§ 32. *Lagrange's Case of the Three-Body Problem*

We cannot resist the temptation to conclude our analysis of relative motion with the proof of a famous principle enounced by Lagrange (Paris Academy, 1772): *The three-body problem can be solved in closed and elementary form if one assumes that the triangle formed by the three celestial bodies always remains similar to itself.* The masses of the three bodies are completely arbitrary.

The proof of this principle will show that

1. The plane through the three mass points is fixed in space.
2. The resultant of the Newtonian forces on each of the three points passes through their common mass center.
3. The triangle formed by them is equilateral.
4. The three points describe conic sections similar to each other, with the common mass center at one focus.

The proof given by Lagrange is rather involved. It can be simplified if, with Laplace, we assume from the start the first conclusion above. Carathéodory[1] has, however, shown that even without this assumption an elementary proof is possible. His starting point is our vector equation (29.4) resolved in orthogonal coordinates. We shall follow his proof with minor modifications.

We consider the plane S which passes through the three points P_1, P_2, P_3 (masses m_1, m_2, m_3) and therefore also through their center of mass O. Without spoiling the generality of the problem we can define the latter as being at rest. S therefore rotates about the fixed point O; this rotation includes a component causing S to turn into itself about its normal at O. Call ω the total angular velocity. We imagine ourselves to be located on a frame fixed in S from which we observe the motion of the points P_k, in a way similar to that in which we observed the motion of Foucault's pendulum

[1] *Sitzber. bayer. Akad. Wiss., München.* 257 (1933).

from the earth. From O we measure the radius vectors $\mathbf{r}_k$ to the points P_k; $\mathbf{v}_k$ and $\frac{d\mathbf{v}_k}{dt}$ are their velocities and accelerations as observed from S. Making use of the vector rule (24.7), we write down the differential equations (29.4) of the motion in the form

(1) $$\frac{d\mathbf{v}_k}{dt} + 2\omega \mathbf{X}\mathbf{v}_k + \omega(\mathbf{r}_k \cdot \omega) - \mathbf{r}_k\omega^2 + \dot{\omega}\mathbf{X}\mathbf{r}_k = \frac{\mathbf{F}_k}{m_k}.$$

$\mathbf{F}_k$ is the vector sum of the Newtonian gravitational forces acting at m_k. Thus, for example,

(2) $$\frac{\mathbf{F}_1}{m_1} = \frac{Gm_2}{|\mathbf{r}_2 - \mathbf{r}_1|^2}\frac{\mathbf{r}_2 - \mathbf{r}_1}{|\mathbf{r}_2 - \mathbf{r}_1|} + \frac{Gm_3}{|\mathbf{r}_3 - \mathbf{r}_1|^2}\frac{\mathbf{r}_3 - \mathbf{r}_1}{|\mathbf{r}_3 - \mathbf{r}_1|}.$$

We fix a Cartesian coordinate system in S, with origin at O, and x, y, arbitrarily oriented, in the plane of S; at O we erect the z-axis perpendicular to S. In Eulerian fashion we resolve ω along these axes,

(3) $$\omega = (\omega_1, \omega_2, \omega_3).$$

Let the component ω_3 (rotation of S into itself) be determined by considering the direction of one of the vectors $\overrightarrow{OP}_k$ fixed in S. But we assumed that the triangle $P_1P_2P_3$ was to remain similar to itself; it follows that each of the other two vectors $\overrightarrow{OP}_k$ as well has a direction fixed in S. We can then write

(4) $$\mathbf{r}_k = \lambda(t)(a_k, b_k, 0),$$

where a_k, b_k are the Cartesian components of P_k at some given initial time. The function $\lambda(t)$ determines the common change in scale of the vectors $\overrightarrow{OP}_k$ and hence also that of the triangle $P_1P_2P_3$: with $\dot{\lambda}$ and $\ddot{\lambda}$ the derivatives of λ, we obtain from (4) that

(4a)
$$\mathbf{v}_k = \dot{\lambda}(t)(a_k, b_k, 0),$$
$$\frac{d\mathbf{v}_k}{dt} = \ddot{\lambda}(t)(a_k, b_k, 0).$$

It further follows that the resultant force $\mathbf{F}_k$ of Eq. (1) has a vanishing z-component, and x- and y-components inversely proportional to λ^2. We shall write this force in abbreviated form as

(5) $$\frac{\mathbf{F}_k}{m_k} = \frac{1}{\lambda^2(t)}(L_k, M_k, 0).$$

Next we write down the z-component of Eq. (1) perpendicular to S,

$$2\dot{\lambda}(\omega_1 b_k - \omega_2 a_k) + \lambda\omega_3(a_k\omega_1 + b_k\omega_2) + \lambda(\dot{\omega}_1 b_k - \dot{\omega}_2 a_k) = 0,$$

or, factoring out $a_k, b_k,$

(6) $\qquad \{-2\dot{\lambda}\omega_2+\lambda(\omega_3\omega_1-\dot{\omega}_2)\}a_k+\{2\dot{\lambda}\omega_1+\lambda(\omega_3\omega_2+\dot{\omega}_1)\}b_k=0.$

The two brackets $\{\,\}$ are functions of t independent of k. Calling them $f(t)$ and $g(t)$, we obtain

(6a) $\qquad\qquad\qquad\qquad \dfrac{f(t)}{g(t)}=-\dfrac{b_k}{a_k}.$

We have, however, assumed that the points P_k form a triangle, i.e., are not collinear. The three ratios b/a must therefore be unequal. In that case we can satisfy (6) only by putting $f=g=0$. Explicitly

(7) $\qquad\qquad \begin{aligned} 2\dot{\lambda}\omega_1 &= -\lambda(\omega_3\omega_2+\dot{\omega}_1),\\ 2\dot{\lambda}\omega_2 &= \lambda(\omega_3\omega_1-\dot{\omega}_2). \end{aligned}$

Multiplication by ω_1 and ω_2 respectively, followed by addition, yields

$$\frac{2\dot{\lambda}}{\lambda}=-\frac{\omega_1\dot{\omega}_1+\omega_2\dot{\omega}_2}{\omega_1{}^2+\omega_2{}^2}$$

and, by quadrature,

(8) $\qquad\qquad \omega_1{}^2+\omega_2{}^2=\dfrac{C}{\lambda^4}, \quad C=\text{constant of integration}.$

We proceed to write the x- and y-components of the differential Eq. (1). They are

$$\ddot{\lambda}a_k-2\omega_3\dot{\lambda}b_k+\omega_1\lambda(a_k\omega_1+b_k\omega_2)-\lambda a_k(\omega_1{}^2+\omega_2{}^2+\omega_3{}^2)-\dot{\omega}_3\lambda b_k=\frac{L_k}{\lambda^2},$$

$$\ddot{\lambda}b_k+2\omega_3\dot{\lambda}a_k+\omega_2\lambda(a_k\omega_1+b_k\omega_2)-\lambda b_k(\omega_1{}^2+\omega_2{}^2+\omega_3{}^2)+\dot{\omega}_3\lambda a_k=\frac{M_k}{\lambda^2},$$

or, arranged in factored form,

(9) $\qquad \begin{aligned} \{\ddot{\lambda}-\lambda(\omega_2{}^2+\omega_3{}^2)\}a_k-\{2\omega_3\dot{\lambda}+\lambda(-\omega_1\omega_2+\dot{\omega}_3)\}b_k &= \frac{L_k}{\lambda^2},\\[2mm] \{2\omega_3\dot{\lambda}+\lambda(\omega_1\omega_2+\dot{\omega}_3)\}a_k+\{\ddot{\lambda}-\lambda(\omega_1{}^2+\omega_3{}^2)\}b_k &= \frac{M_k}{\lambda^2}. \end{aligned}$

The brackets $\{\,\}$ of the first equation and similarly those of the second, when multiplied by λ^2, must therefore each satisfy three linear equations with constant coefficients (independent of t). This is possible only if they are themselves constant. It follows that the difference of the first and fourth brackets and that of the third and second brackets each equal a constant divided by λ^2. We then have

(10) $\qquad\qquad \omega_1{}^2-\omega_2{}^2=\dfrac{A}{\lambda^3}, \quad 2\omega_1\omega_2=\dfrac{B}{\lambda^3}.$

A suitable consolidation gives

$$(\omega_1 \pm i\,\omega_2)^2 = \frac{A \pm iB}{\lambda^3}$$

from which the absolute magnitude

(11) $$\omega_1{}^2 + \omega_2{}^2 = \frac{D}{\lambda^3}, \quad D = (A^2 + B^2)^{\frac{1}{2}}$$

is obtained. A comparison with (8) would lead to

(11a) $$\lambda = \frac{C}{D} = \text{const.}$$

unless both C and D were to vanish. Now according to (10) $\lambda = \text{const.}$ would make both ω_1 and ω_2 constant, so that, from (7), ω_3 would have to be zero. By suitable choice of the coordinates x, y one could even make $\omega_2 = 0$; the first Eq. (9) would then yield $L_k = 0$. In that case the three points P_k would have to be collinear, contrary to our hypothesis.

We must therefore put $C = D = 0$ and obtain from either (8) or (11) that

(12) $$\omega_1 = \omega_2 = 0.$$

This proves statement 1, that *the plane S rotates with angular velocity ω_3 into itself; its normal is fixed in space*.

If we apply the equation of angular momentum to our system we see that the motion of the points m_k within the plane S cannot contribute to the areal velocity constant. This constant is hence directly determined by the angular velocity ω_3 of S. We must have

$$\text{const.} = \omega_3 \sum m_k \,|\mathbf{r}_k|^2 = \omega_3 \lambda^2 \sum m_k (a_k^2 + b_k^2).$$

For this we can write

(12a) $$\lambda^2 \omega_3 = \gamma \qquad (\gamma = \text{constant});$$

it follows that

(12b) $$2\lambda \omega_3 + \lambda^2 \dot{\omega}_3 = 0.$$

By virtue of (12) and (12a, b) Eqs. (9) simplify to

(13) $$\lambda^2 \ddot{\lambda} - \frac{\gamma^2}{\lambda} = \frac{L_k}{a_k} = \frac{M_k}{b_k}.$$

The requirement $\dfrac{L_1}{a_1} = \dfrac{M_1}{b_1}$ contained in them says that the moment of F_1 about O vanishes, for

(14) $$\left| \mathbf{r}_1 \times \mathbf{F}_1 \right| = \frac{1}{\lambda^2}(a_1 M_1 - b_1 L_1) = 0,$$

so that $\mathbf{F}_1$ passes through the mass center O. The same holds for $\mathbf{F}_2$ and $\mathbf{F}_3$. This is our assertion 2 which states that *the resultant of the forces acting at P_k passes through the mass center of the particles m_k.*

We can make use of (2) to write (14) more explicitly. We have at once

(15) $$\frac{\mathbf{r}_1 \times \mathbf{F}_1}{m_1 G} = \frac{m_2 \mathbf{r}_1 \times \mathbf{r}_2}{|\mathbf{r}_2 - \mathbf{r}_1|^3} + \frac{m_3 \mathbf{r}_1 \times \mathbf{r}_3}{|\mathbf{r}_3 - \mathbf{r}_1|^3} = 0.$$

But from the definition of the mass center,

(16) $$m_1 \mathbf{r}_1 + m_2 \mathbf{r}_2 + m_3 \mathbf{r}_3 = 0,$$

and therefore

$$m_2 \mathbf{r}_1 \times \mathbf{r}_2 + m_3 \mathbf{r}_1 \times \mathbf{r}_3 = 0.$$

Substitution into (15) yields

$$m_2 \mathbf{r}_1 \times \mathbf{r}_2 \left(\frac{1}{|\mathbf{r}_2 - \mathbf{r}_1|^3} - \frac{1}{|\mathbf{r}_3 - \mathbf{r}_1|^3} \right) = 0,$$

that is,

(17) $$|\mathbf{r}_2 - \mathbf{r}_1| = |\mathbf{r}_3 - \mathbf{r}_1|.$$

Similarly we find

(17a) $$|\mathbf{r}_3 - \mathbf{r}_2| = |\mathbf{r}_1 - \mathbf{r}_2|, \text{ etc.}$$

We have thus arrived at statement 3: *the triangle is equilateral.*

The quotients $\frac{L_k}{a_k}$ and $\frac{M_k}{b_k}$ occurring in (13) can each be determined. To this end, let us call λs the side of the triangle, where

$$s^2 = (a_2 - a_1)^2 + (b_2 - b_1)^2 = (a_3 - a_2)^2 + (b_3 - b_2)^2 = \ldots$$

According to (2) and (5) we then have

$$\frac{L_1}{a_1} = \frac{G}{s^3 a_1} \{ m_2(a_2 - a_1) + m_3(a_3 - a_1) \}$$

and, in view of (16),

(18) $$\frac{L_1}{a_1} = \frac{G}{s^3} \{ -m_1 - m_2 - m_3 \}.$$

The right member of this equation is symmetric in the m_k and the coordinates a_k, b_k; it therefore represents the value not only of $\frac{L_1}{a_1}$, but of $\frac{L_k}{a_k}$ and also of $\frac{M_k}{b_k}$. Substitution of this value in (13) yields

(19) $$\lambda^2 \ddot{\lambda} - \frac{\gamma^2}{\lambda} = -\frac{G}{s^3}(m_1 + m_2 + m_3).$$

This differential equation in λ describes the motion in time, i.e., the rhythm with which our equilateral triangle alternately expands and contracts.

There is, however, a simpler way to gain insight into this secular motion and at the same time into the form of the trajectories; we abandon the plane S and observe the motion from a plane S' coinciding with S, but fixed in space. In S' the only force acting on the mass point m_k is the resultant force $\mathbf{F}_k$ directed toward the mass center which is at rest; the fictitious forces (Coriolis, centrifugal, etc.) occurring in (1) drop out. From (5) and (18) the magnitude of $\mathbf{F}_k$ is

(20) $$|\mathbf{F}_k| = \frac{m_k}{\lambda^2}(L_k^2 + M_k^2)^{\frac{1}{2}} = -\frac{m_k G}{\lambda^2 s^2}(m_1 + m_2 + m_3)\frac{(a_k^2 + b_k^2)^{\frac{1}{2}}}{s}.$$

The only quantity in the right member that varies in time is λ^2. With the help of (4) it can be expressed in terms of $|\mathbf{r}_k|$,

$$\lambda^2 = \frac{|\mathbf{r}_k|^2}{a_k^2 + b_k^2}.$$

Let us replace λ by this value in (20), define a new mass

(20a) $$m_k' = m_k \frac{(a_k^2 + b_k^2)^{\frac{3}{2}}}{s^3}$$

and the total mass $M = m_1 + m_2 + m_3$. Instead of (20) we then obtain

$$|\mathbf{F}_k| = -\frac{m_k' M G}{|\mathbf{r}_k|^2}.$$

Each of our three mass points hence moves in space independently of the others, as if endowed with a mass m_k', and attracted to a mass M at rest in O in a Newtonian manner. *It therefore describes a conic section with one focus at O.*

In order to be able to say something about the magnitude and mutual position of the three conic sections we must take into account the initial conditions implicit in the state of motion we have postulated. Let us for example consider the instant at which $\lambda = \lambda_{\text{extr}}$ when the distance

(21) $$\lambda_{\text{extr}}(a_k^2 + b_k^2)^{\frac{1}{2}}$$

of all the m_k from O is an extremum. According to (4) the radial velocity in S is then equal to zero; the velocity in S', i.e., in space, is given by the component ω_3 of angular velocity multiplied by the distance (21); the factor $(a_k^2 + b_k^2)^{\frac{1}{2}}$ occurring in this distance is thus a *measure of the similarity* not only of the initial velocities and initial distances from the common mass center, but at the same time of the size of the three conic sections resulting from these initial values. *With this, statement 4 is established.* The positions of the three conic sections are distinguished by the angles which the three radius vectors $\overrightarrow{OP_k}$ form with each other.

In the special case $m_1 = m_2 = m_3$, where the mass center coincides with the intersection of the medians of the equilateral triangle, the conic sections are congruent and displaced by $120°$ with respect to each other.

In addition to this motion in conic sections there is, according to Lagrange, a class of motions expressible in elementary form in which the three bodies are located on a rotating straight line. However we do not want to go into this here.

Let us finally point out that from the specialized three-body problem of Lagrange one can pass to a correspondingly specialized n-body problem. In the case of n equal masses and suitable initial velocities one then obtains n congruent Kepler ellipses, which are displaced by an angle $\frac{2\pi}{n}$ with respect to each other and traversed in the same rhythm. At one time this mode of motion was temporarily advanced for electrons to explain the L-spectra of X-rays [*Physikal. Zeits.* **19**, 297 (1918)].

CHAPTER VI

INTEGRAL VARIATIONAL PRINCIPLES OF MECHANICS AND LAGRANGE'S EQUATIONS FOR GENERALIZED COORDINATES

§ 33. Hamilton's Principle

We have already met a variational principle of mechanics, that of d'Alembert. This principle compares the state of a system at any given arbitrary instant with a neighboring state obtained from it by a virtual displacement. The principles which we are about to consider are *integral principles*. They differ from the former in that we shall be concerned with the successive states of the system during a finite interval of time, or, what amounts to the same thing, over a finite section of the trajectory. These states are then compared with certain corresponding virtual neighboring states.

The different integral principles with their various names are distinguished by the way in which the correspondence between the original states and their neighboring or varied states is established. They all have this in common: the quantity to be varied has the dimensions of *action*. They can therefore all be collected under the name " principles of Least Action[1]."

While power, as we already know, is a quantity of dimensions Energy × Time^{-1}, action has dimensions Energy × Time. An example of this is the elementary quantum of action, or Planck's constant, which we shall encounter in § 45, i.e., the quantity

$$h = 6.624 \cdot 10^{-27} \text{ erg sec.}$$

We shall first deal with *Hamilton's principle*. It differs from that of Maupertuis, to be treated in § 37 (though historically the latter came first), in that here *the time is not varied*. This means that the system arrives at any given point of the actual trajectory, of coordinates x_k, at the same time as at the corresponding point of the varied trajectory, of coordinates $x_k + \delta x_k$. The following statement sums up this property of Hamilton's principle:

(1) $$\delta t = 0.$$

---

[1] In English-speaking countries this usage is not common. We shall hence at once distinguish Hamilton's principle from the principle of least action (sometimes called the Principle of Maupertuis).—TRANSLATOR.

We must remark at this point that when we speak of the trajectory or path of the system, we do not mean the trajectory of a point of the system in a space of three dimensions, but rather a curve in a space of many dimensions, characteristic of the motion of the system as a whole. Thus, in the case of f degrees of freedom, this curve lies in the f-dimensional space of the coordinates $q_1 \ldots q_f$ (cf. p. 48).

In addition to the condition (1) we demand that another restriction be imposed on the variations in Hamilton's principle; the end points O and P of the section of the trajectory under consideration and of its varied neighboring trajectory must coincide in space. Hence we have, for any coordinate x,

(2) $\delta x = 0$ at $t = t_0$ and at $t = t_1$.

The adjoining figure has been drawn to aid in visualizing symbolically, in three dimensions, the relation of the actual path (solid) to the virtual one (dashed). The displacement δq, resulting from the variations of the coordinates δx, is to be completely arbitrary except at the two end points, with the restriction that δq be continuous and differentiable in t. There is a one-to-one correspondence between any point on the real path and one on the varied path, obtained from the former by a displacement δq, and two such points belong to the same time t.

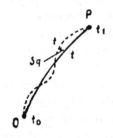

FIG. 51. Variation of the " trajectory " in Hamilton's principle. The time is not varied.

We shall now derive Hamilton's principle. We start out with the form (10.6) of d'Alembert's principle,

(3) $\sum_{k=1}^{n} \{ (m_k \ddot{x}_k - X_k)\delta x_k + (m_k \ddot{y}_k - Y_k)\delta y_k + (m_k \ddot{z}_k - Z_k)\delta z_k \} = 0.$

We therefore consider a system of n discrete mass points which may, however, be coupled by either holonomic or non-holonomic forces of constraint of unspecified nature. It follows that the δx_k, δy_k, δz_k, which must of course satisfy these constraints, are not independent of each other; in the holonomic case of f degrees of freedom only f can be chosen arbitrarily. In the non-holonomic case they are related by differential conditions.

We shall at first take up a purely formal transformation of relation (3), by writing

(4) $\ddot{x}_k \delta x_k = \frac{d}{dt}(\dot{x}_k \delta x_k) - \dot{x}_k \frac{d}{dt}(\delta x_k),$

where we shall at once ask ourselves what the meaning of an expression such as $\frac{d}{dt}(\delta x_k)$ is. For this purpose we compare not only the actual path of the x_k with the virtual path of the $x_k + \delta x_k$, but also the velocity $\dot{x}_k$ along the actual path with the velocity $\dot{x}_k + \delta \dot{x}_k$ along the virtual path *at the same instant t*. The latter velocity is defined by the identity

$$\frac{d}{dt}(x_k + \delta x_k) = \dot{x}_k + \frac{d}{dt}(\delta x_k).$$

We equate these two ways of writing the varied velocity and obtain

(5) $$\frac{d}{dt}(\delta x_k) = \delta \dot{x}_k.$$

Let us introduce this result into (4),

(6) $$\ddot{x}_k \delta x_k = \frac{d}{dt}(\dot{x}_k \delta x_k) - \dot{x}_k \delta \dot{x}_k = \frac{d}{dt}(\dot{x}_k \delta x_k) - \tfrac{1}{2}\delta(\dot{x}_k^2).$$

Similar equations hold of course for the coordinates y_k and z_k. Hence (3) can now be written in the form

(7)
$$\frac{d}{dt}\sum m_k(\dot{x}_k \delta x_k + \dot{y}_k \delta y_k + \dot{z}_k \delta z_k) =$$
$$\sum \frac{m_k}{2}\delta(\dot{x}_k^2 + \dot{y}_k^2 + \dot{z}_k^2) + \sum(X_k \delta x_k + Y_k \delta y_k + Z_k \delta z_k).$$

The second term on the right is nothing else but the virtual work δW, that is, the work done by the external forces in our virtual displacement. On the other hand, the first term on the right is the variation of the kinetic energy T given by

$$T = \sum \frac{m_k}{2}(\dot{x}_k^2 + \dot{y}_k^2 + \dot{z}_k^2)$$

which occurs when we pass from the real to the virtual trajectory. Eq. (7) can therefore be simplified to

(8) $$\frac{d}{dt}\sum m_k(\dot{x}_k \delta x_k + \dot{y}_k \delta y_k + \dot{z}_k \delta z_k) = \delta T + \delta W.$$

Before deriving some further conclusions from this, we shall digress for a moment to make some remarks about the relation (5). Let us write it down once more, in the form

(9) $$\frac{d}{dt}\delta x = \delta \frac{dx}{dt}.$$

If we recall that t is not varied, and that $\delta t = 0$ implies $\delta dt = 0$, we can replace (9) by

(9a) $$\frac{d\delta x}{dt} = \frac{\delta dx}{dt} \quad \text{or also} \quad d\delta x = \delta dx.$$

Eq. (9a), especially in the second form $d\delta = \delta d$, plays a fruitful if somewhat mysterious role in the older calculus of variations of the Euler type. We note that (9a) really says the same thing as the somewhat trivial Eq. (5) relating the time derivative of the virtual displacement to the virtual variation of velocity, except that (9a) contains the two assumptions that the time is not subject to variation and that the virtual displacement is continuous.

We now return to Eq. (8) and integrate it over t from t_0 to t_1. The left-hand side vanishes because of (2) and we are left with

$$(10) \qquad \int_{t_0}^{t_1} (\delta T + \delta W)\,dt = 0.$$

Owing to the type of variation embodied in Hamilton's principle, this can also be rewritten as

$$(11) \qquad \delta \int_{t_0}^{t_1} T\,dt + \int_{t_0}^{t_1} \delta W\,dt = 0.$$

It would, however, be erroneous to replace the latter integral by $\delta \int W\,dt$; for while it is true that the virtual work δW and the amount of work done in dt, i.e., dW, have a well-defined meaning, this is not so for ⁺he work W itself. W is, in general, not a " state variable." It is a state variable only if dW is a perfect differential, that is, if the external forces satisfy those conditions which guarantee the existence of a potential energy V [cf. § 6, (3)]. In that case we can replace

$$\int \delta W\,dt \quad \text{by} \quad -\int \delta V\,dt = -\delta \int V\,dt$$

in Eq. (11), which then takes the classically simple form

$$(12) \qquad \delta \int_{t_0}^{t_1} (T - V)\,dt = 0.$$

This is the equation one usually thinks of when one speaks of *Hamilton's principle*. It is valid, according to the statements of p. 46, *for conservative systems*. We can call equation (11) Hamilton's principle generalized to include non-conservative systems.

We now claim that Eqs. (12) or (11), respectively, contain the sum-total of mechanics, just as does d'Alembert's principle. This emphasizes the special significance of the energy-like expression $T - V$. In mechanics it is called the Lagrangian function (or *Lagrangian*, for short) and takes Eq. (12) into

$$(13) \qquad \delta \int_{t_0}^{t_1} L\,dt = 0 \quad \text{where} \quad L = T - V.$$

In words, *the time integral of the Lagrangian is an extremum.* Helmholtz relied heavily on the variational principle in Hamiltonian form in his last works; he extended it to electrodynamics, and called L the *kinetic potential.* The name " free energy," as opposed to the " total energy " $T+V$, would be equally justified in view of its wide use in thermodynamics.

Hamilton's principle derives special value from the fact that it is totally independent of the choice of coordinates. In fact, T and V (as well as δW) are quantities of immediate physical significance, which can be expressed in any desired set of coordinates. We shall make use of this property in the following section.

Hertz was of the opinion that Hamilton's principle was valid only for holonomic systems. This error was corrected by O. Hölder (Göttinger Nachr. 1896).

Hamilton's principle goes counter to our need for cause-and-effect relationships, as do all the other variational principles involving action integrals. For here the sequence of events is determined not by the present state of the system, but is instead derived under equal consideration of both its past and future states. It seems then that the variation principles are *not causal,* but rather *teleological.* We shall come back to this point in § 37, where we shall deal with the historical origin of the principles. There we shall also briefly touch on the conversion of Hamilton's principle into forms useful in fields of physics other than mechanics.

§ 34. *Lagrange's Equations for Generalized Coordinates*

Let us consider an arbitrary mechanical system. We shall for the present assume that its parts are coupled by holonomic conditions only. The number of degrees of freedom of the system is f. We can then introduce f independent coordinates which determine the position of the system at any given instant. We shall call them, as on p. 49,

(1) $$q_1, q_2, \ldots q_f.$$

These are our position coordinates. To them we add the " velocity coordinates "

(1a) $$\dot{q}_1, \dot{q}_2, \ldots \dot{q}_f.$$

The q_k and $\dot{q}_k$ together completely specify the state of the system at any instant.

Let us be more explicit : let the system be described for the moment by $n > f$ coordinates $x_1, \ldots x_n$, which need not necessarily be Cartesian. Let $n-f$ conditions hold between them, of the form

(2) $$F_k(x_1, x_2, \ldots x_n) = 0, \quad k = f+1, f+2, \ldots n$$

We can then define q_k as some function F_k of $x_1, \ldots x_n$,

(2a) $$F_k(x_1, x_2, \ldots x_n) = q_k, \; k = 1, 2, \ldots f.$$

Let us denote the partial derivatives of F_k with respect to x_i by F_{ik}; a differentiation of (2) and (2a) with respect to t then gives

(2b) $$\sum_{i=1}^{n} F_{ik}(x_1, \ldots x_n) \, \dot{x}_i = \begin{cases} \dot{q}_k, & k = 1, 2, \ldots f \\ 0, & k = f+1, \ldots n. \end{cases}$$

We can calculate from this the $\dot{x}_i$ as linear functions of the $\dot{q}_k$, with coefficients that depend on the $x_1, \ldots x_n$, or, by virtue of (2) and (2a), on the $q_1, \ldots q_f$. The kinetic energy T, a homogeneous quadratic function of the $\dot{x}_i$, just as it would be were it expressed in Cartesian coordinates to begin with, again becomes a homogeneous quadratic function of the $\dot{q}_k$ with coefficients that depend on the q_k. For the present we shall postulate that the potential energy V is a function of the q_k only, without, in principle, excluding the possibility of later making V a function of the $\dot{q}_k$ as well. In this connection we may now complete the definition (33.13) of L by stating that

L is to be regarded as a function of the q_k and $\dot{q}_k$.

For the time being we shall exclude an explicit dependence of L on t.

It is in this sense that we now write down the variation of L, i.e., the difference between the values of L in the virtual varied state $q_k + \delta q_k$, $\dot{q}_k + \delta \dot{q}_k$ and in the original state q_k, $\dot{q}_k$:

(3) $$\delta L = \sum_k \frac{\partial L}{\partial q_k} \delta q_k + \sum_k \frac{\partial L}{\partial \dot{q}_k} \delta \dot{q}_k.$$

This variation is now introduced into Hamilton's principle

(3a) $$\int_{t_0}^{t_1} \delta L \, dt = 0.$$

This form differs from that in (33.13) in that we have written the variation under the integral sign, whereas we had previously put it in front. The two forms are, of course, equivalent by virtue of rule (33.1), which says that t and dt are not varied. In any case, Eq. (3a) corresponds to the formulation (33.10) in which we first encountered the principle.

We now carry out the integration over the time indicated by (3a) on the general term of the second sum of (3). For this purpose we alter the form of this term by a *partial integration*, a procedure which has been characteristic of the whole calculus of variations ever since Euler[2]:

$$(4) \qquad \int_{t_0}^{t_1} \frac{\partial L}{\partial \dot{q}_k} \delta \dot{q}_k \, dt = \int_{t_0}^{t_1} \frac{\partial L}{\partial \dot{q}_k} \frac{d}{dt} \delta q_k \, dt = \frac{\partial L}{\partial \dot{q}_k} \delta q_k \Big|_{t_0}^{t_1} - \int_{t_0}^{t_1} \frac{d}{dt} \frac{\partial L}{\partial \dot{q}_k} \delta q_k \, dt.$$

In the last member of this double equality the first term vanishes because of the conditions laid down in (33.2). The complete expression (3) for δL therefore yields

$$(4a) \qquad \int_{t_0}^{t_1} \delta L \, dt = - \int_{t_0}^{t_1} \sum_k \left(\frac{d}{dt} \frac{\partial L}{\partial \dot{q}_k} - \frac{\partial L}{\partial q_k} \right) \delta q_k \, dt = 0.$$

Now the δq_k are independent of each other. We can therefore make all but one of them zero. This one we can also make zero everywhere along the "trajectory" of Fig. 51 except in the neighborhood of a single point, or, what amounts to the same thing, during a time interval Δt at an arbitrary time t. In order to satisfy (4a) we now require that

$$(5) \qquad \left(\frac{d}{dt} \frac{\partial L}{\partial \dot{q}_k} - \frac{\partial L}{\partial q_k} \right) \int_{\Delta t} \delta q_k \, dt = 0.$$

But Δt is finite, and δq_k does not vanish during the interval Δt. Hence we must have, for any time t and any index k,

$$(6) \qquad \frac{d}{dt} \frac{\partial L}{\partial \dot{q}_k} - \frac{\partial L}{\partial q_k} = 0.$$

These are Lagrange's equations for generalized coordinates, or, as they are also called, *Lagrange's equations of the second kind*, specialized to the case so far considered in which the forces acting on the system have a potential and the internal constraints of the system are holonomic.

If one or the other of these assumptions is dropped, we arrive at an extended form of these equations; let us hence consider two cases.

The first case is that in which the forces are not derivable from a potential. In that case the form (33.11) of Hamilton's principle will have to be our starting point. We think of the virtual work δW of the external forces as expressed in terms of the virtual displacements δq_k, and are led to write

$$(7) \qquad \delta W = \sum Q_k \delta q_k.$$

---

[2] In general we use the term "Euler's equation" of a given variational problem to designate an equation of type (6), and the derivation of (6) from (4) and (5) is typical of the derivation of Euler's equation in any such problem. We can therefore say that Lagrange's equations are the Euler equations for the variational problem characterized by the function L.

We shall call the coefficients Q_k here introduced the *generalized components of force* associated with the coordinates q_k. This is a formal extension of the force concept, which is of course admissible as a mathematical definition. Furthermore it is quite useful. Thus we can now restate the definition of the moment of a force about an axis given in (9.7) as follows: the moment of a force is the generalized force associated with the corresponding angle of rotation. It is clear that the quantities Q_k defined in (7) no longer possess vector character, nor need they in general have the dimensions of dynes any longer. From (7) it is seen that their dimensions depend instead on the dimension of the associated q_k. Thus moments of force must, as we already know, have dimensions of work, hence ergs, for the associated δq_k are angles and therefore dimensionless.

If we now introduce (7) in (33.11) and carry out the transformations indicated by Eqs. (4) and (5), we clearly obtain, in place of (6),

$$(8) \qquad \frac{d}{dt}\frac{\partial T}{\partial \dot{q}_k} - \frac{\partial T}{\partial q_k} = Q_k.$$

We can write this in a somewhat more general form as

$$(8a) \qquad \frac{d}{dt}\frac{\partial L}{\partial \dot{q}_k} - \frac{\partial L}{\partial q_k} = Q_k.$$

This is more general because now we can take into account the case where some of the forces acting are derivable from potentials, others not. We need only write the Q_k corresponding to the latter type of forces on the right side of (8a). The potential energy of the former, on the other hand, can be combined with the kinetic energy T to form the Lagrangian L of (8a).

Eqs. (8a) are then the *Lagrange equations for forces some of which are not derivable from potentials.*

If now we drop the second of the previously stated assumptions, i.e., postulate that the constraints of the system are in part non-holonomic, the introduction of the coordinates q_k is made invalid. For by definition non-holonomic conditions cannot be put in the form (2) and hence cannot be eliminated by proper choice of the q. We are then forced to introduce an excessive number of q, that is, a number greater than the number of degrees of freedom for infinitesimal motion. The latter is $f - r$ where f is the number of degrees of freedom for finite motion and r the number of non-holonomic conditions. These can be written as virtual conditions in a form similar to Eq. (7.4),

$$(9) \qquad \sum_{k=1}^{f} F_{k\mu}(q_1, \ldots q_f)\,\delta q_k = 0, \quad \mu = 1, 2, \ldots r.$$

They imply a restriction on allowable variations δq_k. One takes this restriction into account by multiplying each of the Eqs. (9) by a Lagrangian multiplier λ_μ and then adding it under the integral of (33.13). One obtains, with the F in somewhat abbreviated notation,

$$\int_{t_0}^{t_1} \left(\delta L + \sum_{\mu=1}^{r} \lambda_\mu F_{k\mu} \, \delta q_k \right) dt = 0.$$

The Eulerian transformation proceeds as in (4), where instead of (4a) we obtain

(10) $$\int_{t_0}^{t_1} \sum_k \left(\frac{d}{dt} \frac{\partial L}{\partial \dot{q}_k} - \frac{\partial L}{\partial q_k} - \sum_{\mu=1}^{r} \lambda_\mu F_{k\mu} \right) \delta q_k \, dt.$$

Here the δq_k are no longer independent of each other, but are connected through relations (9). One can, however, argue as on p. 67: of the bracketed () coefficients of δq_k in (10), r can be made to vanish by a suitable choice of the λ_μ. In the remaining sum over k, only $f - r$ of the q_k, all independent of each other, are left. The same line of reasoning as after (5) now forces us to the conclusion that the remaining brackets must vanish, too. We then obtain the complete system of f equations,

(11) $$\frac{d}{dt} \frac{\partial L}{\partial \dot{q}_k} - \frac{\partial L}{\partial q_k} = \sum_{\mu=1}^{r} \lambda_\mu F_{k\mu}.$$

We can designate these as *Lagrange's equations of the mixed type*, since they fall halfway between Lagrange's equations of the first and second kind.

We may mention that this mixed type occurs not only when we are *unable* to eliminate some of the conditions (case of non-holonomic constraints), but also whenever we *do not wish* to eliminate them. For it can happen that we are interested in the force of constraint that a holonomic condition exerts on the system. This force, as it turns out, is represented by the λ_μ associated with the condition in question [just as in Eq. (18.7) dealing with the spherical pendulum], and can be obtained by integration of Eq. (11).

Evidently we can finally combine the types (11) and (8a), for the case that we simultaneously drop both assumptions stated after (6).

Instead of doing this, we shall lastly concern ourselves with the following question: how and under what assumptions can the principle of the conservation of energy be derived from Lagrange's equations (6)?

As already emphasized, above Eq. (3), L is a function of the q and the $\dot{q}_k$; we further require, as earlier, that L *not contain* t *explicitly*. In that case Eq. (3) is valid not only for the virtual changes δq, $\delta \dot{q}$, but also for the secular changes dq, $d\dot{q}$, so that we have

(12)
$$\frac{dL}{dt} = \sum_k \dot{q}_k \frac{\partial L}{\partial q_k} + \sum_k \ddot{q}_k \frac{\partial L}{\partial \dot{q}_k}.$$

On the other hand we emphasized at the same place that T is a homogeneous quadratic function[3] of the $\dot{q}_k$. We can therefore apply the Euler rule

(13)
$$2\,T = \sum_k \dot{q}_k \frac{\partial T}{\partial \dot{q}_k}$$

for homogeneous functions. Differentiation with respect to the time gives

(14)
$$2\frac{dT}{dt} = \sum_k \dot{q}_k \frac{d}{dt} \frac{\partial T}{\partial \dot{q}_k} + \sum_k \ddot{q}_k \frac{\partial T}{\partial \dot{q}_k}.$$

We now subtract (12) from (14). Because $L = T - V$, the left member becomes

$$\frac{dT}{dt} + \frac{dV}{dt}.$$

On the right the second terms cancel *provided V is independent of $\dot{q}_k$*. In that case, by means of Eq. (6), the first terms on the right cancel as well, so that we obtain

(14a)
$$\frac{dT}{dt} + \frac{dV}{dt} = 0$$

from which we conclude that

(15)
$$T + V = E.$$

The law of the conservation of energy is therefore a consequence of Lagrange's equations.

We must now examine the assumptions leading to this important conclusion.

a) From the meaning of T we can say that the kinetic energy is determined by the position and velocity of the system, hence by q and $\dot{q}$; T could depend on t explicitly only as a result of the elimination of the equations

---

[3] Even when this is not the case and L is instead assumed to be any desired function of the q_k and $\dot{q}_k$, a generalized conservation law of the form $H = \sum \frac{\partial L}{\partial \dot{q}_k} \dot{q}_k - L = \mathrm{const.}$ can be given. In Chapter VIII we shall call the function H thus defined the "Hamiltonian"; the conservation law contained in Eq. (15c) is a special case of the above equation.

of constraint, in case the latter depend on t [4]. Now we have already seen on p. 68 that such constraints do work on the system, and therefore upset the conservation of energy. It is then indeed necessary for the validity of the conservation law that T not contain the time explicitly.

b) The assumption that L does not depend explicitly on t therefore reduces to the assumption that V is independent of t. This condition, too, is necessary. Otherwise one would have to add the term

$$- \frac{\partial V}{\partial t}$$

on the right side of Eq. (12). This term would then reappear with opposite sign in the right-hand member of Eq. (14a). Instead of $T + V = $ const. we should then obtain

(15a) $$\frac{d}{dt}(T+V) = \frac{\partial V}{\partial t},$$

that is, the law of the conservation of energy would be invalidated.

c) Suppose that V depends not only on the q_k but also on the $\dot{q}_k$. With the aid of (6) we obtain as the difference of the right members of (14) and (12)

(15b) $$\sum \dot{q}_k \frac{d}{dt}\frac{\partial V}{\partial \dot{q}_k} + \sum \ddot{q}_k \frac{\partial V}{\partial \dot{q}_k} = \frac{d}{dt}\sum \dot{q}_k \frac{\partial V}{\partial \dot{q}_k}.$$

This case does lead to a conservation law, which has, however, the unfamiliar form

(15c) $$T + V - \sum \dot{q}_k \frac{\partial V}{\partial \dot{q}_k} = \text{const.}$$

One more conclusion can be drawn from the above which will be useful to us later. We shall calculate $L - 2T = -(T+V)$, by using the expression (13) for $2T$ and reverting to the assumption that V is a function of only the q_k. We then arrive at

$$(T+V) = L - \sum \dot{q}_k \frac{\partial T}{\partial \dot{q}_k} = L - \sum \dot{q}_k \frac{\partial L}{\partial \dot{q}_k},$$

or

(16) $$T + V = \sum \dot{q}_k \frac{\partial L}{\partial \dot{q}_k} - L.$$

The total energy $T + V$ can be calculated from the expression for the Lagrangian.

---

[4] Sometimes such time-dependent conditions are called rheonomous (fluid) as opposed to time-independent conditions which are characterized as scleronomous (fixed, rigid).

The rather abstract developments of this section will come to life with the examples of the following section. To prepare ourselves for these we shall specialize the two expressions

$$\frac{\partial L}{\partial \dot{q}_k} \quad \text{and} \quad \frac{\partial L}{\partial q_k}$$

occurring in (6) for the simplest case, the motion of an isolated mass point expressed in Cartesian coordinates x, y, z. We have

$$T = \frac{m}{2}(\dot{x}^2 + \dot{y}^2 + \dot{z}^2), \qquad \frac{\partial L}{\partial \dot{x}} = \frac{\partial T}{\partial \dot{x}} = m\dot{x}, \text{ etc.}$$

$$\frac{\partial L}{\partial x} = -\frac{\partial V}{\partial x} = X, \text{ etc.}$$

Since, according to this equation, $\frac{\partial L}{\partial \dot{x}}$ represents the x-coordinate of momen-tum, we shall, quite in general, call $\frac{\partial L}{\partial \dot{q}_k}$ the *component of generalized momentum belonging to* q_k. Since on the other hand $\frac{\partial L}{\partial x}$ furnishes the x-component of the force, we shall label the two terms resulting from $\frac{\partial L}{\partial q}$ as *q-components of generalized force*,

(17)
$$\frac{\partial T}{\partial q} - \frac{\partial V}{\partial q} = \frac{\partial T}{\partial q} - Q.$$

Q is an *external* force as in Eq. (7), whereas $\frac{\partial T}{\partial q}$ is a *fictitious Lagrange* force dependent on the way in which the q coordinate varies with position. In the case of Cartesian coordinates x, y, z where curves of constant q are parallel to each other, a given q_i is independent of the q_k $(k \neq i)$ and the fictitious force vanishes.

§ 35. *Examples Illustrating the Use of Lagrange's Equations*

We have chosen examples which were treated earlier by elementary methods, in order to demonstrate the superiority of the Lagrange formalism.

(1) The Cycloidal Pendulum.

The obvious coordinate q in this case is the angle of rotation of the wheel generating the cycloids in Fig. 26. The Cartesian coordinates expressed in terms of this angle are, according to (17.2),

$$x = a(\phi - \sin \phi), \qquad \dot{x} = a(1 - \cos \phi)\,\dot{\phi}$$

$$y = a(1 + \cos \phi), \qquad \dot{y} = -a \sin \phi\,\dot{\phi}.$$

From these we calculate

$$T = \frac{m}{2}(\dot{x}^2 + \dot{y}^2) = ma^2(1 - \cos\phi)\,\dot{\phi}^2$$

$$V = mgy = mga(1 + \cos\phi)$$

(1) $$L = ma^2(1 - \cos\phi)\dot{\phi}^2 - mga(1 + \cos\phi).$$

This is all we need to know about the geometry and mechanics of our system. The Lagrange formalism automatically takes care of the rest:

$$\frac{\partial L}{\partial\dot{\phi}} = 2ma^2(1 - \cos\phi)\dot{\phi}, \qquad \frac{\partial L}{\partial\phi} = ma^2\sin\phi\,\dot{\phi}^2 + mga\sin\phi$$

$$\frac{d}{dt}\frac{\partial L}{\partial\dot{\phi}} = 2ma^2(1 - \cos\phi)\ddot{\phi} + 2ma^2\sin\phi\,\dot{\phi}^2$$

or, when substituted into the differential equation (6),

$$(1 - \cos\phi)\,\ddot{\phi} + \frac{1}{2}\sin\phi\,\dot{\phi}^2 = \frac{g}{2a}\sin\phi.$$

Introduction of the half angle and division by $2\sin\frac{1}{2}\phi$ simplifies this to

(2) $$\sin\frac{\phi}{2}\,\ddot{\phi} + \frac{1}{2}\cos\frac{\phi}{2}\,\dot{\phi}^2 = \frac{g}{2a}\cos\frac{\phi}{2}.$$

It can easily be verified that the left member equals $-2\frac{d^2}{dt^2}\cos\frac{1}{2}\phi$. Our differential equation (2) is therefore identical to the previous Eq. (17.6), by means of which we were able to prove the rigorously isochronous behavior of the cycloidal pendulum.

(2) The Spherical Pendulum

Here the angles θ and ϕ, polar angle and geographic longitude respectively on the sphere of radius l, are the given coordinates of the mass point. The line element is

$$ds^2 = l^2\,(d\theta^2 + \sin^2\theta\,d\phi^2)$$

so that the kinetic energy becomes

$$T = \frac{m}{2}l^2\,(\dot{\theta}^2 + \sin^2\theta\,\dot{\phi}^2).$$

As in (18.5a) the potential energy is $V = mgl\cos\theta$ and therefore

(3) $$L = \frac{m}{2}l^2\,(\dot{\theta}^2 + \sin^2\theta\,\dot{\phi}^2) - mgl\cos\theta.$$

And now the automatic calculation along the Lagrange pattern sets in. After division by constant factors, the differential equations for θ and ϕ are

(4)
$$\ddot{\theta} - \sin\theta\cos\theta\,\dot{\phi}^2 - \frac{g}{l}\sin\theta = 0$$

$$\frac{d}{dt}(l^2\sin^2\theta\,\dot{\phi}) = 0.$$

The second of these equations is the law of conservation of areal velocity, in agreement with (18.8). Note that we have here avoided the calculation which necessarily preceded this equation in the earlier treatment. With the help of the areal velocity constant C of Eq. (18.8), the first of Eqs. (4) can be written

$$\ddot{\theta} = \frac{C^2}{l^4}\frac{\cos\theta}{\sin^3\theta} + \frac{g}{l}\sin\theta.$$

The second term on the right is equivalent to the gravitational torque $|\mathbf{L}| = mgl\sin\theta$, this being the generalized component of force associated with the angle $q = \theta$ in the sense of (34.7). The first term is a fictitious Lagrange force in the sense of (34.17); the origin of this force is the fact that the lines along which the angle θ is measured on a sphere do not run parallel but diverge from the pole.

It is instructive to apply to this example the extension of Lagrange's equations for which provision was made in Eq. (34.11) by introducing the excess coordinate r together with the θ and ϕ. Now r is of course fixed through the relation $r = l$; nevertheless we are interested in this coordinate because it will give us, by means of the multiplier λ, the pressure of the mass point on the surface of the sphere, or, what amounts to the same thing, the tension in the suspension cord of the pendulum. In order to obtain the pertinent differential equation we need only replace (3) by

(5) $$L = \frac{m}{2}(\dot{r}^2 + r^2\dot{\theta}^2 + r^2\sin^2\theta\,\dot{\phi}^2) - mgr\cos\theta$$

and form a third Lagrange equation to be added to the two of (4),

(6) $$\frac{d}{dt}m\dot{r} - mr\dot{\theta}^2 - mr\sin^2\theta\,\dot{\phi}^2 + mg\cos\theta = \lambda r.$$

We have put the quantity $F_{k\mu}$ occurring in (34.11) equal to r, for in order to obtain agreement with Eq. (18.1) we have written the condition $r = l$ in the form

$$F = \frac{1}{2}(r^2 - l^2) = 0.$$

If we set $r = l$ and $\dot{r} = \ddot{r} = 0$, it follows from (6) that

(7) $$\lambda l = mg \cos \theta - ml(\dot{\theta}^2 + \sin^2\theta \; \dot{\phi}^2).$$

This is in agreement with (18.6) if there we transform the rectangular coordinates to θ, ϕ. Such a calculation is once more avoided by the use of the Lagrange scheme.

(3) The Double Pendulum

Here the two angles ϕ and ψ of Fig. 38 are suitable coordinates q_k. In the notation of § 21 we write

(8)
$$X = L \sin \phi, \quad x = L \sin \phi + l \sin \psi$$
$$Y = L \cos \phi, \quad y = L \cos \phi + l \cos \psi.$$

From these we get the following exact relations:

$$T = \frac{M}{2}(\dot{X}^2 + \dot{Y}^2) + \frac{m}{2}(\dot{x}^2 + \dot{y}^2)$$

$$= \frac{M+m}{2} L^2 \dot{\phi}^2 + \frac{m}{2} l^2 \dot{\psi}^2 + mLl \cos (\phi - \psi) \dot{\phi} \dot{\psi},$$

$$V = -MgY - mgy = -(M+m)gL \cos \phi - mgl \cos \psi.$$

The sign of the last expression is negative because (cf. Fig. 38) Y and y have been taken positive in the direction of the force of gravity. We shall here call $\varLambda$ the Lagrangian formed from $T - V$ since we have used the letter L for the length of the pendulum suspension. We obtain

$$\frac{\partial \varLambda}{\partial \dot{\phi}} = (M+m) L^2 \dot{\phi} + mLl \cos (\phi - \psi) \; \dot{\psi},$$

$$\frac{\partial \varLambda}{\partial \dot{\psi}} = ml^2 \dot{\psi} + mLl \cos (\phi - \psi) \dot{\phi},$$

$$\frac{\partial \varLambda}{\partial \phi} = -(M+m)gL \sin \phi - mLl \sin (\phi - \psi) \; \dot{\phi}\dot{\psi},$$

$$\frac{\partial \varLambda}{\partial \psi} = -mgl \sin \psi + mLl \sin (\phi - \psi) \dot{\phi} \dot{\psi}.$$

In writing down the Lagrange equations from these relations we shall at once go over to small ϕ, ψ. $\dot{\phi}$, $\dot{\psi}$ are quantities of the same magnitude as ϕ, ψ; their squares can therefore be neglected. The equations in question are then

(9)
$$\ddot{\phi} + \frac{g}{L} \phi = -\frac{m}{M+m} \frac{l}{L} \ddot{\psi},$$

$$\ddot{\psi} + \frac{g}{l} \psi = -\frac{L}{l} \ddot{\phi}.$$

These are identical with the Eqs. (21.3); we need merely switch back from coordinate angles ϕ, ψ to coordinate distances X, x by making use of the transformation equations (8) which, for small ϕ, ψ, simplify to

$$\phi = \frac{X}{L}, \quad \psi = \frac{x-X}{l}.$$

The identity is immediate for the second of Eqs. (9) and (21.3); the same is true of the first Eq. (9) and the first Eq. (21.3) provided we introduce for $\ddot{\psi}$ in the right member its value from the second Eq. (9). The discussion of the oscillation process following Eq. (21.3) is hence immediately applicable to our present Eqs. (9) and need not be repeated here.

In concluding we wish to emphasize that in the present purely formal treatment there was no mention whatever of the tension in the pendulum string l; this tension is implicitly contained in the Lagrange equations of motion as an internal reaction of the system, as has already been stressed in the footnote on p. 112.

(4) The Heavy Symmetrical Top

The classical coordinates q_k of this problem are the Eulerian angles θ, ϕ and ψ [θ and ϕ have been introduced already in (25.4) and (26.5a)]. We shall define them and their corresponding angular velocities as follows (cf. Fig. 52):

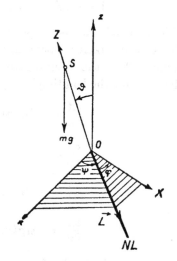

1. θ is the angle between the vertical and the axis of the top; $\dot{\theta}$ is the angular velocity about the line of nodes which is perpendicular to both of these directions.

2. ψ is the angle which the line of nodes makes with a fixed direction in the horizontal plane, for instance the x-axis; $\dot{\psi}$ is the angular velocity about the vertical.

3. ϕ is the angle which the line of nodes makes with a fixed direction in the equatorial plane of the top, for example the X-axis; $\dot{\phi}$ is the angular velocity about the axis of symmetry of the top.

Fig. 52. Definition of the Eulerian angles θ, ϕ, ψ, and their sense. The labeling of the axes is in agreement with the systems of coordinates introduced on p. 139 (z = vertical, Z = axis of top, x = horizontal line fixed in space, X = line in the equatorial plane of the top, fixed in the top).

The $\dot{\theta}$, $\dot{\phi}$, $\dot{\psi}$ are *holonomic* but *curvilinear* components of the angular velocity vector ω, as opposed to the ω_1, ω_2, ω_3 which were *rectilinear* but *non-holonomic* components of rotational velocity. Table (10) below shows the direction cosines between both sets of components. The table also gives the sense of rotation of $\dot{\theta}$, $\dot{\phi}$, $\dot{\psi}$ (rule of right-handed screw):

(10)

	$\dot{\theta}$	$\dot{\phi}$	$\dot{\psi}$
ω_1	$\cos\phi$	0	$\sin\theta\sin\phi$
ω_2	$-\sin\phi$	0	$\sin\theta\cos\phi$
ω_3	0	1	$\cos\theta$

The first two columns follow in an obvious manner from what was said in 1 and 3. In order to understand the third column, note that the projection of the vertically oriented vector $\dot{\psi}$ in the equatorial plane is $\dot{\psi}\sin\theta$; this vector in turn is resolved in the equatorial plane into the two components indicated opposite ω_1 and ω_2, viz., $\dot{\psi}\sin\theta\sin\phi$ and $\dot{\psi}\sin\theta\cos\phi$ respectively.

Notice that our table, unlike those in § 2, can be read only from left to right, not from top to bottom. From its rows we now obtain

(11) $\omega_1 = \cos\phi\,\dot{\theta} + \sin\theta\sin\phi\,\dot{\psi},$
$\omega_2 = -\sin\phi\,\dot{\theta} + \sin\theta\cos\phi\,\dot{\psi},$ (11a) $\omega_1^2 + \omega_2^2 = \dot{\theta}^2 + \sin^2\theta\,\dot{\psi}^2.$
$\omega_3 = \dot{\phi} + \cos\theta\,\dot{\psi}.$

Putting $I_2 = I_1$, the expression (26.17) therefore becomes

(12) $$T = \frac{I_1}{2}(\dot{\theta}^2 + \sin^2\theta\,\dot{\psi}^2) + \frac{I_3}{2}(\dot{\phi} + \cos\theta\,\dot{\psi})^2.$$

By virtue of Eq. (25.6a) for the gravitational potential energy V we have

(13) $$L = T - V = \frac{I_1}{2}(\dot{\theta}^2 + \sin^2\theta\,\dot{\psi}^2) + \frac{I_3}{2}(\dot{\phi} + \cos\theta\,\dot{\psi})^2 - P\cos\theta,$$

$$P = mgs.$$

L is therefore *independent* of the position coordinates ϕ and ψ and depends only on their change with time. We say that ϕ *and* ψ *are cyclic coordinates.* The name has its origin in the dynamic behavior of a rotating wheel (Greek: κυκλοσ) which is determined not by its instantaneous position but only by its speed of revolution. Hence

$$\frac{\partial L}{\partial \phi} = \frac{\partial L}{\partial \psi} = 0.$$

From Lagrange's equations the time derivatives of the quantities

$$\frac{\partial L}{\partial \dot\phi} \quad \text{and} \quad \frac{\partial L}{\partial \dot\psi}$$

must then vanish. At the end of the last section we called these quantities the generalized momenta associated with ϕ and ψ. From now on we shall always designate them by p. Thus we write in general

(14)
$$p_k = \frac{\partial L}{\partial \dot q_k} .$$

We can then assert that if the coordinates q_k are cyclic, *the momenta p_k conjugate to cyclic coordinates are integrals of the motion* (i.e., constants of integration). In our case we already know the significance of these constants from (25.6). We have

(15)
$$p_\phi = M'', \quad p_\psi = M'.$$

Previously, on p. 141, we lacked the expressions of these constants in terms of the position coordinates of the top. These can now be derived by application of the general rule (14):

(16)
$$p_\phi = \frac{\partial L}{\partial \dot\phi} = I_3 (\dot\phi + \cos\theta\,\dot\psi),$$

$$p_\psi = \frac{\partial L}{\partial \dot\psi} = I_1 \sin^2\theta\,\dot\psi + I_3 \cos\theta(\dot\phi + \cos\theta\dot\psi).$$

Combination of (15) and (16) results in

(17)
$$\dot\phi + \cos\theta\dot\psi = \frac{M''}{I_3},$$

$$I_1 \sin^2\theta\dot\psi = M' - M''\cos\theta.$$

Eqs. (17) exhaust the content of two of the Lagrange equations. The third one expresses the rate of change of

(18)
$$p_\theta = \frac{\partial L}{\partial \dot\theta} = I_1 \dot\theta$$

and becomes, if (17) is used to eliminate $\dot\phi$ and $\dot\psi$,

(19)
$$I_1\ddot\theta = \frac{(M' - M''\cos\theta)(M'\cos\theta - M'')}{I_1 \sin^3\theta} + P\sin\theta.$$

The right-hand member, which comes from $\frac{\partial L}{\partial \theta}$, contains not only the gravitational effect familiar to us from (25.4), but in addition a fictitious force which is a consequence of the nature of the coordinate system used, as we know from p. 192.

Eq. (19) has the character of a generalized pendulum equation. We need not be detained with its integration, for we can avail ourselves of the integral of energy

$$(20) \qquad T+V=E$$

which must be identical with the result of a first integration of (19). Let us once more eliminate the quantities $\dot{\phi}$ and $\dot{\psi}$ of Eq. (12) with the help of (17). Then (20) yields

$$(21) \qquad \frac{I_1}{2}\left\{ \dot{\theta}^2 + \left(\frac{M'-M''\cos\theta}{I_1\sin\theta}\right)^2 \right\} + \frac{M''^2}{2I_3} + P\cos\theta = E.$$

Since Eq. (21) contains three constants of integration, namely M', M'', and E, it must be the general integral of first order for the problem of the top. Finally, just as in § 18 for the spherical pendulum, we replace θ and $\dot{\theta}$ by

$$\cos\theta = u; \quad \dot{\theta}\sin\theta = -\dot{u}.$$

We then obtain

$$(22) \qquad \left(\frac{du}{dt}\right)^2 = U(u)$$

where

$$(23) \qquad U(u) = \left(\frac{2E}{I_1} - \frac{M''}{I_1 I_3} - \frac{2P}{I_1}u\right)(1-u^2) - \left(\frac{M'-M''u}{I_1}\right)^2.$$

Since $U(u)$ is a polynomial of third degree in u, the time t must be given by an elliptic integral of the first kind, as in the case of the spherical pendulum:

$$(24) \qquad t = \int^{u} \frac{du}{U^{\frac{1}{2}}}.$$

The azimuth angle ψ is given from Eq. (17) by an elliptic integral of the third kind (cf. p. 100),

$$(25) \qquad \psi = \int^{u} \frac{M'-M''u}{I_1(1-u^2)}\frac{du}{U^{\frac{1}{2}}}.$$

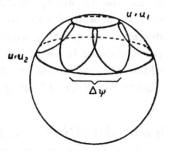

FIG. 53. Trace of the axis of the heavy symmetrical top on a sphere of unit radius.

We can now repeat the considerations following Fig. 29 on p. 99, and arrive at the picture of Fig. 53. The trace of the axis of the top on a unit sphere oscillates back and forth between the two circles of latitude $u=u_2$ and $u=u_1$, which it touches. At the points of tangency, as shown in Fig. 53, the trace may either merely pass by, or make a loop; the loop may in turn degenerate into a cusp. During each oscillation the axis of the top advances by the same azimuth angle $\Delta \psi$, obtained from Eq. (25) by a complete elliptic integral of the third kind, similar to that in (18.15).

In particular, if the top is to describe a regular precession about the vertical, it is necessary that the parallel circles u_1 and u_2 become merged; the curve $U(u)$ of Fig. 29 (p. 99) must then touch the axis of abscissae from below. This shows that the regular precession of the heavy top is a particular form of motion (whereas, in the case of the top under no forces, it is the general form of motion).

If the two roots u_1 and u_2 do not coincide exactly, but only approximately, we still *seem* to have a uniform advance of the axis of the top about the vertical; on closer observation one notices, however, that *small nutations* are superposed on this uniform advance, giving rise to what we called a "pseudo-regular precession". This is the typical phenomenon that one observes in the usual experiments with tops: one first imparts the greatest possible angular momentum to the top about its axis by pulling a string off its rim, and then sets it point down in a socket pan, taking great care not to add a perceptible lateral impulse to the motion.

We explain this behavior as follows: in such an experiment the initial angular momentum **M** is close to the axis of symmetry; this also follows from the Poinsot method for the initial axis of rotation. Hence the axis of rotation describes at first a narrow circuit on the unit sphere of Fig. 43. The parallel circles $u=u_1$, $u=u_2$ touching this circuit are close neighbors and remain close during the entire course of the motion, as can be seen from our general illustration in Fig. 53. The angular momentum and hence also the angular velocity are at first very great; they, too, remain unchanged during the motion apart from frictional losses. The nutations are therefore very rapid and almost invisible. The top seems reluctant to yield to the influence of gravity, instead constantly "sidestepping" in a direction perpendicular to the force of gravitation. It is this paradoxical behavior which has for centuries attracted amateur and professional investigator alike to the theory of the spinning top.

§ 36. *An Alternate Derivation of Lagrange's Equations*

Even though the derivation of Lagrange's equations for generalized coordinates from Hamilton's principle is unsurpassed in clarity and brevity, we feel that it is somewhat artificial. The transformation properties of the various dynamic variables, which form the core of the Lagrange equations, are not brought to light. The following derivation will remedy this situation.

We focus our attention on a system of $\frac{n}{3}$ mass points (n being divisible by 3), subject to arbitrary constraints, chosen holonomic for the sake of simplicity. The number of constraints is equal to $n-f$, where f is the number of degrees of freedom of the system. Our notation will be that of Eq. (34.2). We shall number the coordinates, assumed orthogonal, $x_1, x_2 \ldots x_n$; similarly we let $X_1, X_2 \ldots X_n$ be the components of the external forces. Finally we shall call $\xi_1, \xi_2 \ldots \xi_n$ the components of momentum of our mass points. We would have preferred to name them $p_1, p_2 \ldots p_n$, as agreed to in (35.14); this notation must, however, be reserved for the generalized momenta. We have

(1) $$\xi_i = m_i \dot{x}_i, \quad i = 1, 2 \ldots n,$$

where the m_i are of course equal in groups of three. The motion of our system is described by Lagrange's equations of the first kind (12.9) which are, in the present notation,

(2) $$\frac{d\xi_i}{dt} = X_i + \sum_{\mu=f+1}^{n} \lambda_\mu \frac{\partial F\mu}{\partial x_i}, \quad i = 1, 2 \ldots n.$$

We now introduce the generalized position coordinates $q_1, \ldots q_f$, which can be and are to be chosen in such a way that, just as in (34.2), the $n-f$ conditions $F\mu = 0$ are identically satisfied. Then Eqs. (34.2b) must hold between the old and the new velocity coordinates; we solve these for the $\dot{x}$ and write them as follows:

(3) $$\dot{x}_i = \sum_{k=1}^{f} a_{ik} \dot{q}_k, \quad i = 1, 2 \ldots n.$$

The a_{ik}, called F_{ik} in (34.2b), are functions of the $x_1 \ldots x_n$ and therefore also of the $q_1 \ldots q_f$, as stressed in § 34. We see that whereas the old and new *position* coordinates are connected by an *arbitrary point transformation*, the *velocity* coordinates transform *linearly*, the coefficients depending on the position coordinates.

What is the transformation character of the *components of force*? We shall call the new force components Q_k and define them as in (34.7) by means of the invariance of virtual work, that is

(4) $$\delta W = \sum_{i=1}^{n} X_i \delta x_i = \sum_{k=1}^{f} Q_k \delta q_k.$$

We now pass from virtual to real displacements and from these to the corresponding velocities. By virtue of (3), Eq. (4) becomes

(4a) $$\sum_{k=1}^{f} Q_k \dot{q}_k = \sum_{i=1}^{n} X_i \sum_{k=1}^{f} a_{ik} \dot{q}_k.$$

The $\dot{q}_k$, unlike the $\dot{x}_i$, are independent of each other. Hence their coefficients on the right and left of (4a) must be equal, so that

$$(5) \qquad Q_k = \sum_{i=1}^{n} a_{ik} X_i, \quad k=1, 2 \ldots f.$$

This is the *transpose* of transformation (3); in (3) we sum over the k, in (5) over the i. Written explicitly,

$$\dot{x}_1 = a_{11}\dot{q}_1 + a_{12}\dot{q}_2 \ldots \qquad Q_1 = a_{11}X_1 + a_{21}X_2 + \ldots$$
$$\dot{x}_2 = a_{21}\dot{q}_1 + a_{22}\dot{q}_2 \ldots \qquad Q_2 = a_{12}X_1 + a_{22}X_2 + \ldots$$

The transposition hence consists of an interchange of a_{ik} and a_{ki}. We say that *the components of force transform contravariantly*[5] (*or are " contragredient "*) *to the velocity coordinates.*

The *components of momentum transform like the components of force*, that is, covariantly to them. For we can think of the momenta as those impulsive forces which cause our mass points, initially at rest, to take on the required velocities. If we call the new momenta p_k, they can be expressed in terms of the old ξ_i by means of the relations

$$(6) \qquad p_k = \sum_{i=1}^{n} a_{ik} \xi_i.$$

These are the defining equations for the p_k. The definition is rather clumsy, but can be converted quite readily to a more meaningful form. For this purpose let us consider, as on p. 186, the expressions for the kinetic energy as function of the $\dot{q}$ on the one hand, and as function of the $\dot{x}$ on the other. We shall distinguish the two expressions, wherever necessary, by writing

$$T_{\dot{q}} \quad \text{or} \quad T_{\dot{x}}.$$

We then form

$$(7) \qquad \frac{\partial T_{\dot{q}}}{\partial \dot{q}_k} = \sum_{i=1}^{n} \frac{\partial T_{\dot{x}}}{\partial \dot{x}_i} \left[\frac{\partial \dot{x}_i}{\partial \dot{q}_k} \right].$$

The bracket is to remind us that in differentiating with respect to $\dot{q}_k$ we must keep the q_k as well as all $\dot{q}_i$ $(i \neq k)$ fixed. According to Eq. (3) the term in brackets is just a_{ik}. On the other hand the elementary expression

$$T_{\dot{x}} = \frac{1}{2} \sum m_i \dot{x}_i^2 \quad \text{evidently yields} \quad \frac{\partial T_{\dot{x}}}{\partial \dot{x}_i} = \xi_i.$$

---

[5] In the theory of general relativity it is customary to denote by a superscript (Q^k, p^k) those quantities which, like Q and the p (about to be defined), transform contravariantly (i.e., are " contragredient ") to the $\dot{q}_k$. We believe, however, that this usage, so important in general relativity, can here be dispensed with,

Instead of (7) we then have

$$\text{(8)} \qquad \frac{\partial T_{\dot{q}}}{\partial \dot{q}_k} = \sum_{i=1}^{n} a_{ik}\xi_i.$$

The right member is identical to that of (6). Hence the result:

$$\text{(9)} \qquad p_k = \frac{\partial T_{\dot{q}}}{\partial \dot{q}_k}$$

We can now assume that the external forces are derivable from a potential V independent of the $\dot{q}$, and introduce the Lagrangian $L = T - V$, so that (9) can be rewritten

$$\text{(9a)} \qquad p_k = \frac{\partial L}{\partial \dot{q}_k}.$$

We have thus quite generally justified the definition of the p_k anticipated in (35.14).

We are now in a position to transform the equations of motion (2) to generalized coordinates. To this end we multiply them successively by the different a_{ik} $(k = 1 \ldots n)$ and sum over i. By Eq. (5), the first term in the right member becomes

$$\text{(10)} \qquad Q_k = -\frac{\partial V}{\partial q_k}.$$

In the second term on the right the factor of λ_μ is

$$\text{(11)} \qquad \sum_{i=1}^{n} a_{ik}\frac{\partial F_\mu}{\partial x_i} \quad \text{for} \quad \mu = f+1, \ldots n$$

Now Eq. (3) tells us that

$$\text{(12)} \qquad a_{ik} = \frac{\partial x_i}{\partial q_k}.$$

This becomes evident if one writes (3) in the equivalent form $dx_i = \sum a_{ik}dq_k$, and holds all q except q_k fixed. Instead of (11) we can now also write

$$\sum_{i=1}^{n} \frac{\partial F_\mu}{\partial x_i}\frac{\partial x_i}{\partial q_k} = \frac{\partial F_\mu}{\partial q_k}.$$

But according to (34.2) it is precisely for $\mu = f+1, \ldots n$ that the F_μ have been made identically zero by our choice of the q_k, so that the partial derivatives of the F_μ with respect to the q_k vanish as well. Hence the right member of our equation reduces to (10).

The left member,

$$\sum_i a_{ik}\frac{d\xi_i}{dt},$$

is transformed into

(13) $$\frac{d}{dt}\sum_i a_{ik}\xi_i - \sum_i \xi_i \frac{da_{ik}}{dt} = \frac{dp_k}{dt} - \sum_i \xi_i \frac{d}{dt}\frac{\partial x_i}{\partial q_k},$$

where we have made use of (6) and (12). The last sum can be written in the form

$$\sum m_i \dot{x}_i \frac{\partial \dot{x}_i}{\partial q_k} = \frac{\partial}{\partial q_k}\frac{1}{2}\sum m_i \dot{x}_i^2 = \frac{\partial}{\partial q_k}T_{\dot{q}},$$

where the index $\dot{q}$ of T is to remind us that T must be converted to a function of the q, $\dot{q}$ before the differentiation with respect to q_k is carried out. The right side of (13) will then become

(13a) $$\frac{dp_k}{dt} - \frac{\partial T}{\partial q_k}.$$

Since it is to be equal to (10), we finally obtain

(14) $$\frac{dp_k}{dt} = \frac{\partial T}{\partial q_k} - \frac{\partial V}{\partial q_k} = \frac{\partial L}{\partial q_k}.$$

Referring back to (9a) we see that this is identical with the Lagrange equation in form (34.6), or, if we do not assume the existence of a potential energy, with that in the form (34.8).

We have thus convinced ourselves that we need not have recourse to Hamilton's principle to derive the Lagrange equations; we need merely make a thorough study of the transformation properties of the dynamic variables involved.

§ 37. *The Principle of Least Action*

In the conclusion to § 33 we spoke of the *teleological* character of our integral principles. "Teleological" means "shaped by a purpose," "directed toward an end." "Among all possible motions, Nature chooses that which reaches its goal with the minimum expenditure of action." This statement of the principle of least action may sound somewhat vague, but is completely in keeping with the form given it by its discoverer.

In the formulation of the principle not only teleological but also theological beliefs played a role. Maupertuis recommended his principle with the assertion that it best expressed the wisdom of the Creator. Leibniz, too, must have had such arguments in mind, as shown by the title of his *Theodicée* (justification of God).

Maupertuis published his principle in the year 1747. He was referred to a letter of Leibniz of the year 1707 (the original of this letter has been lost); he nevertheless defended his priority with passion, even throwing his weight as president of the Berlin Academy into the dispute. The principle acquired a mathematically definite form only later in the hands of Euler and especially Lagrange.

In the formulation of the principle of least action given above two things are not clear.

1. What is meant by the word "action"? Clearly not the same entity as in Hamilton's principle, for we are dealing with a formulation which, though related to that of Hamilton, is yet distinct from it.

2. What is meant by the phrase, "all possible motions"? It is quite essential to define precisely the class of all motions to be considered for comparison; only thus shall we be able to choose from this class the real motion as the most purposeful or favorable.

Regarding 1: Leibniz took the product $2T\,dt$ as his element of action. In what is to follow we, too, shall designate by *action integral* the quantity[6]

$$(1) \qquad\qquad S = 2 \int_{t_0}^{t_1} T\,dt.$$

Maupertuis, who, like Descartes, regarded the momentum mv as basic in mechanics, took $mv\,ds$ to be the element of action. It is clear, however, that the definitions of Leibniz and Maupertuis are equivalent in the case of the single mass point, since

$$(2) \qquad\qquad 2T\,dt = mv \cdot v\,dt = mv\,ds.$$

This equality carries over to arbitrary mechanical systems, provided that by action we understand the sum of the $m_k v_k ds_k$ for all the mass points of the system.

Regarding 2: in Hamilton's principle we had restricted the sum total of motions to be compared by means of conditions (1) and (2) of § 33. Here we shall keep (2), but alter (1). Instead of $\delta t = 0$ we shall now require that

$$(3) \qquad\qquad \delta E = 0.$$

We shall therefore compare only trajectories of the same energy E as that of the real trajectory under investigation. This condition implies of course that our principle is now valid only for motions in which energy is conserved,

---

[6] The factor of 2 is of course unimportant as far as the minimum property of S is concerned. It is, however, convenient, especially for the formulations of § 44. Incidentally Leibniz was still in doubt as to whether he should take mv^2 or, as we nowadays do, $\frac{1}{2}mv^2$ as "live force."

i.e., motions caused by forces which have a potential. If we call the potential energy of the real path V, that of the varied paths $V + \delta V$, we must have because of (3),

(4) $$\delta T + \delta V = 0, \quad \delta V = -\delta T, \quad \delta L = \delta T - \delta V = 2\,\delta T.$$

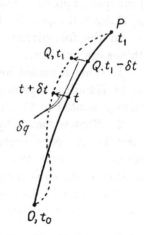

To visualize the change in the state of affairs caused by condition (3) we recall Fig. 51. There two points related by a variation δq belonged to the same time t. This is now no longer the case. The time of the varied point is $t + \delta t$ rather than t (cf. Fig. 54). Hence our varied path does not reach the end point at time $t = t_1$, but, according to the way our figure is drawn, at a later time. On the varied path a point Q is reached at a time $t = t_1$, whereas on the original path the corresponding point (also labeled Q) is reached at an earlier time $t_1 - \delta t_1$.

We now repeat the calculations of § 33. Eqs. (3) and (4) of that section remain valid, but Eq. (5) must be altered because, as stressed there, it is valid only for $\delta t = 0$. We find the condition replacing (5) by forming

Fig. 54. Variation of the "trajectory" in the principle of least action. Since the energy is not varied, point q of the original path and $q + \delta q$ of the varied path belong to different times t and $t + \delta t$. To the endpoint P on the real path is assigned the point Q on the varied one.

(5) $$\delta \dot{x} = \frac{d(x + \delta x)}{d(t + \delta t)} - \frac{dx}{dt}.$$

Let us transform the quotient of differentials on the right by writing

(6) $$\frac{\dfrac{d(x + \delta x)}{dt}}{\dfrac{d(t + \delta t)}{dt}} = \frac{\dfrac{dx}{dt} + \dfrac{d}{dt}\delta x}{1 + \dfrac{d}{dt}\delta t} = \frac{dx}{dt} + \frac{d}{dt}(\delta x) - \dot{x}\frac{d}{dt}(\delta t) + \ldots,$$

where we have neglected products in small quantities of order higher than the first. From (5) we therefore have

$$\delta \dot{x} = \frac{d}{dt}(\delta x) - \dot{x}\frac{d}{dt}(\delta t),$$

or

(7) $$\frac{d}{dt}(\delta x) = \delta \dot{x} + \dot{x}\frac{d}{dt}(\delta t).$$

If we introduce this in (33.4), we have, with index k arbitrary,

(8) $$\ddot{x}_k \delta x_k = \frac{d}{dt}(\dot{x}_k \delta x_k) - \dot{x}_k \delta \dot{x}_k - \dot{x}_k^2 \frac{d}{dt}(\delta t).$$

Eq. (8) is valid for coordinates y and z as well as x. Therefore (33.3), rather than leading to (33.8) as before, yields in this instance

(9) $$\frac{d}{dt}\sum m_k(\dot{x}_k\delta x_k + \dot{y}_k\delta y_k + \dot{z}_k\delta z_k) = \delta T + 2T\frac{d}{dt}(\delta t) + \delta W.$$

Here we make use of (4) to put

(9a) $$\delta W = -\delta V = +\delta T,$$

thereby giving as the right member of (9)

(10) $$2\delta T + 2T\frac{d\delta t}{dt}.$$

Let us now integrate (9) from t_0 to t_1. In this process the left member vanishes because of condition (33.2); we then obtain, using (10),

(11) $$2\int_{t_0}^{t_1}\delta T\,dt + 2\int_{t_0}^{t_1}T\,d\delta t = 0.$$

This, however, is nothing but

(12) $$2\delta\int_{t_0}^{t_1}T\,dt = 0$$

or, recalling (1),

(12a) $$\delta S = 0.$$

This concludes the explicit proof of the principle of least action, as envisaged by Maupertuis.

Let us subject the transition from (11) to (12) to some further scrutiny. In Hamilton's principle the two symbols

$$\delta\int T\,dt \quad \text{and} \quad \int\delta T\,dt$$

could be used interchangeably because of the condition $\delta t = 0$; use of this was made for instance in the transition from Eq. (33.10) to (33.11). From our present viewpoint the expressions are, however, different in character, as comparison of Eqs. (11) and (12) above will show.

In particular, let us consider a motion under no forces. In that case $T = E$, so that, with the help of (3), Eq. (12) gives

(13) $$\delta\int_{t_0}^{t_1}dt = \delta(t_1 - t_0) = 0.$$

This is the *principle of least time* (principle of " earliest arrival ") which Fermat formulated and applied to the refraction of light, after Heron, in ancient times, had treated the reflection of light in a similar fashion.

In the case of a single free mass point we can put $v = $ const. instead of $T = E$ and write in place of (12),

(14) $\delta \int v \, dt = \delta \int ds = 0.$

This is the principle of the " shortest path." It determines the trajectory of a free mass point, for instance on a curved surface or — as in general relativity — in a manifold of arbitrary curvature.   Such a trajectory is called a *geodesic*. We shall come back to this point in § 40.

In his celebrated Königsberg *Vorlesungen über Dynamik* of 1842 (published by Clebsch) Jacobi justified the necessity for completely eliminating the time t from the principle of least action. This is possible, because

$$T = E - V = \frac{1}{2} \sum m_k \, v_k^2 = \frac{1}{2} \frac{\Sigma m_k \, ds_k^2}{dt^2}$$

and therefore

$$dt = \left(\frac{\Sigma m_k \, ds_k^2}{2 (E - V)} \right)^{\frac{1}{2}}.$$

Instead of (12) we can then require that

(15) $\delta \int [2(E - V)]^{\frac{1}{2}} [\Sigma m_k \, ds_k^2]^{\frac{1}{2}} = 0.$

With E fixed, the variation here concerns only the spatial properties of the trajectory of the system; there is no longer any mention of the passage of time during the motion.

Let us come back once more to the teleological aspect of the principles of Hamilton and of least action.   Notice that the " least action " may, under certain circumstances, also be a " greatest action "; for in demanding that $\delta \ldots = 0$ we do not necessarily obtain a minimum, but rather in general only an *extremum*.   We see this most simply in the example of the geodesics on the surface of a sphere, which are arcs of great circles. Suppose that initial point O and endpoint P lie on a specified hemisphere. Then the arc of a great circle connecting them directly is indeed shorter than all arcs lying in planes through O and P but not containing the center of the sphere.   Yet the complementary arc, which proceeds from O to P in the opposite direction, traversing the hemisphere not containing the two endpoints, is also a geodesic; and this line is longer than all other arcs of circles which join O to P over this hemisphere.   We therefore conclude that in general we do not need to think of the integral principles as demonstrating the " purposefulness " of Nature; they merely constitute an unusually impressive mathematical formulation of an extremal property common to the laws of dynamics.

Maupertuis claimed that his principle was generally valid for all laws of nature.   Nowadays we are more inclined to accord this property to Hamilton's principle. We mentioned on p. 185 that Helmholtz made this principle the basis of his studies in electrodynamics.   Since that time integral variational principles of Hamiltonian form have been used in the most diverse fields.

In volume II we shall have direct recourse to this principle in order to gain deeper understanding of the concept of fluid pressure. A special advantage of this procedure will be that we shall obtain not only the differential equations — in this case partial differential equations — of the problem, but also the boundary conditions which the solutions of these equations must satisfy. The same turns out to be true for other problems with continuous mass distributions (capillarity, vibrating membrane, etc.). In many cases it is first necessary to look for the Lagrangian L of the problem at hand before L can be used in the variational principle. Such a case, for instance, is the motion of an electron in a magnetic field; there the force acting is not derivable from a potential V. Relativistic problems form another case; there one should not use the expression for the kinetic energy derived in (4.10) to build the Lagrangian. Instead the expression

$$(16) \qquad m_0 c^2 \int (1 - \beta^2)^{\frac{1}{2}} dt$$

must be used as the kinetic contribution to the action principle. The Eulerian derivation of this term leads directly to the relativistic momentum **p** of (3.19) and therefore also to the law of the velocity-dependent electron mass. In general, especially outside of mechanics, the search for the Lagrange function L which leads (via the variational principle) to given differential laws is an arduous problem for the solution of which there are no universally valid rules. The previously mentioned problem of the electron in a magnetic field was solved in a simple manner by Larmor and Schwarzschild. A separation of L into a kinetic and a potential contribution according to the pattern $L = T - V$ is then in general no longer feasible.

It is to be emphasized that the quantity under the integral of (16) is nothing but the element of *proper time* (2.17), which was recognized by Minkowski as the simplest invariant of the special theory of relativity; Einstein furthermore generalized it in the form of a world line element in the general theory of relativity. In the form (16) Hamilton's principle therefore automatically satisfies the invariance requirement of relativity theory. In this property Planck[7] saw the "most brilliant success which Hamilton's principle has achieved."

---

[7] Cf. the instructive article in *Die Kultur der Gegenwart*, Part III, § III, 1, p. 701 (B. G. Teubner, Leipzig 1915).

CHAPTER VII

DIFFERENTIAL VARIATIONAL PRINCIPLES OF MECHANICS

§ 38. Gauss' Principle of Least Constraint

Gauss was not only a very eminent mathematician, but also an astronomer and geodesist, and, as such, a passionate calculator of numerical results. It was he who founded the method of least squares, which he evolved with successively greater depth in three extensive treatises. If, as happened now and then, he was asked (against his will) to deliver a lecture at the University of Goettingen, his preferred topic was always the method of least squares.

His brief paper of 1829 entitled " On a New General Fundamental Principle of Mechanics "[1] is concluded with the characteristic sentence, " It is quite remarkable that Nature modifies free motions incompatible with the necessary constraints in the same way in which the calculating mathematician uses least squares to bring into agreement results which are based on quantities connected to each other by necessary relations."

Gauss called his new fundamental principle the *principle of least constraint*. He defined the measure of constraint as follows: consider a mass point of the system, and form the product of its mass by the " square of the deviation of this point from free motion." The sum of this product over all mass points of the system defines the constraint. Let us number the mass points and their rectangular coordinates as on p. 66. We then have

$$(1) \qquad Z = \sum_{k=1}^{3n} m_k \left(\ddot{x}_k - \frac{X_k}{m_k} \right)^2$$

as the measure of the constraint of a system of n mass points; for the " free motion " which would occur were the internal constraints neglected is given by

$$\ddot{x}_k = \frac{X_k}{m_k}.$$

Thus the quantity contained in the parenthesis of (1) is indeed the " deviation from free motion " caused by the constraint on the kth mass point.

---

[1] *Crelle's Journal f. Math.* **4**, 232 (1829); Werke **5**, 23.

It can (cf. p. 61) also be called the " lost force " divided by the mass, so that instead of (1) we can write

$$(2) \qquad Z = \sum_{k=1}^{3n} \frac{1}{m_k} (\text{lost force})_k^2.$$

Notice that here the lost forces and reciprocal masses play the same role as the errors and weights in the calculation of errors.

We must now define what is meant by the expression " least constraint," that is, we must indicate what quantities are to be kept fixed and what quantities are to be varied in the calculation of $\delta Z = 0$.

We shall keep *fixed*

a) the instantaneous state of the system, *i.e.*, the position and velocity of each of its mass points. We must therefore put

$$(3) \qquad \delta x_k = 0, \quad \delta \dot{x}_k = 0.$$

b) the constraints to which the system is subject. If we take these in the holonomic form $F_i(x_1, x_2, \ldots) = 0$, we must, in the variation δZ, take into account the secondary condition

$$(4) \qquad \sum_{k=1}^{3n} \frac{\partial F_i}{\partial x_k} \delta x_k = 0, \quad i = 1, 2 \ldots r,$$

where r is the number of conditions, $3n - r = f$ therefore the number of degrees of freedom of the system. Let us differentiate Eq. (4) twice with respect to t. This yields terms in δx, $\delta \dot{x}$ and $\delta \ddot{x}$. Because of (3) we need keep only those in $\delta \ddot{x}$, that is,

$$(4a) \qquad \sum_{k=1}^{3n} \frac{\partial F_i}{\partial x_k} \delta \ddot{x}_k = 0.$$

c) the forces acting on the system and, of course, the masses, so that we have

$$(5) \qquad \delta X_k = 0, \quad \delta m_k = 0.$$

The remaining quantity $\ddot{x}_k$ is then the only one to be *varied*.

Taking the secondary conditions (4a) into account by the method of Lagrange's undetermined multipliers, we obtain from (1)

$$(6) \qquad \delta Z = 2 \sum_{k=1}^{3n} \left\{ m_x \ddot{x}_k - X_k - \sum_{i=1}^{r} \lambda_i \frac{\partial F_i}{\partial x_k} \right\} \delta \ddot{x}_k = 0.$$

Only $f = 3n - r$ of the $\delta \ddot{x}_k$ are independent. As on p. 66 we can, however, choose our λ_i in such a way as to make r of the $\{\ \}$ vanish, so that only

f terms are left in (6). The $\delta \ddot{x}_k$ of these remaining f terms can now be treated as independent. It follows that their f associated $\{\,\}$ must vanish. We therefore arrive at Lagrange's equations of the first kind in the form (12.9).

Clearly the proof extends without change to non-holonomic constraints. Thus we are indeed confronted by a " new general fundamental principle of mechanics," as claimed by Gauss in the title of his paper. This fundamental principle is fully equivalent to d'Alembert's principle. Like the latter it is a *differential* principle in that it deals only with the present behavior of the system, not its future or past behavior. Here we do not need the rules of the calculus of variations, but only those of the ordinary differential calculus in the determination of the maxima and minima.

§ 39. *Hertz's Principle of Least Curvature*

Strictly speaking this principle is but a special case of that of Gauss. Nevertheless Hertz was able to call his principle, if not new, at least completely general; the reason for this is that he succeeded in replacing all forces by means of connections between the system in question and other systems interacting with it (cf. p. 5). Hertz was hence able to restrict himself to *systems under no forces*. In order to give the principle its sought-for geometric interpretation, he found himself obliged, moreover, to assume all masses to be multiples of a *unit mass*, say of atomic origin. The factor m_k in Gauss' expression (38.1) then becomes 1, while X_k becomes 0. It follows that (38.1) goes over to

$$(1) \qquad Z = \sum_{k=1}^{N} \ddot{x}_k^2.$$

Here we have indicated by means of the upper index N of the summation that the number of unit masses of the system to be summed has been augmented in an unspecified manner by a suitable number of unit masses corresponding to the interacting systems coupled to the given system.

Let us change (1) by writing

$$\frac{d^2 x_k}{ds^2} \quad \text{in place of} \quad \ddot{x}_k, \quad \text{where} \quad (2) \ \ ds^2 = \sum_{k=1}^{N} dx_k^2.$$

This is permitted because of the special form of the principle of energy. This principle is a consequence of Lagrange's equations of the first kind, and hence also of the principle of least constraint. For our present specialization the principle of energy can be written

$$\frac{1}{2} \sum_{k=1}^{N} \left(\frac{dx_k}{dt}\right)^2 = E$$

or, more concisely,

$$\left(\frac{ds}{dt}\right)^2 = \text{const.}$$

A division of (1) by the square of this constant yields thus the quantity

(3)
$$K = \sum_{k=1}^{N} \left(\frac{d^2 x_k}{ds^2}\right)^2.$$

Hertz calls ds the *element of line*, $K^{\frac{1}{2}}$ the *curvature* of the trajectory described by the system, and postulates

(4) $\delta K = 0.$

Every free system remains in a state of rest or of uniform motion along a path of least curvature.

The mode of expression (cf. Art. 309 of Hertz's book cited earlier) is chosen so as to recall Newton's formulation of the first law.

The mathematical treatment of postulate (4) follows that of Gauss and, on the basis of the conditions of variation stipulated under (a) and (b) on p. 211, evidently leads to Lagrange's equations of the first kind for a system under no forces (with $m_k = 1$).

What justifies Hertz in calling ds the " line element " and $K^{\frac{1}{2}}$ the " curvature "? Evidently these concepts are to be interpreted in a poly-dimensional sense. We are not in three dimensions, but in an N-dimensional Euclidean space of coordinates $x_1, x_2 \ldots x_N$. In this space the element of line is indeed given by (2). We shall now discuss the cases of two and three dimensions in order to show that the square of the curvature of a trajectory is quite generally given by (3).

According to Eq. (5.10) we have, in the space of coordinates x_1, x_2,

(5)
$$K = \frac{1}{\rho^2} = \left(\frac{\Delta\epsilon}{\Delta s}\right)^2.$$

From Fig. 4b, $\Delta\epsilon$ is the angle between two neighboring tangents to the path whose points of contact with the path are a distance Δs apart. These tangents have direction cosines

(6) $\dfrac{dx_1}{ds}, \quad \dfrac{dx_2}{ds}$ and $\dfrac{dx_1}{ds} + \dfrac{d^2 x_1}{ds^2}\Delta s, \quad \dfrac{dx_2}{ds} + \dfrac{d^2 x_2}{ds^2}\Delta s,$ respectively.

Now these direction cosines are at the same time the coordinates of the two points formed by the intersection of a unit circle about the origin of coordinates with two radii drawn from the origin parallel to the tangents; moreover the angle $\Delta\epsilon$ is measured by the arc of distance between these two points of intersection. According to (6) we therefore have

$$\Delta \epsilon^2 = \left[\left(\frac{d^2 x_1}{ds^2} \right)^2 + \left(\frac{d^2 x_2}{ds^2} \right)^2 \right] \Delta s^2$$

and from (5),

(7) $$K = \left(\frac{d^2 x_1}{ds^2} \right)^2 + \left(\frac{d^2 x_2}{ds^2} \right)^2 .$$

In the space of the three coordinates x_1, x_2, x_3, $\Delta \epsilon$ is once again the angle between neighboring tangents to the three-dimensional trajectories. The unit circle is now replaced by a unit sphere through the center of which parallels to the two tangents are to be drawn. The distance between their points of intersection with the surface of the sphere measures $\Delta \epsilon$ in units of arc:

$$\Delta \epsilon^2 = \left[\left(\frac{d^2 x_1}{ds^2} \right)^2 + \left(\frac{d^2 x_2}{ds^2} \right)^2 + \left(\frac{d^2 x_3}{ds^2} \right)^2 \right] \Delta s^2 .$$

From (5) we thus obtain an expression for K which now has three terms.

The generalization to a space of N dimensions and to the equation (3) of N terms is now obvious.

With this we must conclude our report on the mechanics of Hertz. As mentioned on p. 5, his is an interesting and stimulating idea, carried out with great logic; because of the complicated replacement of forces by connections it has, however, borne little fruit.

§ 40. A Digression on Geodesics

We define as geodesics of an arbitrary curved surface the trajectories of mass points under no forces (hence no friction) constrained to move on the surface. Let the mass of a particle be equal to 1, and the equation of the surface $F(x, y, z) = 0$.

The principle of least action states that these geodesics are also the shortest possible lines or, more generally (cf. p. 208) lines whose lengths are extrema. Since conservation of energy holds, the velocity along the path is constant. By choosing the constant of energy properly we can put the velocity equal to 1 and therefore replace $\frac{d}{dt}$ by $\frac{d}{ds}$.

We obtain the basic definition of geodesics if we describe our trajectories by Lagrange's equations of the first kind. Written vectorially, these are, in our case,

(1) $$\dot{\mathbf{v}} = \lambda \, \mathbf{grad} \, F.$$

$\dot{\mathbf{v}}$ has the direction of the principal normal to the trajectory if, as in our case, $v = \text{const.}$ so that $\dot{v} = 0$ (cf. § 5, beginning of (3)); it follows (cf. same place) that $\dot{\mathbf{v}}$ lies in the osculating plane. $\mathbf{grad} \, F$, on the other hand, has the direction of the normal to the surface, since for any translation (dx, dy, dz) on the surface we have

$$\frac{\partial F}{\partial x}\,dx + \frac{\partial F}{\partial y}\,dy + \frac{\partial F}{\partial z}\,dz = 0,$$

so that the direction

$$\frac{\partial F}{\partial x} : \frac{\partial F}{\partial y} : \frac{\partial F}{\partial z}$$

is indeed normal to that of the displacement. Eq. (1) therefore contains the basic definition of geodesics which states that *the principal normal of a geodesic coincides with the normal to the surface*, or equivalently, *the osculating plane of a geodesic contains the normal to the surface*.

We now appeal to the principle of least curvature. According to it the geodesic has a smaller curvature than neighboring paths; the neighboring paths are, according to conditions (38.3), restricted to pass through the same point with the same tangent as the geodesic at the point considered. We obtain the total class of these neighboring paths by passing through the tangent in question all possible skew planes and determining their intersections with the surface; the plane containing the normal to the surface furnishes the geodesic. According to Hertz's principle these skew sections have a greater curvature than the normal section, or, equivalently, a smaller radius of curvature.

This fact is in agreement with Meusnier's theorem in the differential geometry of surfaces, which states that the radius of curvature of an oblique section equals the projection of the radius of curvature of the normal section on the plane of the oblique section. We thus recognize in Meusnier's theorem a quantitative expression of the general content of the principle of least curvature.

Let us finally apply Lagrange's equations of the second kind to our geodesics. We thereby enter the sphere of thought of Gauss' great treatise of 1827 ("Disquisitiones generales circa superficies curvas"), which, extended to four dimensions, is also the sphere of thought of the general theory of relativity.

While Lagrange introduces arbitrary curvilinear coordinates q, Gauss uses as coordinates on the surface two arbitrary families of curves which cover the surface with a "grid." As customary, we shall call them

(2) $$u = \text{const.}, \quad v = \text{const.}$$

In these coordinates Gauss writes the line element ds in the form

(3) $$ds^2 = E\,du^2 + 2F\,du\,dv + G\,dv^2.$$

The "first differential parameters" E, F and G are to be thought of as functions of u and v. They are connected with the rectangular coordinates x, y, z of the points on the surface by the relations

$$E = \left(\frac{\partial x}{\partial u}\right)^2 + \left(\frac{\partial y}{\partial u}\right)^2 + \left(\frac{\partial z}{\partial u}\right)^2, \quad G = \left(\frac{\partial x}{\partial v}\right)^2 + \left(\frac{\partial y}{\partial v}\right)^2 + \left(\frac{\partial z}{\partial v}\right)^2,$$

$$F = \frac{\partial x}{\partial u}\frac{\partial x}{\partial v} + \frac{\partial y}{\partial u}\frac{\partial y}{\partial v} + \frac{\partial z}{\partial u}\frac{\partial z}{\partial v}.$$

The square of the line element divided by $2dt^2$ is the expression of the kinetic energy T of our (unit) mass point moving on the surface. We can thus transform the Lagrange equations for generalized coordinates to Gaussian notation by forming

$$p_u = \frac{\partial T}{\partial \dot u} = E\dot u + F\dot v$$

$$2\frac{\partial T}{\partial u} = \frac{\partial E}{\partial u}\dot u^2 + 2\frac{\partial F}{\partial u}\dot u\dot v + \frac{\partial G}{\partial u}\dot v^2.$$

If, finally, we put $\frac{d}{ds}$ in place of $\frac{d}{dt}$, the differential equation of the geodesics is, according to the method of Lagrange,

(4) $$\frac{d}{ds}\left(E\frac{du}{ds} + F\frac{dv}{ds}\right) = \frac{1}{2}\left\{\frac{\partial E}{\partial u}\left(\frac{du}{ds}\right)^2 + 2\frac{\partial F}{\partial u}\frac{du}{ds}\frac{dv}{ds} + \frac{\partial G}{\partial u}\left(\frac{dv}{ds}\right)^2\right\}$$

for the u-coordinate. We need not write down the corresponding differential equation for the v-coordinate; by virtue of the principle of energy (in our case $\frac{ds}{dt} = 1$) it must be identical to (4).

Gauss derives Eq. (4) in Art. 18 of the cited treatise by means of the principle of the shortest path. Here we merely wanted to point out the fact that Gauss' method of general surface parameters (2) is equivalent to Lagrange's method of the mechanics of systems. Both methods are invariant with respect to an arbitrary transformation of coordinates and depend only on the intrinsic properties of the surface or of the mechanical system respectively.

CHAPTER VIII

THE THEORY OF HAMILTON
§ 41. Hamilton's Equations

In Lagrange's equations our independent variables were the q_k and $\dot{q}_k$. In Hamilton's equations, which we shall now derive in two different ways, the q_k and p_k are the independent variables; the latter is defined by Eq. (36.9a). Whereas the characteristic function of Lagrange's equations was the " free energy " $T-V$, regarded as function of the q_k and $\dot{q}_k$, in Hamilton's equations the characteristic function is the total energy $T+V$, regarded as function of the q_k and p_k. This function we call the Hamiltonian function or simply the *Hamiltonian*, and we designate it by $H(q, p)$ just as we called the free energy the Lagrangian and designated it by $L(q, \dot{q})$. Between H and L there exists relation (34.16), which we shall write

(1) $$H = \Sigma p_k \dot{q}_k - L,$$

using the definition of the p_k.

Let us at once extend the basis of the theory by recalling the last part of § 37: we shall drop the decomposition of L into a kinetic and a potential contribution and, in addition, permit an explicit dependence on t. According to p. 190, such a dependence may arise if either the equations of constraint or the defining equations for the coordinates contain the time. We then write the Lagrangian in the generalized form

(1a) $$L = L(t, q, \dot{q}).$$

Let us keep Eq. (1) as our definition of the Hamiltonian associated with L,

(1b) $$H = H(t, q, p)$$

although H then loses the meaning of total energy. As before, the p_k are given by the relation

(1c) $$p_k = \frac{\partial L}{\partial \dot{q}_k}.$$

If we take Hamilton's principle

(1d) $$\delta \int_{t_0}^{t} L\, dt = 0$$

as our fundamental principle of mechanics, we obtain Lagrange's equations

217

just as in § 34 — in spite of the new extended meaning of L. For purposes of the following we shall write these equations in the form

(1e)
$$\dot{p}_k = \frac{\partial L}{\partial q_k}.$$

(1) Derivation of Hamilton's Equations from Lagrange's Equations

Let us write down the total differentials of H and L:

(2)
$$dH = \frac{\partial H}{\partial t}dt + \sum \frac{\partial H}{\partial q_k}dq_k + \sum \frac{\partial H}{\partial p_k}dp_k,$$

(2a)
$$dL = \frac{\partial L}{\partial t}dt + \sum \frac{\partial L}{\partial q_k}dq_k + \sum \frac{\partial L}{\partial \dot{q}_k}d\dot{q}_k,$$

and, by means of Lagrange's equations (1e) and the definition (1c) of the p_k, transform dL to

(2b)
$$dL = \frac{\partial L}{\partial t}dt + \sum \dot{p}_k dq_k + \sum p_k d\dot{q}_k.$$

Let us, on the other hand, form the total differential of (1) with the help of (2b):

(3)
$$dH = \sum \dot{q}_k dp_k + \sum p_k d\dot{q}_k - \frac{\partial L}{\partial t}dt - \sum \dot{p}_k dq_k - \sum p_k d\dot{q}_k.$$

Cancelling of the last term on the right against the second term yields

(3a)
$$dH = -\frac{\partial L}{\partial t}dt - \sum \dot{p}_k dq_k + \sum \dot{q}_k dp_k.$$

This expression for dH must, of course, be identical to that of Eq (2). If we equate the coefficients of dt, we obtain

(3b)
$$\frac{\partial H}{\partial t} = -\frac{\partial L}{\partial t}.$$

Comparison of the coefficients of dq_k and dp_k yields

(4)
$$\dot{p}_k = -\frac{\partial H}{\partial q_k}, \quad \dot{q}_k = \frac{\partial H}{\partial p_k}.$$

These relations, exhibiting an amazing symmetry, are " Hamilton's ordinary differential equations " or, for short, *Hamilton's equations*.

Incidentally, they first occurred in the much earlier " Mécanique analytique " of Lagrange (Sec. 5, § 14), where they were, however, derived and put to use only for the special case of small vibrations.

(2) Derivation of Hamilton's Equations from Hamilton's Principle

In the light of (1) we write this principle in the form

$$-\delta \int L\,dt = \delta \int [H(t,\,q,\,p) - \Sigma p_k \dot{q}_k]\,dt$$

(5)

$$= \sum_k \int \left(\frac{\partial H}{\partial q_k}\,\delta q_k + \frac{\partial H}{\partial p_k}\,\delta p_k - \dot{q}_k\,\delta p_k - p_k\,\delta \dot{q}_k \right) dt = 0,$$

where we can transform the last term in the parenthesis by partial integration,

(6)

$$-\int_{t_0}^{t_1} p_k\,\delta \dot{q}_k\,dt = \int_{t_0}^{t_1} \dot{p}_k\,\delta q_k\,dt - p_k\,\delta q_k \Big|_{t_0}^{t_1}.$$

The integrated term vanishes because of the way in which the variation is carried out in Hamilton's principle. Substitution of (6) in (5), followed by collection of terms in δq_k and δp_k, yields

(7)

$$\sum_k \int \left(\left\{ \frac{\partial H}{\partial q_k} + \dot{p}_k \right\}\,\delta q_k + \left\{ \frac{\partial H}{\partial p_k} - \dot{q}_k \right\}\,\delta p_k \right) dt = 0.$$

If it were permitted to treat the δq_k and the δp_k as independent variations, one would be justified in putting the factors of δq_k and δp_k separately equal to 0 for every value of the index k, and so obtain Hamilton's equations (4). This, however, is not allowed; for while q_k and p_k enter in H as independent variables, they are related in time through Eq. (1c), a fact which might conceivably cause our Eq. (7) to be satisfied identically. We notice, however, that a partial differentiation of (1) with respect to p_k (q_k being held constant) causes the second $\{\ \}$ of (7) to vanish identically. We conclude, therefore, that the first $\{\ \}$ must vanish as well.

One of the reasons why we derived Hamilton's equations in the second way is that we wish now to make an important remark connected with it.

We know that Lagrange's equations are invariant under arbitrary "point transformations," i.e., that they keep their form if we replace the q_k by a new set of coordinates Q_k connected with the former by relations of the type

(8)
$$Q_k = f_k(q_1, q_2, \ldots q_f).$$

The associated P_k are then given by

(8a)
$$P_k = \frac{\partial L}{\partial \dot{Q}_k} = \sum_i \frac{\partial L}{\partial \dot{q}_i}\,\frac{\partial \dot{q}_i}{\partial \dot{Q}_k} = \sum_i p_i\,a_{ik},$$

that is, by linear functions of the p_i whose coefficients a_{ik} are functions of the q_k, just as in (36.3).

We shall now show that Hamilton's equations are invariant under the much more general transformations

(9)
$$Q_k = f_k(q,\,p)$$
$$P_k = g_k(q,\,p),$$

where the f_k and g_k are arbitrary functions of the two sets of variables q_k and p_k—arbitrary, that is, to within a restriction to be mentioned below. In particular, the g_k need no longer be linear in the p_k.

Let us suppose that Eqs. (9) are solved for the q, p in terms of the Q, P [we must of course require that Eqs. (9) be so constituted that this is possible] and are substituted in the expression $H(q, p)$. Let us call $\overline{H}$ this new transformed Hamiltonian. We then have

(10) $H(q, p) = \overline{H}(Q, P).$

Let us, moreover, compare the quantity $\Sigma p_k \dot{q}_k$ occurring in (5) with $\Sigma P_k \dot{Q}_k$. It is easy to see that the two expressions would be equal in a transformation (8), (8a). We now require that this equality be maintained in a general transformation (9), apart from an additive term. The latter we require to be a complete time derivative of a function F' of the q and p, or, alternatively, of a function F of the q and Q[1]. We hence put

(11) $\sum p_k \dot{q}_k = \sum P_k \dot{Q}_k + \frac{d}{dt} F(q, Q)$

with arbitrary F. This is the restriction on transformation (9) mentioned above.

In the substitution of Eqs. (10) and (11) in (5) the additional term $\frac{dF}{dt}$ vanishes in the integration and subsequent variation, since δq and δQ vanish at the endpoints; Eq. (5) then retains its earlier form, becoming

$$\delta \int (\overline{H}(Q, P) - \sum P_k \dot{Q}_k) \, dt = 0.$$

Furthermore, nothing is changed in our transformations (6) and (7); we conclude that Hamilton's equations remain valid in the new variables. In complete correspondence to our Eqs. (4) we now have

(12) $\dot{P}_k = -\frac{\partial \overline{H}}{\partial Q_k}, \quad \dot{Q}_k = \frac{\partial \overline{H}}{\partial P_k}.$

Transformations (9), as subjected to restriction (11), are called *canonical transformations* or[2] *contact transformations*. The reason for the latter name

---

[1] If F' is originally given as a function of q and p, we can of course solve for p from the first Eq. (9) and substitute it in F', thus obtaining a new function F of q and Q.

[2] The terms are not entirely synonymous, their difference being one of definition. We need not be detained with this difference, but remark that under suitable conditions either of the two transformations can be shown to be a special case of the other. Cf., for instance, Whittaker, *Analytical Dynamics* (Dover), Chapter XI, or Osgood, *Mechanics* (Macmillan), Chapter XIV.—TRANSLATOR.

is a geometric one. Let us consider a hypersurface in the f-dimensional space of the $q_1, q_2, \ldots q_f$, given by

$$(13) \qquad\qquad z = z(q_1, \ldots q_f);$$

the quantities

$$p_k = \frac{\partial z}{\partial q_k}$$

determine the position of the tangent plane to the hypersurface and can, for this reason, be interpreted as the coordinates of this plane. We require that there exist a condition

$$(14) \qquad\qquad dz = \sum_{k=1}^{f} p_k \, dq_k$$

between the coordinates of the point q_k and those of the plane p_k. This condition insures the "union of lineal elements," i.e., the continuity of the coordinates p_k as we pass from an arbitrary point of coordinates q_k to a neighboring point. Let us now introduce new coordinates Q_k, P_k by means of Eq. (9) and calculate (13) in terms of these new coordinates. Let the result be

$$z = Z(Q, P).$$

We now demand that this new expression again represent a hypersurface touched by the planes of coordinates P at the points determined by the Q. From (14) we must therefore have

$$(15) \qquad\qquad dZ = \sum_{k=1}^{f} P_k \, dQ_k,$$

or, with ρ a factor of proportionality,

$$(16) \qquad\qquad dZ - \sum P_k \, dQ_k = \rho \, (dz - \sum p_k \, dq_k).$$

Thus the contact between the surface and its tangent plane at a given point has been preserved in a transformation of the point. Let us compare condition (16) with Eq. (11), which, when multiplied by dt, can be written

$$(16a) \qquad\qquad \sum p_k \, dq_k = \sum P_k \, dQ_k + dF.$$

If we put $dF = dz - dZ$ in (16a) and $\rho = 1$ in (16), the two conditions are in agreement. This may constitute adequate justification for the name "contact transformation."

In transformations of the generality of Eqs. (9) the meaning of the P_k as components of momentum is obscured. For this reason we prefer to call the P_k, Q_k *canonical variables;* P_k and Q_k are then said to be *canonically conjugate.* Because Hamilton's equations are invariant under transformations (9) [with restriction (11)], they are often called *Hamilton's canonical equations.*

It is to this invariance under canonical transformations that Hamilton's equations owe their special significance in astronomical perturbation theory. They also play an important role in the statistical mechanics of Gibbs, a topic which we shall discuss in Vol. V.

We conclude our treatment of Hamilton's equations with a remark dealing with the principle of energy.

In agreement with Eq. (2) we have, quite generally,

$$\frac{dH}{dt} = \frac{\partial H}{\partial t} + \sum_k \left(\frac{\partial H}{\partial q_k} \dot{q}_k + \frac{\partial H}{\partial p_k} \dot{p}_k \right).$$

According to (4), the parenthesis vanishes for all k. We then obtain, in general,

$$(17) \qquad \frac{dH}{dt} = \frac{\partial H}{\partial t}.$$

If, in particular, H does not depend *explicitly* on t, we arrive at the *conservation law*

$$(18) \qquad \frac{dH}{dt} = 0, \quad H = \text{const.}$$

This law is more general than that of the conservation of energy, for, according to (1) and (1c), it states that

$$(18a) \qquad \sum \frac{\partial L}{\partial \dot{q}_k} \dot{q}_k - L = \text{const.}$$

where L must not depend explicitly on t, but otherwise can be quite arbitrary. It is this conservation law to which we alluded in footnote 3 of Chapter VI. Eq. (18a) leads to the *conservation of energy* if L can be split up into two contributions, a kinetic one homogeneous of second degree in the $\dot{q}_k$, and a potential one independent of the $\dot{q}_k$.

§ 42. Routh's Equations and Cyclic Systems

In Eqs. (10) and (11) of § 34 we considered a " mixed type " of equation resulting from a combination of Lagrange equations of the first and the second kind. We shall now become acquainted with a mixed type of equation arising from a combination of Lagrange's equations of the second kind with those of Hamilton. The new equations bear the name of Routh[3] who, for several decades, dominated the study of mechanics in Cambridge as " coach " and examiner in the " tripos." Somewhat later Helmholtz[4] developed the same equations as the basis of his theory of monocyclic and polycyclic systems, a theory which he intended to use in the solution of the fundamental problems of thermodynamics.

We subdivide the degrees of freedom of the system into two groups. One group, containing $f-r$ degrees of freedom, can be described by Lagrange's position and velocity coordinates

$$q_1, q_2, \ldots q_{f-r}; \quad \dot{q}_1, \dot{q}_2, \ldots \dot{q}_{f-r},$$

and the other, containing r degrees of freedom, is to be represented in terms of Hamilton's canonical variables

$$q_{f-r+1}, q_{f-r+2}, \ldots q_f; \quad p_{f-r+1}, p_{f-r+2}, \ldots p_f.$$

Instead of the Lagrangian L or the Hamiltonian H we now construct a Routh function R, which is to be a function of the $2f$ variables enumerated above and, for the sake of generality, of the time as well:

(1) $R(t, q_1, q_2, \ldots q_f; \dot{q}_1, \dot{q}_2, \ldots \dot{q}_{f-r}, p_{f-r+1}, \ldots p_f).$

R is to be defined by the equation

(2) $$R = \sum_{k=f-r+1}^{f} p_k \dot{q}_k - L(t, q_1, \ldots q_f; \dot{q}_1, \ldots \dot{q}_f).$$

We see that for $r=f$, R transforms to the Hamiltonian (41.1); for $r=0$, where the summation on the right vanishes, it goes over into the Lagrangian (apart from sign). Evidently we could have replaced definition (2) of R by the equivalent condition

(2a) $$R = H(t, q_1, \ldots q_f; p_1, \ldots p_f) - \sum_{k=1}^{f-r} p_k \dot{q}_k.$$

We now proceed as in Eqs. (41.2) to (41.4). We form the total differential of R, on the one hand from (1),

---

[3] In this connection we wish to mention the two volumes of Routh's *Treatise on the Dynamics of a System of Rigid Bodies*; I, Elementary Part, II, Advanced Part. It is a collection of problems of unique variety and richness. Routh first developed his form of the dynamical equations in the prize article *A Treatise of Stability of a Given State of Motion* (1877).

[4] *Berliner Akad.* (1884) and *Crelle's Journal f. Math* **97**.

(3) $\qquad dR = \dfrac{\partial R}{\partial t}\, dt + \sum\limits_{k=1}^{f} \dfrac{\partial R}{\partial q_k}\, dq_k + \sum\limits_{k=1}^{f-r} \dfrac{\partial R}{\partial \dot q_k}\, d\dot q_k + \sum\limits_{k=f-r+1}^{f} \dfrac{\partial R}{\partial p_k}\, dp_k,$

and on the other, from (2),

(3a) $\qquad dR = \sum\limits_{k=f-r+1}^{f} \dot q_k\, dp_k + \sum\limits_{k=f-r+1}^{f} p_k\, d\dot q_k - dL,$

For dL we can use the expression (41.2b), which we shall, for greater clarity, decompose into

(3b) $\qquad dL = \dfrac{\partial L}{\partial t}\, dt + \sum\limits_{k=1}^{f} \dot p_k\, dq_k + \sum\limits_{k=1}^{f-r} p_k\, d\dot q_k + \sum\limits_{k=f-r+1}^{f} p_k\, d\dot q_k.$

Substitution in (3a) causes the last term of (3b) to cancel against the middle term of (3a), so that we are left with

(4) $\qquad dR = -\dfrac{\partial L}{\partial t}\, dt - \sum\limits_{k=1}^{f} \dot p_k\, dq_k - \sum\limits_{k=1}^{f-r} p_k\, d\dot q_k + \sum\limits_{k=f-r+1}^{f} \dot q_k\, dp_k.$

A term-by-term comparison with (3) yields the relation

$$\frac{\partial R}{\partial t} = -\frac{\partial L}{\partial t}$$

and the scheme of equations given below:

	for $k = 1, 2, \ldots f-r$	for $k = f-r+1, f-r+2, \ldots f$
(5)	$\dot p_k = -\dfrac{\partial R}{\partial q_k}$	$\dot p_k = -\dfrac{\partial R}{\partial q_k}$
	$p_k = -\dfrac{\partial R}{\partial \dot q_k}$	$\dot q_k = +\dfrac{\partial R}{\partial p_k}$

The $f-r$ equations on the left are of the Lagrange type with $L = -R$, whereas the r equations on the right are of the Hamiltonian type with $H = R$.

The application of these equations to cyclic systems, which Routh had in mind when he formulated them, proceeds as follows: we assume that the coordinates of the second group are cyclic, so that, from p. 197, they do not occur in the Lagrangian; in that case neither do they occur in the Routh function. The associated p_k are then constant [from the upper equation of the right group of Routh's equations (5) or, as remarked on p. 198, from Lagrange's equations]. We can now replace these constant values of the p_k, and, with the help of Eq. (41.1c), those of the (generally not constant) associated $\dot{q}_k$ in Eq. (2). We thus obtain a Routh function which depends only on the $f-r$ coordinates of the first group of q_k and $\dot{q}_k$. For these coordinates the left group of Eq. (5) above is valid. We have, therefore, reduced the problem to $f-r$ equations of the Lagrange type.

Routh used his method chiefly in the difficult problems of the stability of given states of motion. Let us instead illustrate the method with a reasonably simple example, that of the symmetrical top. The cyclic coordinates of this doubly cyclic problem are the Eulerian angles ϕ and ψ; according to Eqs. (35.15) to (35.17), we have

$$p_\phi \dot{\phi} + p_\psi \dot{\psi} = M''\left(\frac{M''}{I_3} - \cos\theta \frac{M'-M''\cos\theta}{I_1\sin^2\theta}\right) + M'\frac{M'-M''\cos\theta}{I_1\sin^2\theta}$$

$$= \frac{M''^2}{I_3} + \frac{(M'-M''\cos\theta)^2}{I_1\sin^2\theta} \; ;$$

by virtue of (35.13) the Routh function then becomes

$$R = \frac{M''^2}{I_3} + \frac{(M'-M''\cos\theta)^2}{I_1\sin^2\theta} - \frac{I_1}{2}\dot{\theta}^2 - \frac{(M'-M''\cos\theta)^2}{2I_1\sin^2\theta} - \frac{M''^2}{2I_3} + P\cos\theta$$

$$= -\frac{I_1}{2}\dot{\theta}^2 + \Theta(\theta), \quad \Theta = \frac{M''^2}{2I_3} + \frac{(M'-M''\cos\theta)^2}{2I_1\sin^2\theta} + P\cos\theta \; .$$

With $q_k = \theta$, the lower equation in the left group of our present Eqs. (5) then yields

$$p_k = I_1\dot{\theta}$$

and the upper equation of the same group gives

(6) $$I_1\ddot{\theta} = -\frac{\partial\Theta}{\partial\theta},$$

which is, of course, in agreement with the " generalized pendulum equation " (35.19). This example may serve to illustrate the usefulness of Routh's method, particularly for problems more difficult than the one presented.

In 1891 Boltzmann gave a series of lectures on Maxwell's electromagnetic theory at the University of Munich. He devoted his first lectures to the detailed consideration of a doubly cyclic mechanical system in order to illustrate the mutual inductive effect between two electrical circuits. The carefully worked mechanical model, consisting mainly of two pairs of beveled gears with centrifugal governors, is preserved in the museum of our Institute. To us it seems much more complicated than Maxwell's theory which it was intended to illustrate. Hence we shall not use it to clarify this theory, but instead take advantage of it in an exercise on the differential of an automobile, to which it is similar in its essential features.

Let us finally generalize the mathematical formalism which led us from Lagrange's to Hamilton's and to Routh's equations. We consider a function Z of two variables (or two sets of variables) x and y, and let

(7) $$dZ(x,y)=X\,dx+Y\,dy.$$

If we wish to replace x, y by X, Y as independent variables, it is convenient to consider, instead of Z, the "modified function"

(8) $$U(X,Y)=xX+yY-Z(x,y).$$

Indeed a differentiation of (8) at once gives, in view of (7),

(9) $$dU(X,Y)=x\,dX+y\,dY.$$

Eqs. (7) and (9) are identical to the "reciprocity relations"

(10)
$$\frac{\partial Z}{\partial x}=X,\quad \frac{\partial Z}{\partial y}=Y,$$
$$\frac{\partial U}{\partial X}=x,\quad \frac{\partial U}{\partial Y}=y.$$

If, on the other hand, we wish to replace only one of the original variables, say y, by its "canonically conjugate" Y, we shall have to "modify" (8) to

(11) $$V(x,Y)=yY-Z,$$

which yields

(12) $$dV(x,Y)=-X\,dx+y\,dY$$

with the "reciprocity relations"

(13) $$\frac{\partial V}{\partial x}=-X,\quad \frac{\partial V}{\partial Y}=y.$$

The transition from Z to U can be compared with that from Lagrange to Hamilton, that from Z to V with the transition from Lagrange to Routh.

Such a change of independent variables and the attendant modification of the characteristic function is called a *Legendre transformation* and plays an extensive role in analysis. We have mentioned it chiefly in order to be able to refer to it in our study of thermodynamics (Vol. V).

§ 43. The Differential Equations for Non-Holonomic Velocity Parameters

Whereas the differential equations considered so far were all modeled after those of Lagrange for generalized coordinates, the theory of the spinning top brought us in contact with equations of an entirely different, much simpler structure, viz., Euler's Eqs. (26.4) for the angular velocities ω_1,

ω_2 and ω_3. Let us determine what relation they bear to Lagrange's equations.

The difference between the two types stems from the fact that the ω_1, ω_2, ω_3 are not holonomic coordinates like the $\dot\theta$, $\dot\psi$, $\dot\phi$, but linear functions of these which are not integrable with respect to t. The connection between them is given by Eq. (35.11). Let us immediately consider the unsymmetrical top with kinetic energy

(1) $$T = \tfrac{1}{2}(I_1\,\omega_1{}^2 + I_2\,\omega_2{}^2 + I_3\,\omega_3{}^2),$$

and for brevity restrict ourselves to the case of a top under no forces.

We start out with Lagrange's equation for the ϕ-coordinate

(2) $$\frac{d}{dt}\frac{\partial T}{\partial \dot\phi} - \frac{\partial T}{\partial \phi} = 0.$$

According to (35.11)

$$\frac{\partial \omega_1}{\partial \dot\phi} = \frac{\partial \omega_2}{\partial \dot\phi} = 0, \quad \frac{\partial \omega_3}{\partial \dot\phi} = 1,$$

$$\frac{\partial \omega_1}{\partial \phi} = \omega_2, \quad \frac{\partial \omega_2}{\partial \phi} = -\,\omega_1, \quad \frac{\partial \omega_3}{\partial \phi} = 0,$$

so that, in view of (1),

$$\frac{\partial T}{\partial \dot\phi} = I_1\,\omega_1\frac{\partial \omega_1}{\partial \dot\phi} + I_2\,\omega_2\frac{\partial \omega_2}{\partial \dot\phi} + I_3\,\omega_3\frac{\partial \omega_3}{\partial \dot\phi} = I_3\,\omega_3,$$

$$\frac{\partial T}{\partial \phi} = I_1\,\omega_1\frac{\partial \omega_1}{\partial \phi} + I_2\,\omega_2\frac{\partial \omega_2}{\partial \phi} + I_3\,\omega_3\frac{\partial \omega_3}{\partial \phi} = (I_1 - I_2)\,\omega_1\,\omega_2.$$

From (2) we then have

(3) $$I_3\frac{d\omega_3}{dt} = (I_1 - I_2)\,\omega_1\,\omega_2.$$

This is the third Euler Eq. (26.4).

A similar calculation for the θ-coordinate yields

$$\frac{\partial \omega_1}{\partial \dot\theta} = \cos\phi, \quad \frac{\partial \omega_2}{\partial \dot\theta} = -\sin\phi, \quad \frac{\partial \omega_3}{\partial \dot\theta} = 0,$$

$$\frac{\partial \omega_1}{\partial \theta} = \dot\psi\cos\theta\sin\phi, \quad \frac{\partial \omega_2}{\partial \theta} = \dot\psi\cos\theta\cos\phi, \quad \frac{\partial \omega_3}{\partial \theta} = -\dot\psi\sin\theta.$$

From (1) we obtain

$$\frac{\partial T}{\partial \dot\theta} = I_1\,\omega_1\cos\phi - I_2\,\omega_2\sin\phi,$$

$$\frac{\partial T}{\partial \theta} = (I_1\,\omega_1\sin\phi + I_2\,\omega_2\cos\phi)\,\dot\psi\cos\theta - I_3\,\omega_3\,\dot\psi\sin\theta.$$

Lagrange's equation

(4)
$$\frac{d}{dt}\frac{\partial T}{\partial \dot\theta} - \frac{\partial T}{\partial \theta} = 0$$

hence becomes

(5)
$$0 = I_1 \frac{d\omega_1}{dt}\cos\phi - I_1 \frac{d\omega_2}{dt}\sin\phi$$
$$- I_1\,\omega_1\sin\phi\,(\dot\phi + \dot\psi\cos\theta) - I_2\,\omega_2\cos\phi\,(\dot\phi + \dot\psi\cos\theta)$$
$$+ I_3\,\omega_3\,\dot\psi\sin\theta.$$

But according to (35.11),

$$\dot\phi + \dot\psi\cos\theta = \omega_3, \quad \dot\psi\sin\theta = \omega_1\sin\phi + \omega_2\cos\phi,$$

so that the second and third lines of (5) can be written

$$(I_3 - I_1)\,\omega_3\,\omega_1\sin\phi - (I_2 - I_3)\,\omega_2\,\omega_3\cos\phi$$

and, together with the first line,

(6)
$$0 = \left\{ I_1\frac{d\omega_1}{dt} - (I_2 - I_3)\omega_2\omega_3 \right\}\cos\phi - \left\{ I_2\frac{d\omega_2}{dt} - (I_3 - I_1)\omega_3\omega_1 \right\}\sin\phi.$$

Finally the Lagrange Equation

$$\frac{d}{dt}\frac{\partial T}{\partial \dot\psi} - \frac{\partial T}{\partial \psi} = 0$$

becomes, after suitable transformation of variables and in view of (3),

(7)
$$0 = \left\{ I_1\frac{d\omega_1}{dt} - (I_2 - I_3)\omega_2\omega_3 \right\}\sin\phi - \left\{ I_2\frac{d\omega_2}{dt} - (I_3 - I_1)\omega_3\omega_1 \right\}\cos\phi.$$

It follows from (6) and (7) that both { } must necessarily vanish, so that we obtain the first and second Euler equations (26.4).

The transformation which we have carried out for one specific example can be performed quite generally[5] in the case of an arbitrary number of non-holonomic velocity parameters defined as linear (or more general) functions of real velocity coordinates. If, as in the case of the rigid body, the kinetic energy takes an especially simple form when expressed in terms of these parameters, such transformations can be of signal value for the integration of the equations of motion; they can also be useful in that they may satisfy non-holonomic conditions. Boltzmann found it necessary

[5] Cf., in particular, G. Hamel, *Math. Ann.* 59 (1904), and *Sitzungsber. der Berl. Math. Ges.* 37 (1938). Furthermore, *Encykl. d. Math. Wiss.* IV.2, Art. Prange No. 3 and ff.

to introduce the components of momentum corresponding to non-holonomic velocities in the kinetic theory of gases. He called these components
" momentoids."

§ 44. The Hamilton-Jacobi Equation

At the beginning of the previous century the most burning question of
theoretical physics was, " wave theory or corpuscular theory of light? "
The wave theory was founded by Huygens and, at the time mentioned,
found its confirmation in Thomas Young's discovery of the phenomenon
of interference. The corpuscular theory, on the other hand, had Newton's
seemingly authoritative backing. W. R. Hamilton, astronomer and profound
mathematical thinker, was just then engaged in a study of the paths of
light rays in optical instruments. The results of these studies[6] began to
appear in print in 1827, at about the time at which the two greatest advocates
of wave optics, Fraunhofer and Fresnel, died at almost the same early
age. Hamilton's work on general dynamics, the results of which we shall
briefly summarize in this section, came somewhat later, but it is intimately
related to his investigations in ray optics[7].

Let us add parenthetically that as a result of Planck's discovery of the
elementary quantum of action the above-mentioned question must now be
posed differently. We no longer ask, " waves *or* corpuscles? " but state,
" waves *as well as* corpuscles ! " It seems at first sight impossible to reconcile
these apparently contradictory concepts; actually they are complementary
rather than contradictory aspects both of optics and of dynamics. Their
reconciliation, as Schrödinger has recognized, results from a logical extension
of Hamilton's ideas and leads to *wave* or *quantum mechanics*.

Ray optics is the mechanics of light particles; in optically inhomogeneous media the paths of these particles are by no means straight lines,
but are determined by Hamilton's ordinary differential equations or
Hamilton's principle which is equivalent to them. From the viewpoint of
wave optics, on the other hand, the rays of light are given by the orthogonal[8]
trajectories of a system of wave surfaces or wave fronts. According to
Huygens' principle, these wave fronts are parallel surfaces. Hamilton under-

---

[6] Treatises on ray optics, *Trans. Roy. Irish Acad.* 1827, with supplements of 1830
   and 1832. His work on dynamics appeared in the *Trans. Roy. Soc. London* 1834
   and 1835.
[7] In the formulation by Jacobi this connection was lost. It was newly worked out
   in 1891 by F. Klein (*Naturforscher-Ges.* in Halle ; *Ges. Abhandl.*, Vol. II,
   pp. 601, 603).
[8] This is true of optically isotropic media. In anisotropic media such as crystals
   the orthogonality between ray and wave front is no longer an ordinary Euclidean
   one, but a non-Euclidean, generalized tensor orthogonality.

took to represent the family of wave surfaces by a (perforce partial) differential equation and to extend this method to the polydimensional space of the q_k of an arbitrary mechanical system. As we shall see, the family of wave surfaces is then given by $S = $ const., where S is the least action function of Eq. (37.1). The trajectories orthogonal to the surfaces are determined by the equation

(1) $$p_k = \frac{\partial S}{\partial q_k}.$$

(1) Conservative Systems

For the moment we deal with a mechanical system in which energy is conserved and can be resolved into a kinetic part T and a potential part V. T, V and H are hence not explicitly dependent on t.

We start out with Eq. (37.9), and replace δW in the right member by

$$-\delta V = \delta(T - E) = \delta T - \delta E.$$

The right member of (37.9) then becomes

(2) $$2\delta T + 2T \frac{d}{dt} \delta t - \delta E.$$

Next we transform the left member of the same equation to generalized coordinates p, q,

(3) $$\frac{d}{dt} \sum p_k \delta q_k.$$

Equating (3) and (2) yields

(4) $$2\delta T + 2T \frac{d}{dt} \delta t - \delta E = \frac{d}{dt} \sum p_k \delta q_k.$$

We integrate (4) with respect to t between the limits 0 and t to obtain

(5) $$\delta S - t\delta E = \sum p \delta q - \sum p_0 \delta q_0,$$

where S is defined by Eq. (37.1) and p_0 and δq_0 refer to the lower limit $t = 0$ of the integration, p and δq to its upper limit t.

Eq. (5) indicates that we must regard the action integral S as a function of the initial position q_0, the final position q and the energy E, i.e., that we are to use the arbitrarily assigned total energy E as variable in place of the time t:

(6) $$S = S(q, q_0, E).$$

According to (5), the motion as a function of time is then given by

(7) $$t = \frac{\partial S}{\partial \overline{E}},$$

where q_0 and q are held fixed. If, instead, we keep E fixed and vary either q or q_0, (5) yields

$$(8) \qquad p = \frac{\partial S}{\partial q}, \quad p_0 = -\frac{\partial S}{\partial q_0}.$$

The first of these relations is in agreement with our assertion (1); as for the second, we shall soon transform it to a more convenient form.

It must be admitted that we have not gained much in the way of knowledge of the motion as long as S is not known in the form (6). Let us, however, recall the equation of energy

$$H(q, p) = E,$$

where we substitute the value of p from Eq. (8) to obtain

$$(9) \qquad H(q, \frac{\partial S}{\partial q}) = E.$$

We regard (9) as the determining equation for the characteristic function S. It is called "Hamilton's partial differential equation" or the *Hamilton-Jacobi equation* for conservative systems. With T homogeneous of second degree in the p (V can be assumed independent of the p), it is of *second degree and first order*.

Let us suppose that we have found a *complete integral* of this equation, i.e., a solution that contains a number of assignable constants equal to the number of degrees of freedom of the problem. Call these constants

$$\alpha_1, \alpha_2, \ldots \alpha_f.$$

Since S itself does not occur in (9), it is determined by (9) only up to an additive constant. One of the above constants of integration, say α_1, is, therefore, in excess and can be replaced by an additive constant which remains unassigned. We may replace α_1 by our energy parameter E, so that the complete integral can be written in the form

$$(10) \qquad S = S(q, E, \alpha_2, \alpha_3, \ldots \alpha_f) + \text{const.}$$

The classic method used to arrive at such a complete solution is that of *separation of variables*—a method often, but not always, applicable. We shall deal with this method in § 46. In § 45 we shall show how Eq. (10) leads to a knowledge of the motion of the system.

(2) Dissipative Systems

We shall now adopt the general viewpoint that the Lagrangian L and hence the Hamiltonian H depend on t. In general it is then impossible to decompose L and H into T and V; if, in particular, a potential energy V does exist, it will have to depend on the time. This case is important for the perturbation problems of astronomy and quantum mechanics. There exists then no energy principle, hence no total energy constant E. It follows that we cannot use the principle of least action, but must revert to Hamilton's principle. Consequently we define a characteristic function S^*, given by the integral occurring in Hamilton's principle,

$$(11) \qquad S^* = \int_{t_0}^{t} L\, dt,$$

and regard S^* as function of the initial and final positions and of the *time of travel* t,

$$(12) \qquad S^* = S^* (q, q_0, t).$$

This is to be compared to Eq. (6) where the constant total energy E (nonexistent in the present case) took the place of t.

Let us now form $\dfrac{dS^*}{dt}$, first by means of (11),

$$(13) \qquad \frac{dS^*}{dt} = L,$$

next by means of (12),

$$(14) \qquad \frac{dS^*}{dt} = \sum \frac{\partial S^*}{\partial q_k} \dot{q}_k + \frac{\partial S^*}{\partial t} = \sum p_k \dot{q}_k + \frac{\partial S^*}{\partial t}.$$

The relation analogous to (8) used here,

$$(15) \qquad p_k = \frac{\partial S^*}{\partial q_k},$$

is easily verified. Merely differentiate (11) with respect to q_k and recall Eq. (41.1e).

In view of the general definition (41.1) of H, the comparison of (13) and (14) now yields

$$(16) \qquad \frac{\partial S^*}{\partial t} + H = 0;$$

from Eq. (15) we have, therefore,

$$(17) \qquad \frac{\partial S^*}{\partial t} + H\left(q, \frac{\partial S^*}{\partial q}, t\right) = 0.$$

This is the *Hamilton-Jacobi equation in general form*. It includes our earlier Eq. (9) as a special case. To show this, let us assume, as in (a), that H is independent of t. From (17) it follows that S^* is linear in t. Hence we put

$$S^* = at + b$$

and conclude from (16) that $-a=H$, i.e., equal to the energy constant E which now exists; b proves to be identical to our former characteristic function S. Thus Eq. (17) reduces to the special form (9) in this case.

The remarks made in (a) concerning the integration of (9) apply equally well to the more general Eq. (17). The complete integral of the latter now contains $f+1$ constants, one of which is again additive. Instead of (10) we can then write

(18) $$S^*=S^*(q, t, \alpha_1, \alpha_2, \ldots \alpha_f)+\text{const.}$$

§ 45. *Jacobi's Rule on the Integration of the Hamilton-Jacobi Equation*

We stated in connection with Eqs. (44.8) that the second of these did not lend itself to ready integration. The reason for this is that we integrated our partial differential equation, not in the form (44.6), but in the forms (44.10) and (44.18), respectively. In Eq. (44.7),

(1) $$t-\frac{\partial S}{\partial E},$$

on the other hand, we had obtained an equation describing, very directly, *the motion in time.* We shall now prove that if we differentiate S with respect to the constants of integration $\alpha_2, \alpha_3, \ldots \alpha_f$ instead of E, we obtain equations

(2) $$\beta_k=\frac{\partial S}{\partial \alpha_k}, \quad k=2, 3, \ldots f$$

which describe *the geometric configuration of the path of the system, provided that we regard the β_k as a second set of constants of integration.* This is Jacobi's rule for the case (a). In the case (b) it takes the even simpler form

(3) $$\beta_k=\frac{\partial S^*}{\partial \alpha_k}, \quad k=1, 2, \ldots f.$$

Here we have f equations of uniform structure which give both the *temporal and the spatial course of the motion of the system.*

We can introduce the same simplicity into case (a) by formally writing

(3a) $$\beta_1=\frac{\partial S}{\partial \alpha_1},$$

where we have put $t=\beta_1$ and $E=\alpha_1$.

We shall restrict our proof to case (a). Let us recall the definition (41.11) of a contact transformation, which we shall write, for purposes of the following,

(4) $$dF(q, Q) = \sum p_k dq_k - \sum P_k dQ_k.$$

Let us compare this with the total differential of the characteristic function (44.10),

$$dS(q, E, \alpha) = \sum_{k=1}^{f} \frac{\partial S}{\partial q_k} dq_k + \frac{\partial S}{\partial E} dE + \sum_{k=2}^{f} \frac{\partial S}{\partial \alpha_k} d\alpha_k,$$

which becomes, with substitution from (44.8) and (2), (3a) of this section,

(5) $$dS(q, \alpha) = \sum_{k=1}^{f} p_k \, dq_k + \sum_{k=1}^{f} \beta_k \, d\alpha_k.$$

This equation agrees with (4) if we identify

(6) $$F \text{ with } S, \quad Q_k \text{ with } \alpha_k, \quad P_k \text{ with } -\beta_k.$$

Now we know that, by means of a transformation q_k, $p_k \rightarrow Q_k$, P_k satisfying condition (4), we pass from Hamilton's equations (41.4)

$$\dot{p}_k = -\frac{\partial H}{\partial q_k}, \qquad \dot{q}_k = \frac{\partial H}{\partial p_k}$$

to Eqs. (41.12),

$$\dot{P}_k = -\frac{\partial \overline{H}}{\partial Q_k}, \qquad \dot{Q}_k = \frac{\partial \overline{H}}{\partial P_k}.$$

In view of (6) these become, in our case,

(7) $$-\dot{\beta}_k = -\frac{\partial \overline{H}}{\partial \alpha_k}, \qquad \dot{\alpha}_k = -\frac{\partial \overline{H}}{\partial \beta_k}.$$

But from (41.10),

$$\overline{H}(Q, P) = H(q, p),$$

or, by virtue of (6),

$$\overline{H}(\alpha, -\beta) = E = \alpha_1.$$

It follows that

(9) $$\frac{\partial \overline{H}}{\partial \alpha_k} = \begin{cases} 1 \ for \ k=1, \\ 0 \ for \ k>1; \end{cases} \qquad \frac{\partial \overline{H}}{\partial \beta_k} = \begin{cases} 0 \ for \ k=1, \\ 0 \ for \ k>1. \end{cases}$$

Thus Eqs. (7) become

(10) $$\dot{\beta}_k = \begin{cases} 1 \ for \ k=1, \\ 0 \ for \ k>1; \end{cases} \qquad \dot{\alpha}_k = \begin{cases} 0 \ for \ k=1, \\ 0 \ for \ k>1. \end{cases}$$

These equations tell us nothing new regarding the α_k; they merely confirm that the α_k are constants of integration. The same can be said of the equation for β_1; from $\dot{\beta}_1 = 1$ we simply have $\beta_1 = t$ (to within an additive constant of no importance), nothing new in view of Eq. (3a). Eqs. (10) for β_k with $k > 1$, on the other hand, furnish the proof of Jacobi's rule; they state that, like the α_k, the β_k are integration constants.

The proof can be extended without important changes to the case (b) provided we make the definition of a contact transformation somewhat more general. Since we do not need this result in the following, we shall not be detained by it.

§ 46.   Classical and Quantum-Theoretical Treatment of the Kepler Problem

In this section we wish to show how the Hamilton-Jacobi method of integration leads unambiguously and directly to the solution of the planetary problem of astronomy. We shall, furthermore, discover with surprise that the same method is made to order for the requirements of atomic physics and leads in a natural way to the (older) quantum theory.

We begin with the Lagrangian of the two-body problem with fixed sun M, expressed in polar coordinates,

$$(1) \qquad L = \frac{m}{2}(\dot{r}^2 + r^2\dot{\phi}^2) + G\frac{mM}{r}$$

from which we calculate the momenta

$$(1a) \qquad p_r = m\dot{r}, \quad p_\phi = mr^2\dot{\phi}.$$

Substitution of these in (1) and a change of sign in the potential energy yield the Hamiltonian

$$(1b) \qquad H = \frac{1}{2m}\left(p_r^2 + \frac{1}{r^2}p_\phi^2\right) - G\frac{mM}{r}$$

and, from (44.9), the Hamilton-Jacobi equation

$$(2) \qquad \left(\frac{\partial S}{\partial r}\right)^2 + \frac{1}{r^2}\left(\frac{\partial S}{\partial \phi}\right)^2 = 2m\left(E + G\frac{mM}{r}\right).$$

Let us apply in this example the method of "separation of variables" mentioned on p. 231.

We try a solution of the differential equation (2) of the form

$$(3) \qquad S = R + \Phi$$

in which R depends only on r and Φ only on ϕ. If we replace the right member of (2) by the general function $f(r, \phi)$, we obtain

(3a)
$$\left(\frac{dR}{dr}\right)^2 + \frac{1}{r^2}\left(\frac{d\Phi}{d\phi}\right)^2 = f(r, \phi).$$

In general, such a relation does not hold. If, however, f is independent of ϕ, as in our case, we need merely put $\frac{d\Phi}{d\phi}$ equal to a constant, say C (called the " separation constant "). R is then determined by the equation

(4)
$$\left(\frac{dR}{dr}\right)^2 = f(r) - \frac{C^2}{r^2},$$

which is solved by quadrature, yielding a complete integral. The assumption that f is independent of ϕ is evidently equivalent to the fact that in our case ϕ is cyclic, that is, it does not occur explicitly in the differential equation. We see that the method of separation of variables is based on special symmetry properties of the given differential equation, symmetry properties which are often, though by no means always, realized.

We now follow the general pattern of § 45, put $C = \alpha_2$ and separate (2) into

(5)
$$\frac{\partial S}{\partial \phi} = \alpha_2$$

(6)
$$\frac{\partial S}{\partial r} = \left[2m\left(E + G\frac{mM}{r}\right) - \frac{\alpha_2^2}{r^2}\right]^{\frac{1}{2}}.$$

Eq. (5) is the law of conservation of angular momentum, that is, Kepler's second law; the separation constant α_2 is the constant angular momentum, essentially identical to the areal velocity constant used in Eq. (6.2). Eq. (6) gives the variable radial momentum.

To calculate the characteristic function S, we integrate (5) and (6) and form (3). Replacing E by α_1, we obtain

(7)
$$S = \int_{r_0}^{r}\left[2m\left(\alpha_1 + G\frac{mM}{r}\right) - \frac{\alpha_2^2}{r^2}\right]^{\frac{1}{2}} dr + \alpha_2 \phi + \text{const.}$$

The lower limit of integration can be chosen arbitrarily since it merely affects the magnitude of the additive constant.

Let us, for the present, focus our attention on the geometric trajectory, i.e., on Kepler's first law. To do this we follow (45.2) and form

(8)
$$\beta_2 = \frac{\partial S}{\partial \alpha_2} = -\alpha_2 \int_{r_0}^{r}\left[2m\left(\alpha_1 + G\frac{mM}{r}\right) - \frac{\alpha_2^2}{r^2}\right]^{-\frac{1}{2}}\frac{dr}{r^2} + \phi.$$

It is evidently convenient to introduce $s = \frac{1}{r}$ instead of r as variable of integration and to rewrite (8)

(9)
$$\beta_2 - \phi = \alpha_2 \int_{s_0}^{s} \left[2m \left(\alpha_1 + GmMs \right) - \alpha_2^2 s^2 \right]^{-\frac{1}{2}} ds$$

$$= \int_{s_0}^{s} \frac{ds}{[(s - s_{min})(s_{max} - s)]^{\frac{1}{2}}}.$$

Here s_{min} and s_{max} are the reciprocals of the distances from sun to aphelion and perihelion. Comparison of the two integrals yields

(10)
$$s_{min}\, s_{max} = - \frac{2m\alpha_1}{\alpha_2^2}$$

$$s_{min} + s_{max} = \frac{2Gm^2 M}{\alpha_2^2}.$$

Now we wish to obtain (9) in convenient trigonometric form; the transformation

(11)
$$s = \frac{s_{min} + s_{max}}{2} + \frac{s_{max} - s_{min}}{2} u,$$

suggests itself. It takes $s = s_{max}$ into $u = +1$ and $s = s_{min}$ into $u = -1$. From (9) we then have

(12)
$$\beta_2 - \phi = \int_{u_0}^{u} \frac{du}{(1 - u^2)^{\frac{1}{2}}}$$

and, making the assignable lower limit of integration equal to 1,

(13)
$$\phi - \beta_2 = \cos^{-1} u, \quad u = \cos(\phi - \beta_2).$$

Finally we return from u to s via (11) and take note that, according to p. 42, Fig. 7,

$$s_{min} = \frac{1}{a(1 + \epsilon)}, \quad s_{max} = \frac{1}{a(1 - \epsilon)},$$

so

$$s = \frac{1}{a(1 - \epsilon^2)} + \frac{\epsilon}{a(1 - \epsilon^2)} u.$$

From (13) we then obtain the equation of an ellipse in the familiar form

(14)
$$s = \frac{1}{r} = \frac{1 + \epsilon \cos(\phi - \beta_2)}{a(1 - \epsilon^2)},$$

where the constant β_2 can be absorbed into the definition of ϕ.

For experimental reasons the astronomer is interested not so much in the geometrical form of the trajectory as in the motion as a function of time. Here again the Hamilton-Jacobi method gives the answer in the most systematic fashion, namely, by means of Eq. (45.1),

$$t = \frac{\partial S}{\partial E} = \frac{\partial S}{\partial \alpha_1},$$

from which we obtain, after substitution of the variable s,

$$(15) \qquad t = -\frac{m}{\alpha_2} \int_{s_0}^{s} \frac{ds}{s^2 \left[(s - s_{\min})(s_{\max} - s)\right]^{\frac{1}{2}}}.$$

With this equation we complete our former treatment in § 6, where the position of the planet as a function of time was left undetermined. With the help of the " eccentric anomaly " of Problem I.16 as the new variable of integration [its symbol u should not be confused with the auxiliary u in Eq. (11)] equation (15) can be solved by elementary integration and leads directly to the celebrated Kepler equation

$$nt = u - \epsilon \sin u$$

mentioned in the cited problem.

It is well-known that two- and several-body problems play a central role in modern atomic physics as well. In the hydrogen atom the electron moves about the nucleus, the proton, like a planet about the sun. Here, too, the Hamilton-Jacobi method has proved of surprising value. It literally shows us the point at which *quantum numbers* must be introduced.

In the older quantum theory, whenever the kth degree of freedom was separable from the remaining ones, one defined a *phase integral* (also called " action variable ") of the kth degree of freedom given by

$$(16) \qquad J_k = \int p_k \, dq_k.$$

The integral was to be taken over the whole range of values of the variable q_k. One then asked that J_k be an integral multiple of Planck's elementary quantum of action (cf. p. 181),

$$(16a) \qquad J_k = n_k h.$$

With p_k in (16) expressed in terms of the characteristic function S, one obtains

$$(17) \qquad \int \frac{\partial S}{\partial q_k} dq_k = \Delta S_k = n_k h.$$

ΔS_k is the kth " modulus of periodicity " of the function S, i.e., the change suffered by S when q_k runs through a complete cycle of its values.

The electron of a hydrogen atom has coordinates $q_1 = \phi$ and $q_2 = r$. The differential equation (2) for S and its solution (7) can be transferred directly from astronomy to atomic physics, provided we replace the gravitational potential energy by the Coulomb energy $-\dfrac{e^2}{r}$.

Since the range of the coordinate ϕ extends from 0 to 2π, we obtain from (7) and (17)

$$(18) \qquad \Delta S_\phi = 2\pi\,\alpha_2 = n_\phi h.$$

n_ϕ is the *azimuthal quantum number*; α_2, as we know, is identical to the azimuthal moment of momentum p_ϕ.

The range of values of the r-coordinates extend from $r_{\min}$ to $r_{\max}$ and back. Eqs. (7) and (17) therefore yield

$$(19) \qquad \Delta S_r = 2 \int_{r_{\min}}^{r_{\max}} \left[2m\left(E - \frac{e^2}{r} \right) - \frac{n_\phi^2 h^2}{4\pi^2 r^2} \right]^{\frac{1}{2}} dr = n_r h.$$

n_r is the *radial quantum number*. The quadrature is best performed by complex integration in the r-plane; once this is done, (19) becomes

$$(20) \qquad -n_\phi h + 2\pi i\,\frac{me^2}{(2mE)^{\frac{1}{2}}} = n_r h.$$

The energy of the hydrogen electron in the quantum state $n = n_r + n_\phi$ is, therefore,

$$(21) \qquad E = -\frac{2\pi^2 m\,e^4}{h^2 n^2}.$$

It is negative because the energy was set equal to zero for infinite electron-proton distance (see the above expression for the potential energy).

Eq. (21), together with Bohr's postulate of the radiation of energy in quantum jumps, led to the first understanding of the hydrogen spectrum (the so-called Balmer series) and from there to the modern theory of spectral lines in general.

Present-day developments of atomic theory have gone beyond the description of electronic orbits here presented. As mentioned at the beginning of § 44, investigations following Hamilton's line of thought have resulted in a more profound wave-mechanical conception of atomic processes.

PROBLEMS

Chapter I

I.1. *Elastic collision*[1]. n equal masses M are placed in contact with each other along a straight line. Two masses M, both having velocity v, collide with the row of n masses from the left. Evidently the laws of momentum and energy are satisfied if the two masses on the left transfer their velocities to the last two masses on the right. Show that these laws cannot be satisfied if only one mass is expelled on the right, or if the two last masses on the right are set in motion with different velocities v_1, v_2.

I.2. *Elastic collision with unequal masses.* Let the last mass m on the right be smaller than the remaining masses. Let a mass M collide from the left with velocity v_0. Show from the principles of energy and momentum that it is impossible for m to be the only mass set in motion. If it is assumed that only two masses are set in motion, what must be their velocities?

I.3. *Elastic collision with unequal masses.* Let the last mass M' on the right be greater than the remaining ones. Make the same assumptions as in Problem 2, taking notice, however, that the next-to-last mass on the right transfers its momentum toward the left. What is the velocity of M' and of the first mass M at the left end of the row? What happens if M' is very large?

I.4. *Inelastic collision between an electron and an atom.* An electron m, of velocity v, collides centrally with an atom M initially at rest. The atom is excited and is raised from its ground state to an energy level E units above it. What is the minimum initial velocity v_0 that the electron must have?

You will find one quadratic equation each for the final velocities v of the electron and V of the atom. The minimum value v_0 results from the requirement that the radical occurring in the solutions for v and V be real. The value of v_0 is somewhat higher than would be expected if only the conservation of energy were called into play, although the difference is not observable because the ratio $M/m \geqslant 2000$ is very high.

---

[1] It is essential that the student carry out the experiments described in Problems I.1 to I.3 himself. This can be done with coins on a smooth support, with elastic spheres so suspended on strings that they touch in the position of rest, or finally with marbles in a trough.

If the colliding particle is of the same, or approximately the same, mass as the struck one, the required minimum energy is about twice that expected from conservation of energy alone.

I.5. *Rocket to the moon.* A rocket with continuous exhaust shoots vertically upward. Let the exhaust velocity be a relative to the rocket and $\mu = -\dot{m}$ be the mass expelled per second, and assume both constant in time. Assume that the motion occurs under constant gravitational acceleration g, friction being neglected. Set up the equation of motion and integrate it under the assumption that the initial velocity of the rocket on the surface of the earth is zero. If $\mu = \frac{1}{100}$ of the initial mass m_0 and $a = 2000$ meter· sec^{-1}, what height has the rocket reached at $t = 10, 30, 50$ sec.?

I.6. *Water drop falling through saturated atmosphere.* A spherical water droplet falls, without friction and under the influence of gravity, through an atmosphere saturated with water vapor. Let its initial radius ($t=0$) be c, its initial velocity, v_0. As a result of condensation the water drop experiences a continuous increase in mass proportional to its surface; as will be shown, its radius then increases linearly with time. Integrate the differential equation of the motion by introducing r instead of t as independent variable. Show that for $c=0$ the velocity increases linearly with time.

I.7. *Falling chain.* A chain lies pushed together at the edge of a table, except for a piece which hangs over it, initially at rest. The links of the chain start moving one at a time; neglect friction. The energy written in the usual form is here no longer an integral of the motion. Instead, the impulsive (Carnot) energy loss must be taken into account in writing the balance of energy.

I.8. *Falling rope.* A rope of length l slides over the edge of a table. Initially a piece x_0 of it hangs without motion over the side of the table. Let x be the length of rope hanging vertically at time t. The rope is assumed to be perfectly flexible. Show that the principle of energy in the form $T + V = $ const. gives an integral of the motion.

I.9. *Acceleration of moon due to earth's attraction.* The moon's distance to the earth is about 60 earth radii. Assume that the lunar orbit is a circle, once traversed in 27 days, 7 hours and 43 minutes. From this the acceleration of the moon toward the earth (centripetal acceleration) can be calculated. Comparison of this value with that from Newton's law of gravitation gave the first confirmation of this law.

I.10. *The torque as a vector quantity.* Consider a rectangular coordinate system (x, y, z), with $\mathbf{r}$ the radius vector of the point of application of a force $\mathbf{F}$. We now pass to a second coordinate system (x', y', z'), obtained from the former by rotation. Show that the moment of force $\mathbf{F}$ about the origin of the first coordinate system transforms like a vector, i.e., like

$r = (x, y, z)$. To prove this one must assume that both coordinate systems are of similar sense (both right-handed or both left-handed).

I.11. *The hodograph of planetary motion.* From Eq. (6.5) with $A = 0$, the hodograph of planetary motion is given by

$$\xi = \dot{x} = -\frac{GM}{C}\sin\phi, \quad \eta = \dot{y} = +\frac{GM}{C}\cos\phi + B,$$

where M is the mass of the sun, C the angular momentum constant, ϕ the true anomaly (cf. Fig. 6). Show that the trajectory is a hyperbola or ellipse, depending on whether the " pole " $\xi = \eta = 0$ of the hodograph is excluded from or included by the hodograph. Also describe the limiting cases of parabola and circle in terms of the position of this pole.

I.12. *Parallel beam of electrons passing through the field of an ion. Envelope of the trajectories.* A source located at infinity shoots off electrons (charge e, mass m) along parallel paths with constant initial velocity v_0. An ionized atom A (charge E, mass M) is fixed at the origin. If e and E have the same sign, what area surrounding A is never touched by the electrons?

Take the y-axis as the direction of the incident particles; treat the problem as plane. It will be easiest to write the trajectory of an electron in polar coordinates with A as pole of the coordinate system and focus of the hyperbolic trajectory. The boundary of the above-mentioned area is obtained as the envelope of the electronic trajectories. Because of $M \gg m$ one can consider A to be at rest.

Show that if e and E have opposite signs, the envelope of trajectories seems to yield the same boundary, but that it is now devoid of physical meaning.

I.13. *Elliptical trajectory under the influence of a central force directly proportional to the distance.* Consider a mass m under the influence of a force directed toward a fixed point O (center of force)

$$\mathbf{F} = -k\mathbf{r}$$

($\mathbf{r} = \overrightarrow{Om}$, $k = $ const.). Show that the following three laws hold for the motion of m:

1. m describes an ellipse with O as center.
2. The radius vector $\mathbf{r}$ sweeps out equal areas in equal times.
3. The period T is independent of the shape of the ellipse, depending only on the force law, i.e., the values of k and m.

I.14. *Nuclear disintegration of lithium* (Kirchner, Bayer. Akad. 1933). If a hydrogen nucleus (proton, mass m) collides with velocity v_p with a nucleus of Li^7 (lithium of atomic weight 7), the latter splits into two α-particles (mass $m_\alpha = 4\, m_p$). These two α- particles fly off in almost (but not exactly) diametrically opposite directions. Assume that the α-particles fly off symmetrically with respect to the line of collision, and with equal velocities, and calculate the angle 2ϕ between them. Notice that in addition to the kinetic energy E_p of the proton, there occurs another energy E liberated as a result of the mass defect, which is greatly in excess of E_p and is likewise transmitted to the two α- particles. Thus the final answer for $\cos\phi$ contains not only m_p and m_α, but also the kinetic energy E_p of the proton, and E.

In the units customary in atomic physics, $E = 14 \cdot 10^6$ ev (electron-volt). In an experiment $E_p = 0.2 \cdot 10^6$ ev; what are the values of v_α and 2ϕ?

I.15. *Central collisions between neutrons and atomic nuclei; effect of a block of paraffin.* Neutrons are slowed down but little by a lead plate 50 cm thick; a layer of paraffin about 20 cm deep, on the other hand, absorbs them completely. This can easily be understood if one remembers that in a central collision the kinetic energy of the neutron (mass $m = 1$) is completely transferred to one of the hydrogen nuclei of the paraffin (proton mass $M_1 = 1$); whereas the amount of energy transferred in a central collision with a lead nucleus (mass $M_2 = 206$) is hardly worth mentioning. Draw a curve showing the kinetic energy which the initially motionless atomic nucleus (of mass M) receives from the neutron (mass m) in a central collision, as a function of the ratio M/m.

I.16. *Kepler's equation.* The secular variation of the motion of the planet in its orbit is determined, in differential form, by the principle of angular momentum. In order to obtain the secular variation in integral form, one can, following Kepler, proceed as follows (Fig. 55).

Draw a circle about the center of the ellipse with the major axis as diameter. We now associate a point K on the circle with the planet located at point E of the ellipse at the time t. If we take the principal axes of the ellipse as coordinate axes, point K has the same abscissa as E. Whereas E is given by its polar coordinates r, ϕ (pole S), K is determined by polar coordinates a, u (pole M). Thus with the *true anomaly* ϕ we associate

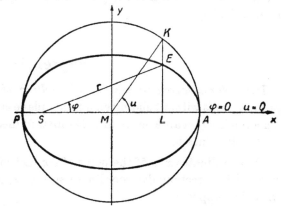

FIG. 55. Kepler's construction of the eccentric anomaly u and its connection with the true anomaly ϕ.

the *eccentric anomaly* u (as in the text, we count both from aphelion in the direction of motion, at variance with astronomical usage, where the anomalies are counted from perihelion, though, of course, also in the direction of motion of the planet).

The coordinates x and y of the planet E can be expressed, on the one hand, in terms of r and ϕ, and on the other, in terms of one of the semi-axes of the ellipse and the eccentric anomaly u, so that, with K given, E is also given. The course of the motion of point K on the circle is then determined by the celebrated Kepler equation

$$nt = (u - \epsilon \sin u).$$

Here ϵ is the eccentricity of the elliptical trajectory, and $n = \left(\dfrac{GM}{a^3}\right)^{\frac{1}{2}} = \dfrac{C}{ab}$, where a, b are the semi-axes, G the gravitational constant, M the mass of the sun, C the areal velocity constant.

In order to derive Kepler's equation, start out with the polar equation of the ellipse, referred to S as pole and the ray SA (aphelion) as polar axis

$$r = \frac{p}{1 - \epsilon \cos \phi}$$

where p is the " parameter " $a(1 - \epsilon^2)$. Now use the transformation relations referred to above to introduce u in place of ϕ, and obtain the equation

$$r = a(1 + \epsilon \cos u).$$

Differentiation of these two equations, elimination of r and ϕ, introduction of the principle of angular momentum and of Eq. (6.8) finally yield Kepler's equation by an integration, provided we stipulate that at $t = 0$ the planet is at aphelion.

Chapter II

II.1. *Non-holonomic conditions of a rolling wheel.* A sharp-edged wheel, of radius a, rolls without sliding on a rough plane support (think, for example, of a hoop rolling on an even road). Its instantaneous position is determined by assigning values to

1. coordinates x, y of the point of contact of the wheel with the support, referred to a rectangular coordinate system x, y, z whose xy-plane coincides with the support;

2. angle θ between axle of wheel and z-axis;

3. angle ψ formed between the tangent to the wheel (intersection of the plane of the wheel with that of the support) and the x-axis;

4. angle ϕ that the radius toward the instantaneous point of contact of the wheel makes with an arbitrary fixed radius, this angle to be counted positive, say, in the direction of rotation.

In finite motion the wheel therefore has five degrees of freedom. The mobility of the wheel is, however, restricted by the condition of pure rolling (without slipping) brought about by the static friction between wheel and support; indeed, with the wheel moving along its instantaneous direction, the distance δs moved along the direction of the tangent must equal $a\,\delta\phi$. By projecting this equation on the coordinate axes we then obtain the conditions of constraint

(1) $$\delta x = a\cos\psi\,\delta\phi, \quad \delta y = a\sin\psi\,\delta\phi$$

which the displacements δx, δy and $\delta\phi$ must satisfy.

Hence the rolling wheel has only three degrees of freedom in infinitesimal motion.

Show that conditions (1) cannot be reduced to equations between the coordinates themselves. To do this, show that the existence of an equation $f(x, y, \phi, \psi) = 0$ [θ does not occur in (1)] is incompatible with conditions (1).

II.2. *Approximate design of a flywheel for a double-acting one-cylinder steam engine* (cf. also § 9 (4)). A double-acting piston engine is one in which steam is introduced alternately on both sides of the piston, so that work is done during both strokes of a cycle.

Let us assume, for simplicity, that the steam pressure remains constant during each stroke (full pressure or Diesel cycle), and let us, moreover, suppose that the connecting rod is of infinite length. The variable torque as a function of crank angle ϕ transmitted from the piston to the crank shaft is then given by

$$L = L_0 \sin\phi$$

for the half-cycle in which the crank moves from the backward to the dead forward position [cf. Eq. (9.5)]. Here L_0 is a constant; ϕ is counted in the sense of rotation from the dead backward position. During the second half-cycle, from forward to dead backward position, under the same assumptions as made above (viz., 1. double-acting engine, 2. operation under full pressure, 3. infinite connecting rod), the torque changes according to the same law, provided ϕ is now measured in the sense of rotation from the dead forward position.

Let the load on the engine be given by the constant torque W, corresponding to a power of N *HP* with n r.p.m. Thus the driving torque L is variable, while the load torque W is constant. As a result the angular velocity of the engine fluctuates between a maximum value $\omega_{\max}$ and a minimum value $\omega_{\min}$, its mean value ω_m being given approximately by

$$\omega_m = \frac{\omega \max + \omega \min}{2}.$$

It is the purpose of the flywheel to prevent the relative fluctuation, that is the degree δ of unbalance of the engine, given by

$$\delta = \frac{\omega \max - \omega \min}{\omega_m},$$

from exceeding a given limit. How great must the moment of inertia of the flywheel be if the inertial effect of the moving masses (piston, piston rod, cross head, connecting rod and crank) is neglected?

II.3. *Centrifugal force under increased rotation of the earth.* How fast must the earth rotate and how long would the day be in order that centrifugal force and gravity just cancel at the equator?

II.4. *Poggendorff's experiment.* From one end of the beam of a balance we suspend a weightless pulley which can rotate without friction. A string U passes over the pulley; on one side the string carries the weight P, on the other, the weight $P+p$, where p is a small additional weight, just as in Atwood's machine. Initially p is fastened to the axle of the pulley by means of a thread u. On the other side of the balance these weights are suitably equilibrated in a pan. The thread u is then burned.

(a) With what acceleration do the weights P and $P+p$ rise and fall, respectively?

(b) Is the beam of the balance displaced in this process?

(c) What is the tension in the string U?

II.5. *Accelerated inclined plane.* An inclined plane is moved in a vertical direction according to a given dependence on the time. Investigate the motion of a body of mass m sliding down the plane without friction; in particular, consider the case that the inclined plane is moved with the constant accelerations $+g$ and $-g$.

II.6. *Products of inertia for the uniform rotation of an unsymmetrical body about an axis.* An unsymmetrical body rotates uniformly about an axis whose ends rest in bearings A and B. What reactions $\mathbf{A}$ and $\mathbf{B}$ must be exerted by the bearings? Calculate them from d'Alembert's principle; show that they result from the total centrifugal force applied at the center of gravity and from the resultant moment of the centrifugal forces acting on the individual mass elements.

From p. 55 we know the reactions caused by the weight of the body alone; we can therefore omit their effect here.

II.7. *Theory of the Yo-yo.* A disk-shaped body of mass M and moment of inertia I is provided with a deep symmetrical groove in the median plane perpendicular to its axis. A string is wound on the shaft of radius r in the groove. The loose end of the string is held in the hand. One then lets the body fall, with the string taut at all times. As the body descends, it acquires a rotational acceleration until the string is unwound. Then follows a transition stage, not to be considered in detail here, the result of which is that the body shifts from one side of the string to the other. The string now winds around the shaft in the opposite sense, and the body climbs with rotational deceleration, and so on. What is the string tension

(a) in descent?

(b) in ascent?

Assume that r is so small compared to the distance of the axis from the loose end of the string that the string can at all times be considered to be vertical.

II.8. *Particle moving on the surface of a sphere.* A mass point moves on the outside of the upper half of a sphere. Let its initial position z_0 and initial velocity v_0 be arbitrary, except that the latter is to be tangential to the surface of the sphere. The motion is to be frictionless, occurring solely under the influence of gravity. At what height does the mass point leave the sphere?

Chapter III

III.1. *Spherical pendulum with infinitesimal oscillations.* In general, the nodal points of the trajectory of a spherical pendulum advance during the course of the motion. For sufficiently small oscillations, however, the nodal points must be fixed, for we are then dealing with an harmonic elliptical motion. Estimate in what order the advance $\Delta\phi$ of the nodal points vanishes with vanishing area of the ellipse.

III.2. *Position of the resonance peak of forced, damped oscillations.* In a forced oscillation with damping the maximum amplitude of oscillation lies, not at $\omega = \omega_0$ as in the case of no damping, but at a value below ω_0 (cf. Fig. 33) depending on the amount of damping.

Find for what value of ω $|C|$ is a maximum.

[Show that the maximum of the velocity amplitude $|C|\omega$ (or of the time average of the kinetic energy) occurs exactly at $\omega = \omega_0$.]

III.3. *The galvanometer.* A galvanometer is connected through a switch with a direct-current source of constant *EMF E*. At time $t=0$, the switch is closed. After a sufficiently long time the galvanometer deflection reaches its final value α_∞. What is its motion between the initial position of rest, $\alpha=0$, $\dot{\alpha}=0$, and the final position, $\alpha=\alpha_\infty$?

Three effects have to be taken into account. First, an external torque proportional to the electric current and hence to the *EMF* acts on the galvanometer of moment of inertia I. Second, there acts a damping torque proportional to the angular velocity, which tends to slow down the motion. Third, the torsion in the suspension acts as a restoring torque and is proportional to the deflection α. Let ρ be the factor of proportionality of the damping torque, ω_0^2 that of the restoring torque.

Distinguish and explain graphically the three cases

(a) weak damping $(\rho < \omega_0)$,

(b) aperiodic (" critical ") damping $(\rho = \omega_0)$,

(c) strong damping $(\rho > \omega_0)$.

III.4. *Pendulum under forced motion of its point of suspension.*

(a) A particle is suspended from an inextensible string and oscillates without damping under the influence of gravity The point of suspension is moved along a straight horizontal line according to some given law of displacement $\xi = f(t)$.

What are the equations of motion of the system, the mass of the string being neglected? Derive them either from d'Alembert's principle or from Lagrange's equations of the first kind.

The equations of motion become considerably simplified if we pass to small oscillations, i.e., retain terms of only the first order.

If we make the additional assumption that the displacements of the point of suspension are harmonic in time, the equations of motion can easily be integrated. As the pendulum is set moving, say by a motion of the point of suspension, its proper frequency is excited; the amplitude of this proper frequency is gradually damped out (though we shall neglect damping in the analysis), thus leading to a steady state of oscillation with the same frequency as that forced on the point of suspension. Show that when the motion has thus become stationary, suspension point and mass m move in the same sense below the resonance frequency, in opposite senses above it.

(b) Make a similar analysis of the case in which the point of suspension is subjected to a vertical displacement η, with special emphasis on the case that the acceleration of the point is constant. What is the period of oscillation if the point of suspension is displaced with accelerations $+g$ and $-g$?

III.5. *Practical arrangement of coupled pendulums, sketched in Fig. 56.* Between two fixed supports A and B is stretched a weightless, flexible and elastic wire. Its tension S is regulated by an adjustable weight G attached to the loose end of the wire hanging over the angle iron B. Two pendulums are suspended *bifilarly* at points C and D which divide the segment AB into three, let us say, equal parts. The bifilar suspensions, indicated as simple suspensions in the sketch, enable the pendulums to swing out with good accuracy in a transverse direction, i.e., normal to the plane of the drawing. By increasing G the coupling between the two pendulums is made weaker (not stronger, as might at first be thought !). In what is to follow, we shall assume the coupling to be weak, which means that S is large compared with the weight of the pendulum bobs. We further suppose the angles of deflection ϕ_1 and ϕ_2 of the pendulums with

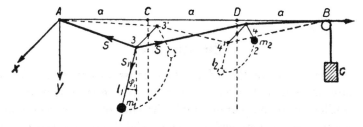

FIG. 56. Wire $ACDB$ is held taut by the weight G. It is deformed into $A34B$ or, for the opposite deflection, into $A3'4'B$, the deflection being caused not only by the gravitational action on masses m_1 and m_2, but also by the inertial effects of the pendulums. The latter are labeled 1 and 2, are of lengths l_1 and l_2, and suspended bifilarly, so that they swing out normally to the plane of the drawing (the bifilar suspensions are not shown in the figure). ϕ_1 and ϕ_2 are the instantaneous deflections from the vertical.

respect to the vertical to be small. (Refer to Fig. 56 for notation; $3'$ and $4'$ are the deflections of the points of suspension C and D symmetrically opposite to 3 and 4.) These angles are then approximated by

$$\sin\phi_1 = \phi_1 = \frac{x_1 - x_3}{l_1}, \quad \cos\phi_1 = 1$$

$$\sin\phi_2 = \phi_2 = \frac{x_2 - x_4}{l_2}, \quad \cos\phi_2 = 1.$$

With neglect of the y-component of the small oscillations we have for m_1 and similarly for m_2,

(1) $\qquad m_1 g = S_1 \cos\phi_1 = S_1 \qquad\qquad m_2 g = S_2 \cos\phi_2 = S_2$

(2) $\quad m_1\ddot{x}_1 = -S_1\sin\phi_1 = \frac{m_1 g}{l_1}(x_3 - x_1), \qquad m_2\ddot{x}_2 = -S_2\sin\phi_2 = \frac{m_2 g}{l_2}(x_4 - x_2).$

At the points of suspension C and D, S_1 and S_2 respectively must, at any instant, be in equilibrium with the tension S; the latter is altered negligibly little by S_1 and S_2. This yields two more conditions between x_1, x_2, x_3 and x_4. We can solve these for x_3 and x_4 and substitute them in (2). We then obtain the simultaneous differential equations of coupled pendulums: Verify that these are indeed in agreement with Eqs. (20.10).

III.6. *The oscillation quencher.* A system capable of oscillating in the x-direction (mass M, proportionality constant of the restoring force, K) is coupled by means of a spring (constant k) to a mass m, in such a way that m, too, can oscillate in the x-direction. We require that when an external force $P_x = c \cos \omega t$ acts on the mass M, this mass M stay at rest. What conditions must be satisfied by the system (m, k)?

Chapter IV

IV.1. *Moments of inertia of a plane mass distribution.* Prove that for any plane mass distribution the moment of inertia about the " polar " axis (perpendicular to the plane) equals the sum of the moments of inertia about two mutually perpendicular " equatorial " axes (in the plane of the mass distribution, intersecting in the polar axis). Specialize the foregoing to a circular disk.

IV.2. *Rotation of a top about its principal axes.* According to Fig. 46a, b, the rotations of an unsymmetrical top about the axes of the largest and smallest moments of inertia are stable, that about the axis of the intermediate moment of inertia is unstable. Prove this analytically. Start out with Euler's equations of motion and put the angular velocity about the axis of rotation $\omega_1 = \text{const.} = \omega_0$. Angular velocities ω_2 and ω_3 about the other two principal axes are initially zero, but, due to a perturbation, acquire values different from zero. If we suppose the perturbation small, the first Euler equation states that to a first approximation ω_1 remains unchanged, $= \omega_0$. From the other two equations one obtains a system of two linear differential equations of first order in ω_2 and ω_3. Put $\omega_2 = a e^{\lambda t}$ and $\omega_3 = b e^{\lambda t}$ with arbitrary constants a and b, and substitute in the two equations. The discussion of the resulting quadratic equation for λ yields the proof of the above assertion.

IV.3. *High and low shots in a billiard game. Follow shot and draw shot.* A billiard ball is hit with horizontal cue in its median plane, i.e., without " English ." At what height h above the center must the cue hit the ball so that pure rolling (no slipping) will ensue? Work out a theory of balls struck high and low, taking into account the kinetic friction between ball and cloth. By how much does the velocity of the center of mass grow in a high shot during the total time friction is acting, and by how much does it decrease in a low shot? What time elapses before only pure rolling remains?

The same method can be used to explain the phenomena taking place on collision with another ball, i.e., follow and draw shots.

IV.4. *Parabolic motion of a billiard ball.* How must a ball be struck so that the initial motion of its center of gravity and the axis of rotation are not normal to each other? Show that the direction of the force of friction is constant as long as the ball slides. What is the trajectory of the center of the ball? After what time does pure rolling ensue?

Chapter V

V.1. *Relative motion in a plane.* A plane rotates with variable angular velocity ω about its normal at one of its points O.

What forces in addition to the centrifugal force must be applied to a mass point so that its equations of motion in the rotating plane take on the same form as in the inertial frame of a spatially fixed plane? It will be convenient to introduce complex variables $x+iy$ in the spatially fixed plane, $\xi+i\eta$ in the rotating plane.

V.2. *Motion of a particle on a rotating straight line.* A mass point moves without friction on a straight line which in its turn rotates with constant angular velocity ω about a fixed horizontal axis intersecting, and perpendicular to, the straight line. Calculate the motion on the rotating straight line as a function of time, and show that the force of constraint (guiding force) and the component of the gravitational attraction along this force just balance the Coriolis force.

V.3. *The sleigh as the simplest example of a non-holonomic system* [after C. Carathéodory, *Z. angew. Math. Mech.* **13**, 71 (1933)]. The sleigh is regarded as a rigid plane system with three degrees of freedom in finite motion, one degree in infinitesimal motion. (Cf. the rolling wheel in Problem II.1, which had five degrees of freedom in finite motion, three in infinitesimal motion.)

Neglect the sliding friction on the snow, or, alternatively, think of it as permanently compensated by the pull of a horse. One must, however, take into account the friction F exerted laterally by the snow tracks against the runners of the sleigh, for it prevents any lateral motion of these. Let this friction be concentrated at one point of application O.

An $\xi\eta$-system is fixed in the sleigh. The ξ-axis runs horizontally along the center line of the runners and passes through the center of mass G with coordinates $\xi=a$, $\eta=0$, and the η-axis passes horizontally through the point of application of F. In the horizontal plane of the snow we fix an xy-system. Let ϕ be the angle between the axes of ξ and x, $\omega=\dot{\phi}$ the instantaneous angular velocity of the sleigh about the vertical; M is the mass of the sleigh, I its moment of inertia about the vertical through the mass center; u, v are the components of the velocity of the point O ($\xi=\eta=0$) along ξ and η.

(a) Derive the three simultaneous differential equations for the quantities u, v, ω with F as external force, using the method of complex variables of Problem V.1.

(b) Simplify them by introducing the non-holonomic condition $v=0$ and determine F from them.

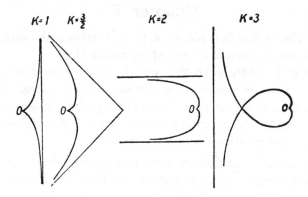

$K=1$ $K=\frac{3}{2}$ $K=2$ $K=3$

FIG. 57.   Trajectory of a sleigh for various values of
k, after Carathéodory.

(c) Integrate them by introducing, instead of ϕ, an auxiliary angle proportional to ϕ.

(d) Verify that the kinetic energy of the sleigh is constant (because F does no work).

(e) Show that, with suitable choice of the time scale, the trajectory of point O in the xy-plane possesses a cusp at $t=0$ and asymptotically approaches straight lines for $t=\pm\infty$, as shown by the curves of Fig. 57 borrowed from Carathéodory.

Chapter VI

VI.1. *Example illustrating Hamilton's principle.*   Calculate the value of Hamilton's integral between the limits $t=0$ and $t=t_1$

(a) for the real motion of a falling particle, $z=\frac{1}{2}gt^2$;

(b) for two fictitious motions $z=ct$ and $z=at^3$, where the constants c and a must be so determined that initial and end positions coincide with those of the real path, in agreement with the rules of variation in Hamilton's principle. Show that the integral has a smaller value for the real motion (a) than for the fictitious ones (b).

VI.2. *Once more the relative motion in a plane and the motion on a rotating straight line.* Treat Problems V.1 and V.2 by means of the Lagrange method.

VI.3. *Once more the free fall on the rotating earth and Foucault's pendulum.* Verify that these problems, too, can be treated by Lagrange's method without knowledge of the laws of relative motion. This procedure is interesting and simpler in thought than that of Chapter V; it does, however, require a careful inspection of the numerous small terms occurring; only *after* the differentiations $\frac{d}{dt}\frac{\partial}{\partial\dot{q}}$ and $\frac{\partial}{\partial q}$ have been carried out should the usual approximations of large terrestrial radius and small angular velocity be made; until then, *all* terms must be carried.

Start out with ordinary spherical polar coordinates r, θ, ψ, where r is measured from the earth's center. Next compare these with coordinates ξ, η, ζ introduced in Fig. 49. Let R be the earth's radius, θ_0, ψ_0 the coordinates of the projection on the earth of the initial position of the freely falling body or of the point of suspension of the pendulum. We then have the following relations between coordinates r, θ, ψ and ξ, η, ζ of the falling or oscillating particle m,

(1) $$\xi = R(\theta - \theta_0), \quad \eta = R\sin\theta(\psi - \psi_0), \quad \zeta = r - R,$$

with

(2) $$\psi_0 = \omega t, \quad \theta_0 = \frac{\pi}{2} - \phi = \text{colatitude}.$$

From this

$$\dot{\xi} = R\dot{\theta}, \quad \dot{\eta} = R\sin\theta(\dot{\psi}-\omega)+\frac{\cos\theta}{\sin\theta}\eta\dot{\theta}, \quad \dot{\zeta} = \dot{r}$$

and, conversely,

(3) $$r\dot{\theta} = \left(1+\frac{\zeta}{R}\right)\dot{\xi},$$

$$r\sin\theta\,\dot{\psi} = \left(1+\frac{\zeta}{R}\right)\dot{\eta} + \omega R\left(1+\frac{\zeta}{R}\right)\sin\theta - \frac{\cos\theta}{\sin\theta}\left(1+\frac{\zeta}{R}\right)\frac{\eta}{R}\dot{\xi}, \quad \dot{r} = \dot{\zeta},$$

where the angle θ occurring on the right must, according to (1), be regarded as a function of ξ.

These values are to be replaced in the expression

$$T = \frac{m}{2}(\dot{r}^2 + r^2\dot{\theta}^2 + r^2\sin^2\theta\,\dot{\psi}^2)$$

for the kinetic energy, which becomes, as a result, a function of $\dot{\xi}$, $\dot{\eta}$, $\dot{\zeta}$, ξ, η and ζ. If we denote the terms later to be dropped by . . . , we can, for instance, calculate from T

(4) $$\frac{\partial T}{\partial \dot{\xi}} = m\left(1 + \frac{\zeta}{R}\right)^2 \dot{\xi} - m\frac{\cos\theta}{\sin\theta}\left(1 + \frac{\zeta}{R}\right)\frac{\eta}{R}\left\{\cdots + \omega R\left(1 + \frac{\zeta}{R}\right)\sin\theta + \cdots\right\}$$

(5) $$\frac{d}{dt}\frac{\partial T}{\partial \dot{\xi}} = m\ddot{\xi} - m\omega\cos\theta\,\dot{\eta} + \cdots$$

(6) $$\frac{\partial T}{\partial \xi} = \frac{1}{R}\frac{\partial T}{\partial \theta} = +m\omega\cos\theta\,\dot{\eta} + \cdots$$

As our potential energy we can take

(7) $$V = mg(r - R) = mg\zeta.$$

Verify that in this manner we obtain Eqs. (30.5) for the free fall and Eqs. (31.2) for Foucault's pendulum, from which follow the results developed earlier.

VI.4. *" Wobbling " of a cylinder rolling on a plane support.* A circular cylinder of radius a has an inhomogeneous mass distribution, so that the center of mass G has the distance s from the axis of the cylinder. The cylinder rolls on a horizontal plane under the influence of gravity. Let m be the mass of the cylinder, I its moment of inertia about an axis through the mass center parallel to its axis of symmetry. Investigate the motion with the help of Lagrange's method, introducing the angle ϕ rotated through as generalized coordinate q. In calculating the kinetic energy, put the point of reference

(a) in the center of mass,

(b) in the geometrical center,

of the cylinder and verify that in both cases the same differential equation in ϕ results.

Show by means of the " method of small oscillations " that the equilibrium of the cylinder is stable with G in the lowest, and unstable with G in the highest position.

VI.5. *Differential of an automobile.* If the driven wheels of an automobile are not to slide, they must be able to turn at different speeds on a curve. This is achieved by the differential (Fig. 58). The engine drives the driving wheel (Ω) in which axle A is fixed. Two bevel gears (ω) sit on A in such a way that they can rotate independently about A. They, in turn, are in mesh with the pair of bevel gears (ω_1, ω_2) on which they may roll (cf. Fig. 58, left) as A turns.

The axle of the rear wheels of an automobile is cut at the center (Fig. 58, right). Fixed to the left end of its right half is the bevel gear (ω_1), to the right end of its left half, the bevel gear (ω_2). The two halves of the rear

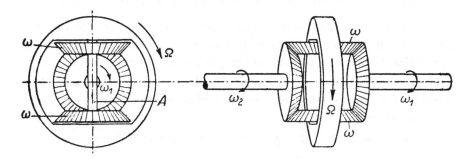

FIG. 58. The differential of an automobile, at the same time a model (after Boltzmann) for the induction effect of two coupled circuits. Left: view along rear axle of vehicle. Right : side view of this axle.

axle are therefore coupled by the differential in such a way that they can turn with different angular velocities.

Set up the kinematic relations between angular velocities Ω, ω, ω_1 and ω_2. Next make use of the principle of virtual work to derive the condition of equilibrium between the driving torque L acting on (Ω) and the torques L_1 and L_2 acting on (ω_1) and (ω_2).

What is the equation of motion of the system? Let I_1 and I_2 be the moments of inertia of (ω_1) and (ω_2), I that of the pair of gears (ω) about the axis of A, I' that of (ω) about the axis of the driving wheel. Neglect the contribution of (Ω) to I'.

If one rear wheel is accelerated, for instance by decreasing friction, the other wheel is retarded, even if driving torque and frictional torque remain equal there.

HINTS FOR SOLVING THE PROBLEMS

Almost all numerical calculations occurring in these problems can be carried out to sufficient accuracy by means of a slide rule. Let us call express attention to this useful tool for quick approximate calculations.

I.1. The proof that $v_1 = v_2 = v$ can be derived either algebraically or geometrically. In the latter case use v_1 and v_2 as rectangular coordinates in a plane diagram.

I.2. The velocities of the expelled masses are, respectively,

$$\frac{2M}{M+m}v_0 \quad \text{and} \quad \frac{M-m}{M+m}v_0.$$

I.3. Here we obtain the formulas of I.2 with a change of sign.

I.4. Verify that the quadratic equation for V leads to the same minimum value v_0 as that for v.

I.5. The differential equation to be integrated is

$$m\dot{v} - \mu a = -mg.$$

With the independent variable $m = m_0 - \mu t$ instead of t one obtains

$$v = -a\ln\left(1 - \frac{\mu}{m_0}t\right) - gt$$

and, by an additional integration ($z =$ height above earth's surface),

(1)
$$z = \frac{am_0}{\mu}\left\{\left(1 - \frac{\mu}{m_0}t\right)\ln\left(1 - \frac{\mu}{m_0}t\right) + \frac{\mu}{m_0}t\right\} - \tfrac{1}{2}gt^2.$$

For small t we obtain, by neglecting higher terms in t,

(2)
$$z = \left(\frac{\mu a}{m_0} - g\right)\frac{t^2}{2}.$$

The numerical calculation with equation (1) yields

$t=$	10	30	50 sec
$z=$	0.54	5.65	18.4 km

I.6. Since water has specific gravity 1, the mass of the drop is $m = \frac{4\pi}{3}r^3$, i.e., $dm = 4\pi r^2 dr$. In condensation, on the other hand, with α a factor of proportionality, $dm = 4\pi r^2 \alpha\, dt$; it follows that $dr = \alpha\, dt$. In terms of r the differential equation is then

$$\alpha\frac{d}{dr}(r^3v)=r^3g.$$

By virtue of the initial condition $v=v_0$ for $r=c$ its solution is

$$v=\frac{g}{\alpha}\frac{r}{4}+\frac{c^3}{r^3}\left(v_0-\frac{g}{\alpha}\frac{c}{4}\right);$$

for $c=0$ and $v_0=0$ we have, respectively,

$$v=\frac{g}{\alpha}\frac{r}{4},\qquad v=\frac{g}{\alpha}\frac{r}{4}\left(1-\frac{c^4}{r^4}\right).$$

I.7. Let x be the instantaneous length of chain hanging down. With the mass of the chain per unit length put equal to 1, the equation of motion is

$$\frac{d}{dt}(x\dot{x})=x\ddot{x}+\dot{x}^2=gx.$$

Since its integration is somewhat difficult — after the substitution $x=u^{\frac{1}{2}}$ it leads to an elliptic integral — we shall be satisfied with expressing the quantities $\dot{T}$, $\dot{V}$ and $\dot{Q}$ (Carnot energy loss per unit time) in terms of x, $\dot{x}$ and $\ddot{x}$ and showing that by the equation of motion we have

$$\dot{T}+\dot{V}+\dot{Q}=0,\quad\text{and hence,}\quad \dot{T}+\dot{V}\neq 0.$$

I.8. Our equation of motion is $l\ddot{x}=gx$. This linear differential equation with constant coefficients has a solution of the form (3.24b). The validity of the principle of energy can be read off either in differential form from the equation of motion, or in integrated form from its solution

$$x=a(e^{\alpha t}+e^{-\alpha t}),\qquad \alpha^2=\frac{g}{l},\qquad a=\frac{x_0}{2}.$$

I.9. The numerical data given in the problem permit the calculation of the centripetal acceleration of the moon in m·sec^{-2}. For the radius r of the earth we can take the original definition of the meter, $r=\frac{2}{\pi}10^7\text{m}$.

The law of gravitation, on the other hand, yields $\frac{g}{60^2}$ as the centripetal acceleration after the gravitational constant G has been eliminated by means of g as on p. 20. The two numerical values thus obtained are in satisfactory agreement.

I.10. Set up the transformation equations for the coordinates as in (2.5), but with $\alpha_0=\beta_0=\gamma_0=0$. The components of the transformed moment **L** are found to be linear expressions of the components of **L** with coefficients equal to the cofactors of the transformation scheme. For the latter we have the relations

$$\rho\gamma_1 = \begin{vmatrix} \alpha_2 & \alpha_3 \\ \beta_2 & \beta_3 \end{vmatrix}, \qquad \rho\gamma_2 = \begin{vmatrix} \alpha_3 & \alpha_1 \\ \beta_3 & \beta_1 \end{vmatrix}, \cdots$$

to be proved from the orthogonality conditions. Here $\rho = \pm 1$, according as the transformed system has the same sense (" unimodular transformation ") as the original system, or the opposite sense.

I.11. From Eqs. (6.8) we have [according to Fig. 7 and Eq. (6.5), B is negative]

$$\epsilon = \frac{-B}{\dfrac{GM}{C}} = \frac{|B|}{\dfrac{GM}{C}}.$$

It follows that for the ellipse $(\epsilon < 1)$ $\dfrac{GM}{C} > |B|$, for the hyperbola $(\epsilon > 1)$ $\dfrac{GM}{C} < |B|$. Now $R = \dfrac{GM}{C}$ is the radius of the hodograph circle, $|B|$ the distance of the center from the pole. From this the assertion made in the statement of the problem follows at once.

The table below, in which

$$v_0 = \frac{GM}{C} + |B|$$

signifies the magnitude of the planet's velocity at perihelion, shows how the limiting cases of the circle and the parabola fit into the scheme.

Planetary Trajectory	ϵ	$|B|$	Hodograph	v_0
circle	$=0$	$=0$	center at pole	$= \dfrac{GM}{C}$
ellipse	<1	$<R$	hodograph includes pole	$< \dfrac{2GM}{C}$
parabola	$=1$	$=R$	hodograph passes through pole	$= \dfrac{2GM}{C}$
hyperbola	>1	$>R$	hodograph excludes pole	$> \dfrac{2GM}{C}$

I.12. In the differential equations (6.4) we must replace GM by $\pm \dfrac{eE}{m}$, where the upper sign (attraction) corresponds to the case of the positive ion, the lower sign (repulsion) to that of the negative ion. Note that here $\dot{x} = 0$, $\dot{y} = -v_0$, and the meaning of ϕ is the same as in Fig. 6, so that Eqs. (6.5) give for $\phi = \dfrac{\pi}{2}$,

$$A = \pm \frac{eE}{m} C, \quad B = -v_0.$$

Eq. (6.6) then becomes

(1)
$$\frac{1}{r} = \pm \frac{eE}{m_0 C^2}(1-\sin\phi) - \frac{v_0}{C}\cos\phi.$$

C changes from trajectory to trajectory with the distance of the firing direction from the y-axis. It follows that (1) represents a family of curves. To obtain the envelope of this family, differentiate Eq. (1) with respect to C, then eliminate C from this and the original equation to obtain

(2)
$$x^2 = p^2 - 2py, \quad p = \pm\frac{4eE}{m_0 v_0^2}.$$

Note that any electron path consists of only one branch of a hyperbola, whereas (1) represents both branches; verify — most simply by a sketch of the corresponding families of curves — that Eq. (2) is the envelope of the actual electron paths only in the case of repulsion.

I.13. It will be easiest to use the method of harmonic oscillations of § 3, (4). It is, however, instructive to check that the methods of § 6 also lead to the desired end.

I.14. The nuclear reaction treated here is not an elastic collision; nor is it an inelastic one. It is, so to speak, a " superelastic " one, in that the nuclear binding energy E is to be added to the primary energy E_p. The kinetic energy of the α-particles can be calculated in the classical form $E_\alpha = \frac{1}{2}m_\alpha v_\alpha^2$.

Elimination of v_α from the equations of energy and momentum then yields Kirchner's result for the symmetrical case,

$$\cos\phi = \left(\frac{m_p}{2m_\alpha}\frac{E_p}{E+E_p}\right)^{\frac{1}{2}}$$

1 ev is that energy which a potential drop of 1 volt ($=10^8$ electromagnetic units of potential) imparts to the electronic charge e ($=1.6\cdot10^{-20}$ electromagnetic units of charge), so that 1 ev$=1.6\cdot10^{-12}$ erg.

The mass of the proton is $m_p = 1.65\cdot10^{-24}$ g, that of the α-particle, is hence $m_\alpha = 6.6\cdot10^{-24}$ g. The latter is needed in order to pass from E_α, at first expressed in ev and then converted to erg, to the velocity v_α. The value of v_α thus found shows that the classical form for E_α was justified and that the relativity correction of Eq. (4.11) is negligible.

I.15. In the second Eq. (3.27) we put $V_0 = 0$ and, say, $v_0 = 1$, so that one can immediately calculate the kinetic energy $\frac{1}{2}MV^2$ of the struck particle after the collision as function of $x = \frac{M}{m}$; in particular one finds it to be a maximum for $x = 1$ and to be small — only 1.9 % of the maximum value — for $x = 206$.

Proceeding from such considerations, Fermi in 1935 worked out his method for the production of " thermal " neutrons, i.e., slow neutrons of uniform velocity, which by frequent collisions have reached equilibrium with the protons of thermal energy contained in the paraffin.

I.16. The coordinates of E are

(1a)
$$x = ML = a \cos u$$
$$= SL - SM = r \cos \phi - \epsilon a$$

(1b)
$$y = EL = r \sin \phi = b \sin u.$$

Write the polar equation of the ellipse in r, ϕ in the form

(1)
$$r = \epsilon r \cos \phi + p, \quad p = a(1 - \epsilon^2).$$

Substitute the value of $r \cos \phi$ from (1a) to obtain

(2)
$$r = \epsilon(a \cos u + \epsilon a) + a(1 - \epsilon^2) = a(1 + \epsilon \cos u).$$

A differentiation of (2) gives

(3)
$$dr = - \epsilon a \sin u\, du.$$

A differentiation of (1) gives

$$\epsilon \sin \phi\, d\phi = - p \frac{dr}{r^2}.$$

From this

(4)
$$\frac{-p}{\epsilon \sin \phi} \dot{r} = r^2 \dot{\phi} = C \quad (C = \text{areal velocity constant}).$$

Eq. (4) is transformed by (1b) and (3) into

$$\frac{pa}{b} r \dot{u} = C.$$

Finally replace r from (2), to arrive at the differential equation

(5) $(1 + \epsilon \cos u)\, du = n\, dt$ (6) $n = \dfrac{Cb}{pa^2}.$

Integration of (5) yields

$$u - \epsilon \sin u = nt.$$

The integration constant vanishes because we agreed to measure time in such a way that for $u = 0$ we have $t = 0$. nt is called the *mean anomaly* and, like the other anomalies, is measured from perihelion in astronomy. The name comes from the fact that by means of Eq. (6.9) the right member of (6) can be transformed to $\dfrac{2\pi}{T}$.

II.1. Reduce

$$\delta f = \frac{\partial f}{\partial x}\delta x + \frac{\partial f}{\partial y}\delta y + \frac{\partial f}{\partial \phi}\delta\phi + \frac{\partial f}{\partial \psi}\delta\psi$$

by means of condition (1) of the problem, to obtain for the right member

$$\left(\frac{\partial f}{\partial x}a\cos\psi + \frac{\partial f}{\partial y}a\sin\psi + \frac{\partial f}{\partial \phi}\right)\delta\phi + \frac{\partial f}{\partial \psi}\delta\psi.$$

Now $\delta\phi$ and $\delta\psi$ can individually be put $=0$, so that

(2) $\qquad \dfrac{\partial f}{\partial \psi} = 0$ $\qquad$ and $\qquad$ (3) $\qquad a\dfrac{\partial f}{\partial x}\cos\psi + a\dfrac{\partial f}{\partial y}\sin\psi + \dfrac{\partial f}{\partial \phi} = 0.$

The latter equation is valid for all ψ and can therefore be differentiated with respect to ψ. With the help of (2) this gives

(4) $$-a\frac{\partial f}{\partial x}\sin\psi + a\frac{\partial f}{\partial y}\cos\psi = 0$$

and, after a second differentiation with respect to ψ,

(5) $$a\frac{\partial f}{\partial x}\cos\psi + a\frac{\partial f}{\partial y}\sin\psi = 0.$$

From (4) and (5) it follows that

(6) $$\frac{\partial f}{\partial x} = \frac{\partial f}{\partial y} = 0.$$

According to (3) we must then also have

(7) $$\frac{\partial f}{\partial \phi} = 0.$$

(2), (6) and (7) show that there does not exist a condition $f=0$ dependent on x, y, ϕ, ψ, i.e., that our system is non-holonomic. Proof of G. Hamel, "Elementare Mechanik," 2nd Ed., Leipzig 1922.

II.2. Draw the work diagram of the engine, that is, the L-curve and the W-line with the crank angle from 0 to π as abscissa. Note that the areas enclosed between the L-curve and the abscissa and the W-line and the abscissa must be equal. This yields a relation between L_0 and W. The angles ϕ_2 and ϕ_1 belonging to ω_{max} and ω_{min} are the points of intersection of the L- and W-curves in the diagram, $\sin\phi_1 = \sin\phi_2 = \dfrac{2}{\pi}$, $\phi_2 = \pi - \phi_1$, $\phi_1 = 39°\ 33' = 0.69$ radians. Determine the difference in the kinetic energy of the flywheel between angles ϕ_2 and ϕ_1 and express it in terms of

I, ω_m and δ. The equation of energy written for the same interval yields the magnitude of the required I in the form

$$I = \frac{W}{\delta\,\omega_m^2}\,(\pi\cos\phi_1 - \pi + 2\phi_1) = \frac{0.66}{\delta\,\omega_m^2}\,W.$$

With

$$N = \frac{W\omega}{75}\,HP \quad \text{and} \quad n = \frac{60}{2\pi}\,\omega \text{ r.p.m.}$$

one obtains

$$I \cong 43,400\,\frac{N}{\delta n^3}\,kg\!\cdot\!m\!\cdot\!\sec^2$$

in the practical system of units.

II.3. For the magnitude of the earth's radius see Problem I.9. For the numerical calculation of the length of the day put $(8\pi)^{\frac{1}{2}} = 5$.

II.4. (a). If one thinks of the beam as held fixed in position, one need only consider the equilibrium of gravity and inertial forces at the pulley in a virtual rotation $\delta\phi$ of the pulley (torque equation). From this one obtains the acceleration $\ddot{x}$ of the weights as a small fraction of g.

(b). Add a virtual rotation of the beam to the foregoing. Here the moments of the inertial forces about the fulcrum of the balance beam enter. One finds that equilibrium does not prevail. The beam is deflected downward on the side of the pan as long as the weight p is falling. In estimating the excess weight the diameter of the pulley may be neglected in comparison with the length of the balance beam. Another procedure using the same approximation is to compare the load on the pan with the load due to weights and inertial forces on the other side of the beam.

II.5. Let the equation of the inclined plane be

(1) $$F(z, x, t) = z - a\,x - \phi(t) = 0.$$

$a = \tan\alpha$ determines the constant inclination α of the plane to the horizontal; $\phi(t)$ is its intersection with the z-axis which varies with time. Lagrange's equations of the first kind (12.9a) give

(2) $$\ddot{x} = -\lambda a, \quad \ddot{z} = \lambda - g.$$

In order to determine λ, differentiate (1) twice with respect to t,

(3) $$\ddot{z} - a\ddot{x} = \ddot{\phi}(t).$$

Substitution of (2) in (3) yields λ; the integration of (2) can now easily be carried out. With initial conditions $\dot{x}=\dot{z}=0$, $x=x_0$, $z=z_0$ at $t=0$ one obtains

$$x=x_0-\frac{a}{1+a^2}\Big(\phi(t)-\phi(0)-\dot{\phi}(0)t+g\frac{t^2}{2}\Big),$$

$$z=z_0+\frac{1}{1+a^2}\Big(\phi(t)-\phi(0)-\dot{\phi}(0)t-ga^2\frac{t^2}{2}\Big).$$

From this we have for $\ddot{\phi}=+g$

$$x=x_0-g\frac{t^2}{2}\sin 2\alpha,\quad z=z_0+g\frac{t^2}{2}\cos 2\alpha$$

and for $\ddot{\phi}=-g$,

$$x=x_0,\quad z=z_0-g\frac{t^2}{2},$$

as for a free fall. $\lambda=0$ only under the last assumption; otherwise λ acts as a pressure against the sliding body and hence does work.

The problem can be solved by means of d'Alembert's principle without introducing λ. Since the time is not to be varied (cf. p. 68), we have, from (1), $\delta z=a\,\delta x$ for the virtual displacements. From d'Alembert's principle it follows that

$$\ddot{x}+(g+\ddot{z})a=0,$$

which, together with (3), allows one to calculate $\ddot{x}$ and $\ddot{z}$ directly. This example illustrates that d'Alembert's method leads to a solution more directly and simply than do Lagrange's equations; the latter, on the other hand, have the advantage that the forces of constraint are quantitatively determined.

II.6. In § 11, (1), d'Alembert's principle was used to derive the equation of acceleration of a system rotating under the influence of an external torque. We introduced a virtual rotation $\delta\phi$ about the axis of rotation, which we shall here take as our x-axis. Only the tangential inertial forces were relevant, since the normal ones, the centrifugal forces, did no work in the rotation $\delta\phi$.

Here we ask for the forces exerted on bearings A and B in a uniform rotation, or, instead, for their reactions **A** and **B**. It is precisely the centrifugal forces which are relevant, whereas the tangential inertial forces drop out in a uniform rotation. If we introduce the virtual translations δy, δz, the virtual work becomes equal to the product of δy and δz by the sum of the y- and z-components of the centrifugal forces acting on the single mass elements, these forces being

$$dm\,y\omega^2, \quad dm\,z\omega^2:$$

An integration yields the inertial components Y and Z of the ordinary swinging motion of the total mass m, which are to be thought of as applied at the center of mass.

Next we introduce virtual rotations $\delta\phi_y$ and $\delta\phi_z$ about the y- and z-axis respectively. The virtual work done in these is given by

$$-\delta\phi_y \int dm\,x\,z\,\omega^2 \quad \text{and} \quad \delta\phi_z \int dm\,x\,y\,\omega^2.$$

They correspond to the torques

$$L_y = -I_{xz}\omega^2 \quad \text{and} \quad L_z = I_{xy}\omega^2.$$

To determine the bearing reactions **A** and **B**, fix the origin of coordinate system xyz, say, at the bearing A, designate by l the distance between the two bearings and by η and ζ the coordinates of the mass center along the y- and z-directions. We then obtain the two component equations

(1)
$$A_y + B_y = -m\eta\omega^2,$$
$$A_z + B_z = -m\zeta\omega^2$$

and the two moment equations

(2)
$$lB_z = -I_{xz}\omega^2,$$
$$lB_y = -I_{xy}\omega^2$$

for the determination of the four unknowns A_y, A_z and B_y, B_z.

Clearly these periodically varying reactions in the bearings are undesirable from the engineering standpoint. To avoid them it is not only necessary that the center of mass lie on the axis of rotation $\eta = \zeta = 0$, Eq. (1), but also that the axis of rotation be a principal axis of the mass distribution, $I_{xz} = I_{xy} = 0$, Eq. (2); see, in this connection, Ch. IV, § 22, near Eq. (15a). The fulfillment of this second condition is just as important as that of the first. The fulfillment of both conditions is called the " balancing " of the rotating body.

II.7. Let S be the tension of the string, z the portion of its length that is unwound at any given instant. We have during (a):

$$I\dot{\omega} = Sr, \quad S = m(g - \ddot{z}).$$

$\dot{z}$ and $\ddot{z}$ are positive; because of $\dot{z} = r\omega$,

(1) $$\ddot{z} = r\dot{\omega} = \frac{Sr^2}{I},$$

(2) $$S = \frac{mg}{1 + \frac{mr^2}{I}};$$

during (b):

The rotation ω retains its sense. The torque of the string tension acts against ω. $\dot{z}$ becomes negative and we have

(3) $$\dot{z} = -r\omega, \quad \ddot{z} = -r\dot{\omega} = +\frac{Sr^2}{I},$$

(4) $$S = \frac{mg}{1 + \frac{mr^2}{I}}.$$

In both cases (a) and (b) the string tension is the same, and constant in time; it is smaller than the weight of the rotating body.

In the transition stage between (a) and (b) one perceives a very noticeable pull in the hand which corresponds to the transition from positive momentum $m\dot{z}$ to negative. During this interval S becomes greater than that given by Eq. (2).

II.8. The condition that the particle jump off is, according to (18.7),

$$\lambda = 0 \quad \text{or} \quad R_n = 0$$

so that, from (18.6),

(1) $$mg\frac{z}{l} = -\frac{m}{l}(x\ddot{x} + y\ddot{y} + z\ddot{z}).$$

Now for every path on the sphere

$$x\dot{x} + y\dot{y} + z\dot{z} = 0, \quad \text{i.e.,} \quad x\ddot{x} + y\ddot{y} + z\ddot{z} = -(\dot{x}^2 + \dot{y}^2 + \dot{z}^2) = -v^2,$$

so that, instead of (1), we can write

(2) $$\frac{mgz}{l} = \frac{mv^2}{l}.$$

The right side does not equal the centrifugal force along the path, since the path is not a geodesic in our case. In agreement with Meusnier's theorem of § 40 it is equal to the projection of this centrifugal force on the normal to the spherical surface.

From the equation of energy

(3) $$v^2 = v_0^2 - 2g(z - z_0).$$

Eq. (2) can therefore be rewritten in terms of the initial values v_0, z_0:

(4) $$3z = 2z_0 + \frac{v_0^2}{g} = 2(z_0 + h_0),$$

where $h_0 = \frac{1}{2}\frac{v_0^2}{g}$ is the height of free fall corresponding to velocity v_0.

III.1. For a pendulum hanging almost vertically coordinates x and y are small quantities of the first order; z equals $-l$ to quantities of the second order. For this reason the third Eq. (18.2) gives, to quantities of the second order,

(1) $$\lambda = -\frac{mg}{l}$$

and the first two Eqs. (18.2) define, as in Problem I.13, a harmonic elliptical motion of circular frequency

(2) $$\frac{2\pi}{T} = \left(\frac{g}{l}\right)^{\frac{1}{2}}.$$

For the areal velocity constant of the elliptical motion we have

(3) $$C = \frac{2\pi ab}{T} = \left(\frac{g}{l}\right)^{\frac{1}{2}} ab \to 0,$$

and for the energy constant (initial state $\theta_0 = \epsilon$, $\dot{\theta}_0 = 0$)

(4) $$E = T + V = mgl\left(-1 + \frac{\epsilon^2}{2}\right).$$

With $u = \eta - 1$ one has, from (18.11),

$$U = -\frac{4g}{l}\left(\eta - \frac{\epsilon^2}{2}\right)\eta - \frac{C^2}{l^4} = \frac{4g}{l}(\eta_1 - \eta)(\eta - \eta_2),$$

$$\eta_{1,2} = \frac{\epsilon^2}{4} \pm \left(\frac{\epsilon^4}{16} - \frac{C^2}{4gl^3}\right)^{\frac{1}{2}}.$$

From (18.15) we then have

(5) $$2\pi + \Delta\phi = \frac{C}{l(lg)^{\frac{1}{2}}} \int_{\eta_2}^{\eta_1} \frac{d\eta}{\eta[(\eta_1 - \eta)(\eta - \eta_2)]^{\frac{1}{2}}}.$$

A substitution modelled after Eq. (46.11) transforms the integral of (5) into the known integral

$$\int_0^\pi \frac{dv}{A + B\cos v} = \frac{\pi}{(A^2 - B^2)^{\frac{1}{2}}}, \quad A = \frac{\epsilon^2}{4}, \quad B = \left(\frac{\epsilon^2}{16} - \frac{C^2}{4gl^3}\right)^{\frac{1}{2}}.$$

Thereupon (5) yields $\Delta\phi = 0$, which was to be proved.

III.2. The first assertion of the problem is proved immediately by differentiation of Eq. (19.10) for $|C|$ with respect to ω; the second assertion is similarly proved by differentiation of $|C|\omega$ with respect to ω.

III.3. Let us designate the proportionality factors of the damping torque and the restoring torque by $2\rho I$ and $\omega_0^2 I$ respectively. One then obtains Eq. (19.9) as the equation of motion of the galvanometer, with the difference that the right member is now a constant C and α replaces x in the notation. Fit the constants a and b of the general solution

$$\alpha = C + e^{-\rho t}\,(a\cos\,[(\omega_0^2 - \rho^2)^{\frac{1}{2}}t] + b\sin\,[(\omega_0^2 - \rho^2)^{\frac{1}{2}}t]$$

to the conditions $\alpha = \dot{\alpha} = 0$ at $t = 0$, and the constant C to the condition $\alpha \to \alpha_\infty$ as $t \to \infty$.

In the case (a) one obtains a transient motion with decreasing oscillations, in the case (c), a monotonic transient motion towards the final position. Case (b) should be treated as limiting case of either (a) or (c); in it we arrive at a secular term containing t as a factor.

III.4. In part (a) of the problem d'Alembert's principle ($x, y =$ coordinates of the oscillating mass point, y positive upward) demands

(1) $$\ddot{x}\,\delta x + (\ddot{y} + g)\delta y = 0.$$

The equation of constraint is

(2) $$(x - \xi)^2 + y^2 = l^2.$$

Its variation (t, and hence also ξ, being held fixed) gives

(3) $$(x - \xi)\delta x + y\,\delta y = 0.$$

Combination of (1) and (3) results in

(4) $$y\ddot{x} - (x - \xi)\,(\ddot{y} + g) = 0.$$

Differentiating (2) twice with respect to t yields a second equation for $\ddot{x}$ and $\ddot{y}$ which, together with (4), furnishes the exact differential equation of the problem.

When passing to small vibrations, one must remember that $x - \xi$ is a small quantity of first order so that, according to (2), $y = -l$ to small quantities of second order. $\dot{y}$ and $\ddot{y}$ are then also small quantities of second order, so that (4) becomes

(5) $$l\ddot{x} + (x - \xi)\,g = 0.$$

With $x - \xi = u$ one obtains the inhomogeneous pendulum equation

(6) $$\ddot{u} + \frac{g}{l}u = -\ddot{\xi},$$

showing that $-m\ddot{\xi}$ acts as driving force. The integration is performed as on p. 101. The phase relation between the motion of the point of suspension and that of the mass point emphasized in the text of the problem corresponds to Fig. 31. It will be instructive to make an experiment with a string whose lower end carries a weight, and whose upper end is moved horizontally to and fro by the hand. For fast motion of the hand (case above resonance) the out-of-phase motion of the two points can be very clearly observed.

Using the method of Lagrange's equations of the first kind, from Lagrange's equation for y one finds $\lambda = -\frac{g}{l}$ up to small quantities of second order, and from the x-equation one obtains Eq. (5).

In part (b) of the problem Eq. (1) remains valid. Condition (2) becomes

(7) $$x^2 + (y - \eta)^2 = l^2.$$

Its variation yields, instead of (4),

(8) $$(y - \eta)\ddot{x} - x(\ddot{y} + g) = 0.$$

If x is treated as a small quantity of first order, (7) gives to quantities of second order

(9) $$y - \eta = -l, \quad \ddot{y} = \ddot{\eta}.$$

By this, (8) becomes

(10) $$\ddot{x} + \frac{\ddot{\eta} + g}{l} x = 0.$$

The same follows from Lagrange's equations of the first kind, since the y-equation yields the value

(11) $$\lambda = -\frac{\ddot{\eta} + g}{l}$$

if approximation (9) is used, so that the x-equation becomes identical to (10).

If the point of suspension is moved upward with constant acceleration $+g$ it follows that the force of gravity seems doubled; if the point is moved downward with $-g$, it seems to be anulled. This points to an equivalence between gravity and acceleration, which, together with the equality of the gravitational and inertial mass (p. 19), formed the foundation of Einstein's theory of gravitation.

III.5. Equilibrium of the tensions at points C and D (necessary because the wire is weightless !) demands that

$$(3) \qquad S_1\frac{x_1-x_3}{l_1}=S\frac{x_3}{a}+S\frac{x_3-x_4}{a}, \quad S_2\frac{x_2-x_4}{l_2}=S\frac{x_4}{a}+S\frac{x_4-x_3}{a},$$

so that, from Eq. (1) of the problem, with $\sigma_1=\frac{m_1g}{S}\frac{a}{l_1}$, $\sigma_2=\frac{m_2g}{S}\frac{a}{l_2}$,

$$(4) \qquad \sigma_1x_1=(2+\sigma_1)x_3-x_4$$
$$\sigma_2x_2=(2+\sigma_2)x_4-x_3.$$

We have presupposed weak coupling, so that σ_1 and σ_2 are small numbers; they can be cancelled in the right members of (4). Solving for x_3, x_4 then yields

$$(5) \qquad x_3=\frac{2}{3}\sigma_1x_1+\frac{1}{3}\sigma_2x_2$$
$$x_4=\frac{2}{3}\sigma_2x_2+\frac{1}{3}\sigma_1x_1$$

and substitution in (2) gives

$$(6) \qquad \ddot{x}_1+\frac{g}{l_1}(1-\sigma_1)x_1=\frac{1}{3}\frac{g}{l_1}(\sigma_2x_2-\sigma_1x_1)$$
$$\ddot{x}_2+\frac{g}{l_2}(1-\sigma_2)x_2=\frac{1}{3}\frac{g}{l_2}(\sigma_1x_1-\sigma_2x_2).$$

These simultaneous differential equations are to be treated just like (20.10). The meaning, for our problem, of the quantities ω_1, ω_2, k_1, k_2 introduced there can be found by comparison with Eq. (6) above.

III.6. The effect of m on M is represented by $k(X-x)$, that of M on m by $k(x-X)$. In the two resulting simultaneous differential equations for X and x put $X=0$. It will be found that the condition required — that only m take part in the oscillation — is given by the resonance requirement that the circular frequency of the proper oscillation of system (m, k) be equal to the circular frequency ω of the external force.

Such an arrangement is used in engineering practice as an " oscillation quencher." It may thus be used on a crank shaft with a flywheel rotating with constant angular velocity ω; there the quencher is a device capable of variable rotation; its purpose is to absorb the oscillations of the crank with which it is coupled. In such a case the angle rotated through takes the place of coordinate x of our problem.

IV.1. Moments of inertia of plane mass distributions are important for the torsion and bending of beams in elasticity theory (Vol. II). Because of $r^2=x^2+y^2$ we have

$$I_p=\int r^2dm=\int x^2dm+\int y^2dm=I_x+I_y.$$

In problems of elasticity the mass is to be thought of as uniformly distributed with density 1 over the cross section of the beam, so that $dm = dS =$ element of area. For a circular disk of radius a and area $S = \pi a^2$ one then obtains

$$I_p = \int r^2 dS = 2\pi \int_0^a r^3 dr = \tfrac{1}{2} S a^2 \text{ and hence, } I_x = I_y = \tfrac{1}{4} S a^2.$$

IV.2. We leave the ratio of the magnitudes of the three principal moments of inertia arbitrary to the very last; we thus embrace, in one and the same calculation, the three cases in which A is the greatest, smallest, and intermediate principal moment of inertia.

IV.3. The impulse Z imparts to the ball (radius a) both a translational and a rotational momentum,

(1) $\qquad\qquad Mv = Z \qquad$ and $\qquad$ (2) $\quad I\omega = Zh,$

where h is the height above the center at which the horizontally held cue strikes the ball. The axis of ω is perpendicular to the median plane. The peripheral velocity u at the lowest point lies in the median plane and equals $a\,\omega$. This is true not only at $t = 0$ (time of the impact), but also for $t > 0$. According to (11.12a), $I = \tfrac{2}{5} M a^2$, so that, for $t = 0$, by Eqs. (2) and (1),

(3) $\qquad\qquad\qquad \tfrac{2}{5} M a u = Z h = M v h.$

$v = u$ means pure rolling, and from (3) requires $h = \tfrac{2}{5} a$. Notice that we have counted u positive in the direction opposite to v. For high shots $h > \tfrac{2}{5} a$ the sliding velocity $u - v$ of the point of contact between ball and cloth is > 0 and opposed to v; friction is therefore directed along v and of magnitude $\mu M g$. Its moment about the center, $\mu M g a$, acts against the rotation ω.

For low shots the friction is directed in the opposite way. In general, we can associate the upper sign with a high shot, the lower sign with a low shot, and write for $t > 0$,

(4) $\qquad\qquad\qquad\qquad \dot{v} = \pm \mu g,$

(5) $\qquad\qquad\qquad\qquad \dot{u} = \mp \tfrac{5}{2} \mu g.$

Discussion by means of graph: draw v and u as ordinates against t as abscissa; both are represented by straight lines which intersect in the case of high shots as well as that of low shots. At the point of intersection

$u=v$ pure rolling takes place. From then on the graphs of u and v run on in coincidence along a horizontal straight line. The abscissa of the intersection is

(6)
$$\tau = \pm \frac{5h-2a}{7a}\ \frac{Z}{\mu g M}.$$

Note that for a low shot the first numerator is negative since h lies between $-a$ and $\frac{2}{5}a$; the negative sign of the right member of (6) is therefore only a formal one. The increase or decrease of velocity for high and low shots respectively is given by $\Delta v = \pm \mu g \tau$. The final velocity of pure rolling becomes

$$v+\Delta v = \frac{5}{7}\frac{h+a}{a}\frac{Z}{M}$$

i.e., proportional to the height $h+a$ of the point of impact above the cloth.

Theory of the follow shot. The ball struck high meets a second ball in a central collision during the time interval $t < \tau$ in which $u > v$. Let u_0 and v_0 be the values of u and v at the moment of impact. v_0 is transferred to the second ball. According to (4), the first ball is then accelerated from $v=0$. From (5), its u decreases from u_0 on down. A new graph shows that there is an intersection at which pure rolling begins to take place. Abscissa of the point of intersection and velocity of pure rolling are, respectively,

(7)
$$\tau_1 = \frac{2}{7}\frac{u_0}{\mu g}, \quad v_1 = \mu g \tau_1 = \frac{2}{7}u_0.$$

Theory of the draw shot. Again the driven ball meets a second one in the interval $t < \tau$, where now, however, $u < v$. For an extremely low shot, which we shall presuppose, u is, as a matter of fact, negative, that is, has the same direction as v. Let u_0 and v_0 be the values of u and v just before impact. v_0 is again transmitted to the second ball. From (4), the first ball is accelerated from $v=0$ in the negative sense: it runs backward. Eq. (5) tells us that u increases from its negative initial value u_0 toward positive values, i.e., its absolute value decreases. The two straight lines of v and u intersect (new diagram); the abscissa of the point of intersection and the final velocity of pure rolling now become

(8)
$$\tau_2 = \frac{2}{7}\frac{|u_0|}{\mu g}, \quad |v_2| = \frac{2}{7}|u_0|.$$

IV.4. The cue is no longer held horizontally as in IV.3, but forms an angle with the horizontal plane; evidently the cue must hit the ball at a point of the upper hemisphere, as in our earlier " high shots." Put the x-axis along the horizontal component of the impulse and the z-axis along the vertical. The components of the impulse $\mathbf{Z}$ become $(Z_x, 0, Z_z)$ and the components of the impulsive torque $\mathbf{N}$ referred to the center of the ball (which is also the origin of the xyz-system),

$$N_x = yZ_z, \quad N_y = zZ_x - xZ_z, \quad N_z = -yZ_x.$$

Here x, y and z are the coordinates of the point of impact of cue and ball. From the N_x, N_y we obtain the angular velocities

$$\omega_x = \frac{5}{2} \frac{N_x}{Ma^2}, \qquad \omega_y = \frac{5}{2} \frac{N_y}{Ma^2}.$$

The associated peripheral velocities at the lowest point P of the ball are

(1) $u_x = -a\omega_y, \quad u_y = +a\omega_x.$

N_z and ω_z do not interest us; they do not generate any sliding at P, but merely a " boring " friction to be neglected. Let the sliding motion at the cloth have components

(2) $v_x - u_x = -\rho \cos \alpha, \qquad v_y - u_y = -\rho \sin \alpha.$

It creates a friction R making an angle $\pi + \alpha$ with the x-axis and having magnitude $\mu g M$. Its influence on the translation and rotation for $t > 0$ is determined by

$$M\dot{v}_x = R_x, \qquad M\dot{v}_y = R_y,$$

$$I\dot{\omega}_x = aR_y, \qquad I\dot{\omega}_y = -aR_x.$$

It follows that

$$\dot{v}_x = -\mu g \cos \alpha, \qquad \dot{v}_y = -\mu g \sin \alpha$$

and, by virtue of (1) and (2),

(4) $\dot{u}_y = -\frac{5}{2}\mu g \sin \alpha, \qquad \dot{u}_x = -\frac{5}{2}\mu g \cos \alpha;$

$$\dot{v}_x - \dot{u}_x = -\frac{d}{dt}(\rho \cos \alpha) = -\frac{7}{2}\mu g \cos \alpha,$$
(5)
$$\dot{v}_y - \dot{u}_y = -\frac{d}{dt}(\rho \sin \alpha) = -\frac{7}{2}\mu g \sin \alpha.$$

Solution for $\dot{\alpha}$ and $\dot{\rho}$ from the last two members of Eqs. (5) gives

   1. $\dot{\alpha} = 0$. The friction has constant direction; since it also has constant magnitude, the path of point P in the horizontal plane becomes a parabola.

The axis of the parabola is parallel to the initial direction α of the sliding motion, which can be gathered from the components of Z and N.

2. $\dot{\rho} = -\frac{7}{2}\mu g$, $\rho = \rho_0 - \frac{7}{2}\mu g t$, $\rho = 0$ for $t = \tau = \frac{2}{7}\frac{\rho_0}{\mu g}$. ρ_0 is the initial magnitude of sliding velocity which can likewise be determined from Z and N. For $t > \tau$ the sliding and friction are permanently 0. The ball pursues a straight course tangent to the parabola.

V.1. Let ϕ be the instantaneous angle by which the rotating plane has turned with respect to the fixed plane. We put

(1) $$x + iy = (\xi + i\eta)e^{i\phi}.$$

Two differentiations with respect to t give, with $\dot{\phi} = \omega$,

(2) $$\ddot{x} + i\ddot{y} = \{\ddot{\xi} + i\ddot{\eta} + 2i\omega(\dot{\xi} + i\dot{\eta}) + i\dot{\omega}(\xi + i\eta) - \omega^2(\xi + i\eta)\}e^{i\phi}.$$

$\xi + i\eta$ is the (complex) vector $\mathbf{r}$ as observed from the rotating plane, $\dot{\xi} + i\dot{\eta} = \dot{\mathbf{r}}$ its velocity observed from the same plane, etc. Since $i(\dot{\xi} + i\dot{\eta}) = (\dot{\xi} + i\dot{\eta})e^{i\frac{\pi}{2}}$ is a vector perpendicular to the latter, we can write,

(3) $$2i\omega(\dot{\xi} + i\dot{\eta}) = 2\boldsymbol{\omega} \times \dot{\mathbf{r}}, \quad i\dot{\omega}(\xi + i\eta) = \dot{\boldsymbol{\omega}} \times \mathbf{r},$$

where $\boldsymbol{\omega}$ is of course directed along the normal to the complex plane. As on p. 165, let us call $\mathbf{w}$ the velocity $\dot{x} + i\dot{y}$, as observed from the fixed plane; we shall, however, retain the designation of superscript dots for the time derivatives referred to the rotating plane, as written in Eq. (3) above. Eq. (2) then transforms to the following equation analogous to (29.4):

(4) $$\dot{\mathbf{w}} = \{\ddot{\mathbf{r}} + 2\boldsymbol{\omega} \times \dot{\mathbf{r}} + \dot{\boldsymbol{\omega}} \times \mathbf{r} - \omega^2 \mathbf{r}\}e^{i\phi}.$$

If $\mathbf{F} = F_x + iF_y$ is the force referred to the fixed plane, $\boldsymbol{\Phi} = F_\xi + iF_\eta$ that referred to the rotating one, we have, from (1), $\mathbf{F} = \boldsymbol{\Phi}e^{i\phi}$, so that

(5) $$\boldsymbol{\Phi} = \mathbf{F}e^{-i\phi}.$$

In the light of (4) and (5) we then have from $m\dot{\mathbf{w}} = \mathbf{F}$ that

(6) $$m\{\ddot{\mathbf{r}} + 2\boldsymbol{\omega} \times \dot{\mathbf{r}} + \dot{\boldsymbol{\omega}} \times \mathbf{r} - \omega^2 \mathbf{r}\} = \boldsymbol{\Phi}.$$

With this we have determined the additional forces required in the problem. In particular, one identifies the second term on the left with the Coriolis force.

We have intentionally treated this problem in complex notation in order to emphasize that two-dimensional vectors are best represented by complex variables.

V.2. Let us choose the plane in which the straight line rotates as the xy-plane; x-axis horizontal, y-axis vertically upward. Let $\phi = \omega t$ be the angle of the straight line with the x-axis. One can reduce the problem to the foregoing one by associating with the rotating straight line a vertical $\xi\eta$-plane in which the line is fixed. This $\xi\eta$-plane must then rotate in the xy-plane with constant angular velocity ω. It is convenient to put the ξ-axis along the rotating straight line. In order to keep the mass point on the ξ-axis one must exert on it a force of constraint in the η-direction. Our external force Φ is therefore the sum of the force of constraint, which we shall call mb, and the force of gravity mg. From Eq. (5) of the foregoing problem the contribution of the latter to Φ is $-img e^{-i\phi}$. Summing, we then have

$$\Phi = \Phi_\xi + i\Phi_\eta = -mg\sin\omega t - img\cos\omega t + imb.$$

In Eq. (6) of the previous problem one can put $\mathbf{r} = \xi$ and, by virtue of (3) ibid., $2\omega \mathbf{x} \dot{\mathbf{r}} = 2i\omega\dot{\xi}$; further one must put $\dot{\omega} = 0$. One obtains

$$(1) \qquad \ddot{\xi} + 2i\omega\dot{\xi} - \omega^2\xi = -mg\sin\omega t + i(b - g\cos\omega t).$$

Its real part gives

$$(2) \qquad \ddot{\xi} - \omega^2\xi = -g\sin\omega t,$$

a differential equation with solution

$$(3) \qquad r = A\cosh\omega t + B\sinh\omega t + \frac{g}{2\omega^2}\sin\omega t.$$

If one puts the imaginary part of (1) equal to zero, one obtains the relation between force of constraint, gravity and Coriolis force given in the problem, viz.,

$$(4) \qquad b = g\cos\omega t + 2\omega\dot{\xi}.$$

V.3. (a) Let $x_0 + iy_0$ determine the position of O in the xy-plane. We then have

$$(1) \qquad \dot{x}_0 + i\dot{y}_0 = (u + iv)e^{i\phi}$$
$$\ddot{x}_0 + i\ddot{y}_0 = \{\dot{u} + i\dot{v} + i\omega(u + iv)\}e^{i\phi}.$$

Let $x + iy$ determine the position of G in the xy-plane. We have

$$(1) \qquad x + iy = x_0 + iy_0 + ae^{i\phi}$$
$$\dot{x} + i\dot{y} = (u + iv + i\omega a)e^{i\phi}$$
$$(2) \qquad \ddot{x} + i\ddot{y} = [\dot{u} + i\dot{v} + i\dot{\omega}a + i\omega(u + iv) - \omega^2 a]e^{i\phi}.$$

In the xy-plane there corresponds to the external force **R** the complex quantity

(2')
$$\mathbf{F} = \mathbf{R}\,ie^{i\phi}.$$

From (2) and (2') the Second Law, $\ddot{x} + i\ddot{y} = \dfrac{\mathbf{F}}{M}$ leads to

$$\dot{u} + i\dot{v} + i\dot{\omega}a + i\omega(u + iv) - \omega^2 a = i\frac{R}{M}$$

or, resolved into components,

(3)
$$\dot{u} - \omega v - \omega^2 a = 0,$$

(4)
$$\dot{v} + \dot{\omega}a + \omega u = \frac{R}{M}.$$

In addition we have, from the law of angular momentum,

(5)
$$I\dot{\omega} = -Ra.$$

   (b) Conditions $v = 0$, $\dot{v} = 0$ simplify (3) and (4) to

(3') $\dot{u} - \omega^2 a = 0,$ (4') $\dot{\omega}a + \omega u = \dfrac{R}{M}.$

Elimination of R from (4') and (5) gives

(6)
$$\dot{\omega}a\left(1 + \frac{I}{Ma^2}\right) + \omega u = 0.$$

Now put $I = Mb^2$ (b = radius of gyration) and

(7)
$$k^2 = 1 + \frac{b^2}{a^2} > 1,$$

which transforms (6) into

(6')
$$k^2\dot{\omega}a + \omega u = 0.$$

After integration of the simultaneous Eqs. (3') and (6') R is determined by (4') or (5).

   (c) Elimination of u from (3') and (6') yields

(8)
$$k^2\frac{d}{dt}\frac{\dot{\omega}}{\omega} = -\omega^2.$$

After multiplication by $\dfrac{\dot{\omega}}{\omega}$ this equation becomes integrable and furnishes

(9) $k^2\left(\dfrac{\dot{\omega}}{\omega}\right)^2 = k^2c^2 - \omega^2,$ (9') $k\dot{\omega} = \omega(k^2c^2 - \omega^2)^{\frac{1}{2}},$

where c is a constant of integration. One gets rid of the square root by putting

(10)
$$\omega = kc\cos\psi.$$

With suitable choice of the sign of the square root, (9') becomes

(10') $$\dot{\psi} = c \cos \psi$$

or

$$c \, dt = \frac{d\psi}{\cos \psi} , \qquad (11) \qquad ct = \tfrac{1}{2} \ln \frac{1 + \sin \psi}{1 - \sin \psi} .$$

We have thus determined ψ as a function of t. We can now express all quantities in terms of ψ; ω from (10), u and R from (6') and (4'):

(12) $\quad u = a k^2 c \sin \psi,$ $\qquad\qquad$ (12') $\quad R = \dfrac{M}{2} a k (k^2 - 1) c^2 \sin 2\psi.$

This completes the integration.

Because of $\omega = \dot{\phi}$, comparison of (10) and (10') finally yields the relation $\dot{\psi} = \dfrac{\dot{\phi}}{k}$. Our auxiliary angle ψ is hence proportional to the angle of rotation ϕ,

(13) $$\psi = \frac{\phi}{k} ,$$

since the constant of integration can be made zero by suitable choice of the arbitrary direction of the x-axis.

(d) From (1'), for $v = 0$,

$$|\dot{x} + i\dot{y}|^2 = \dot{x}^2 + \dot{y}^2 = u^2 + \omega^2 a^2,$$

(14)
$$T = \frac{M}{2}(\dot{x}^2 + \dot{y}^2) + \frac{I}{2}\omega^2 = \frac{M}{2}(u^2 + \omega^2 a^2) + \frac{M}{2}(k^2 - 1)a^2\omega^2$$
$$= \frac{M}{2}(u^2 + k^2 a^2 \omega^2).$$

From (10) and (12) this equals

(15) $$T = \frac{M}{2} a^2 k^4 c^2 (\sin^2 \psi + \cos^2 \psi) = \text{Const.}$$

(e) From (1) and (12)

$$\dot{x}_0 = a k^2 c \sin \psi \cos \phi, \quad \dot{y}_0 = a k^2 c \sin \psi \sin \phi,$$

so that, by virtue of (10') and (13),

(16) $$\frac{dx_0}{d\phi} = a k \tan \psi \cos \phi, \qquad \frac{dy_0}{d\phi} = a k \tan \psi \sin \phi.$$

Eq. (11) tells us that

$$\text{for } \psi = 0, \quad t = 0$$

$$\text{for } \psi = \pm \frac{\pi}{2}, \quad t = \pm \infty.$$

The whole trajectory takes place between $-\dfrac{\pi}{2} < \psi < +\dfrac{\pi}{2}$, $-k\dfrac{\pi}{2} < \phi < +k\dfrac{\pi}{2}$.

At $t=0$ a cusp occurs; for, according to (16) with $\psi=0$, $\phi=0$,

$$\frac{dx_0}{d\phi}=\frac{dy_0}{d\phi}=\frac{d^2y_0}{d\phi^2}=0; \quad \text{on the other hand,} \quad \frac{d^2x_0}{d\phi^2} \text{ and } \frac{d^3y_0}{d\phi^3} \neq 0;$$

the cusp has tangents parallel to the x-axis on both of its branches.

For $t=\pm\infty$ the path becomes asymptotic, for ϕ becomes stationary: from (16), quite generally,

$$\frac{dx_0}{d\phi}=\frac{dy_0}{d\phi}=\pm\infty.$$

In addition (16) yields

$$\frac{dy_0}{dx_0}=\tan\phi=\pm\tan k\frac{\pi}{2},$$

so that the asymptotes are situated symmetrically with respect to the x-axis, with angles $\pm k\dfrac{\pi}{2}$ as shown by Fig. 57 of p. 252 for $k=1, \dfrac{3}{2}, 2, 3$.

VI.1.   With z taken positive in the sense of fall, i.e. downward, $V=-mgz$. Initial position $z=0$ for $t=0$ lies above the final position $z=z_1$ at $t=t_1$.

(a) For $z=\frac{1}{2}gt^2$ we obtain

$$\int L dt = \int_0^{t_1}\left[\frac{m}{2}(gt)^2+mg\cdot\frac{g}{2}t^2\right]dt=\frac{1}{3}mg^2t_1^3.$$

(b) For $z=ct$ we must choose c in such a way that for $t=t_1$

$$z=z_1=g\frac{t_1^2}{2}; \quad \text{we therefore have } c=\frac{gt_1}{2}.$$

With this value we find

$$\int L dt = \int_0^{t_1}\left[\frac{m}{2}\left(\frac{gt_1}{2}\right)^2+mg\frac{gt_1}{2}t\right]dt=\frac{3}{8}mg^2t_1^3.$$

For $z=at^3$, $a=\dfrac{1}{2}\dfrac{g}{t_1}$, on the other hand,

$$\int L dt = \int_0^{t_1}\left[\frac{m}{2}\left(\frac{3g}{2t_1}\right)^2t^4+mg\frac{g}{2t_1}t^3\right]dt=\frac{7}{20}mg^2t_1^2.$$

Whereas in Hamilton's principle we compare paths differing only by infinitesimal amounts, here the trajectories of (b) in the phase space of the q, $\dot{q}$ (here $z, \dot{z}$) differ by finite amounts from the real motion (a). Nevertheless even now the value of Hamilton's integral is smaller for (a) than for (b), as

$$\frac{1}{3} < \frac{3}{8} \quad \text{and} \quad \frac{1}{3} < \frac{7}{20}.$$

This is true here even for arbitrary lengths of path, which need not be the case as a general rule (cf. p. 208).

VI.2. As in Problem V.1, let ξ and η be the coordinates fixed in the rotating plane; let $\mathbf{u} = (\dot\xi, \dot\eta)$ be the velocity measured with respect to this plane. The velocity relative to the fixed plane is then

$$\mathbf{w} = \mathbf{u} + \mathbf{v}, \quad \mathbf{v} = \boldsymbol\omega \times \mathbf{r}$$

[cf., for instance, the first line of the table of p. 139]. Resolution into components gives

$$w_\xi = \dot\xi - \omega\eta, \quad w_\eta = \dot\eta + \omega\xi,$$

$$|\mathbf{w}|^2 = \dot\xi^2 + \dot\eta^2 + 2\omega(\xi\dot\eta - \eta\dot\xi) + \omega^2(\xi^2 + \eta^2).$$

From this it follows, with $T = \frac{1}{2}m|\mathbf{w}|^2$, that

$$\frac{d}{dt}\frac{\partial T}{\partial \dot\xi} = m\frac{d}{dt}(\dot\xi - \omega\eta) = m(\ddot\xi - \omega\dot\eta - \dot\omega\eta)$$

$$\frac{d}{dt}\frac{\partial T}{\partial \dot\eta} = m\frac{d}{dt}(\dot\eta + \omega\xi) = m(\ddot\eta + \omega\dot\xi + \dot\omega\xi)$$

$$\frac{\partial T}{\partial \xi} = m(\omega\dot\eta + \omega^2\xi), \quad \frac{\partial T}{\partial \eta} = m(-\omega\dot\xi + \omega^2\eta).$$

Let Φ_ξ, Φ_η be the components of the external force $\mathbf{F}$ with respect to the moving axes ξ, η; we then obtain Lagrange's equations

$$m(\ddot\xi - 2\omega\dot\eta - \dot\omega\eta - \omega^2\xi) = \Phi_\xi,$$

$$m(\ddot\eta + 2\omega\dot\xi + \dot\omega\xi - \omega^2\eta) = \Phi_\eta.$$

This is in exact agreement with Eq. (6) of Problem V.1, provided we resolve the latter into its components.

In the guiding on a rotating straight rod treated in Problem V.2 we have

$$v^2 = \frac{dr^2 + r^2 d\phi^2}{dt^2} = \dot r^2 + r^2\omega^2, \quad L = \frac{m}{2}(\dot r^2 + r^2\omega^2) - mgr\sin\omega t;$$

$$\frac{d}{dt}\frac{\partial L}{\partial \dot r} = m\ddot r, \quad \frac{\partial L}{\partial r} = mr\omega^2 - mg\sin\omega t.$$

The Lagrange equation resulting from this is identical to Eq. (2) of V.2. It immediately leads to the solution (3) of that problem. With the present method we need not speak of Coriolis and similar forces, though, on the other hand, we do not learn anything about the force of constraint.

VI.3. The terms left out in Eq. (4) of the problem and indicated by ... are

$$\left(1+\frac{\zeta}{R}\right)\dot\eta \quad \text{and} \quad -\frac{\eta}{R}\left(1+\frac{\zeta}{R}\right)\frac{\cos\theta}{\sin\theta}\dot\xi.$$

After multiplication by the factor of {} they would give, on differentiation with respect to t, terms of second or higher order in the ξ, η, ζ or their derivatives. In connection with the differentiated equations (5) and (6) we should remark that terms of second order such as $\zeta\ddot\xi$, $\dot\zeta\dot\xi$, etc., have, of course, been omitted. It is worth noting that through this omission the radius R of the earth disappears from the results. In the complete Eq. (6) we would actually obtain, in addition to the term written down, a term in ω^2, viz.,

$$R\sin\theta\cos\theta\,\omega^2,$$

which evidently represents the ξ-component of the ordinary centrifugal force ; the corresponding ζ-component would occur in $\frac{\partial T}{\partial\zeta}$. These terms must, however, be omitted because they have already been included in the effective gravitational acceleration g, Eq. (30.1).

In the case of Foucault's pendulum one should evidently use not the ordinary form (34.6) of Lagrange's equations, but the mixed type (34.11), coupled with the equation of constraint (31.1).

Incidentally, note that due to the definition of η and ψ_0 in (1) and (2) our problem belongs to the class of problems dependent on the time discussed on p. 217.

VI.4. The center of mass describes a " curtate " cycloid in a plane normal to the axis of the cylinder. Its parametric equations in terms of the angle of rotation ϕ are obtained from Eq. (17.1) for a " common " cycloid by replacing a of Eq. (17.1) in part by s,

$$\xi=a\phi-s\sin\phi, \qquad \dot\xi=(a-s\cos\phi)\dot\phi,$$

$$\eta-a-s\cos\phi, \qquad \dot\eta=s\sin\phi\dot\phi.$$

(a) If we take the mass center as reference point O, we have

$$T_{\text{transl}}=\frac{m}{2}(\dot\xi^2+\dot\eta^2)=\frac{m}{2}(a^2+s^2-2as\cos\phi)\dot\phi^2,$$

$$T_{\text{rot}}=\frac{I}{2}\dot\phi^2,\quad T_m=0,\quad V=mg\eta=mg(a-s\cos\phi).$$

Notice that $\omega=\dot\phi$ is originally the angular velocity of the cylinder about its axis of symmetry, but that, according to (23.8), it is also the angular velocity about a parallel axis through the mass center. Putting $I = mb^2$ ($b =$ radius of gyration) and $c^2 = a^2 + s^2 + b^2$, we have

(1) $$L = T_{\text{transl}} + T_{\text{rot}} - V = \frac{m}{2}(c^2 - 2as\cos\phi)\dot\phi^2 - mg(a - s\cos\phi)$$

$$\frac{1}{m}\frac{d}{dt}\frac{\partial L}{\partial \dot\phi} = (c^2 - 2as\cos\phi)\ddot\phi + 2as\sin\phi\,\dot\phi^2$$

$$\frac{1}{m}\frac{\partial L}{\partial \phi} = as\sin\phi\,\dot\phi^2 - gs\sin\phi.$$

Hence the equation of motion is

(2) $$(c^2 - 2as\cos\phi)\ddot\phi + as\sin\phi\,\dot\phi^2 + gs\sin\phi = 0.$$

(b) If we choose the center of the cross section through the mass center as reference point O, the latter moves horizontally with velocity $a\dot\phi$; with $I' = I + ms^2$ [cf. (16.8)] we have

$$T_{\text{transl}} = \frac{m}{2}a^2\dot\phi^2, \quad T_{\text{rot}} = \frac{I'}{2}\dot\phi^2, \quad V \text{ as above,}$$

but now T_m is not 0; from Eq. (22.11) it is given by

$$T_m = -ma\dot\phi^2 s\cos\phi.$$

As a result

(3) $$L = T_{\text{transl}} + T_{\text{rot}} + T_m - V = \frac{m}{2}(c^2 - 2as\cos\phi)\dot\phi^2 - mg(a - s\cos\phi),$$

which is in agreement with (1), so that we obtain, once more, the equation of motion (2). For small oscillations about $\phi = 0$ it yields

$$\ddot\phi + \frac{g}{l_1}\phi = 0, \qquad l_1 = \frac{c^2 - 2as}{s} = \frac{(a-s)^2 + b^2}{s} \ldots : \text{ stability;}$$

for small oscillations about $\phi = \pi$, on the other hand, with $\psi = \pi + \phi$,

$$\ddot\psi - \frac{g}{l_2}\psi = 0, \qquad l_2 = \frac{c^2 + 2as}{s} = \frac{(a+s)^2 + b^2}{s} \ldots : \text{ instability.}$$

VI.5. 1. *Relations between the angular velocities.* The derivation of these relations is simplest if one remembers that at the points at which the bevel gears (ω) are in mesh with gear (ω_1) on the one hand and gear (ω_2) on the other, the peripheral velocities must, at any instant, be equal. Gears (ω) rotate about axle A with angular velocity ω; in addition, this axle rotates together with (ω) about the common geometric axis of (Ω), (ω_1) and (ω_2) with angular velocity Ω. If r, r_1 and r_2 are the mean radii of bevel gears (ω), (ω_1), (ω_2), we must have at point of contact (ω, ω_1)

$$r\omega + r_1\Omega = r_1\omega_1$$

and at point of contact (ω, ω_2)

$$-r\,\omega + r_2\,\Omega = r_2\,\omega_2.$$

With $r_1 = r_2$ we obtain from this the relations

(1)
$$2\Omega = \omega_1 + \omega_2$$

$$2\omega = \frac{r_1}{r}(\omega_1 - \omega_2).$$

Of course these relations can also be derived by introducing virtual rotations.

2. *Relations between the torques.* The virtual work of L must always equal the sum of the virtual work of L_1 and L_2, i.e.,

$$L\Omega\,\delta t = L_1\,\omega_1\,\delta t + L_2\,\omega_2\,\delta t.$$

We now replace Ω in terms of ω_1 and ω_2 by means of (1) to arrive at

$$\left(\frac{L}{2} - L_1\right)\omega_1 + \left(\frac{L}{2} - L_2\right)\omega_2 = 0.$$

This is possible for arbitrary $\omega_1,\ \omega_2$ only if

(2)
$$\tfrac{1}{2}L = L_1 = L_2.$$

It is seen that the driving torque of the engine is transferred in equal amounts to each rear wheel at all times, no matter what the values of angular velocities ω_1 and ω_2.

3. *Equation of motion of the system.* Here it will be found simplest to use Lagrange's equations of the second kind. We have

$$T = \tfrac{1}{2}\left(I_1\,\omega_1^2 + I_2\,\omega_2^2 + I\omega^2 + I'\,\Omega^2\right).$$

We replace ω and Ω by their expressions in terms of ω_1 and ω_2 and introduce abbreviations

$$L_{11} = I_1 + \frac{I'}{4} + \frac{I}{4}\frac{r_1^2}{r^2},$$

$$L_{22} = I_2 + \frac{I'}{4} + \frac{I}{4}\frac{r_1^2}{r^2},$$

$$L_{12} = L_{21} = \frac{I'}{4} - \frac{I}{4}\frac{r_1^2}{r^2}.$$

Lagrange's equations then become

(3)
$$\frac{d}{dt}(L_{11}\,\omega_1 + L_{12}\,\omega_2) = \frac{L}{2} - W_1,$$

$$\frac{d}{dt}(L_{21}\,\omega_1 + L_{22}\,\omega_2) = \frac{L}{2} - W_2.$$

W_1 and W_2 are two resisting torques acting at the two rear wheels; they have their origin in the static friction at the ground and may, if one wishes, include the other resistances (air, etc.).

If L, W_1 and W_2 are given as functions of the time, one can calculate the parentheses in the left members of (3) as time integrals of the right members, so that ω_1 and ω_2 become known functions of the time.

Averaged over the time, the right members of (3) are equal to zero, so that ω_1 and ω_2 are constant. If, however, the resistance acting on one wheel is decreased, which happens, for instance, if the wheel jumps off a bump in the road and momentarily turns in the air ($W=0$), this wheel is accelerated, whereas the other is decelerated.

4. *Analogy to electrodynamics.* Eqs. (3) are so written that they remind one of the interaction of two inductively coupled currents (see the remarks on p. 225 concerning Boltzmann). If we identify the L_{ij} with the coefficients of induction of the two circuits, ω_1 and ω_2 with the currents flowing in them, the left members of (3) are the electrodynamic induction effects. $\frac{1}{2}L$ corresponds to the " impressed EMF " acting in the circuits, and

$$T = \tfrac{1}{2} L_{11} \omega_1^2 + L_{12} \omega_1 \omega_2 + \tfrac{1}{2} L_{22} \omega_2^2$$

is the total magnetic field energy. According to p. 197, one calls cyclic systems those whose Lagrangian contains only the derivatives of the coordinates with respect to time (here $\omega_1 = \dot{\phi}_1$, $\omega_2 = \dot{\phi}_2$). They therefore constitute the mechanical analogue of stationary electric currents. Both the differential mechanism and the symmetrical top are doubly cyclic systems.

Index

A

Acceleration, moment of, 35
  normal, 33, 36
  resolution along Cartesian
    coordinates, 32
  tangential, 33, 36
Action, 181
  integral, 185, 205, 230, 238
  principle of least, 204, 230
  quantum of, 181, 229, 238
Air resistance, 21
d'Alembert, 53, 60
d'Alembert's principle, 59, 61
Angular acceleration, 62, 142
  velocity, 62, 120, 130
Angular momentum, definition of, 35
  conservation of, 73, 79, 114
  law of, 71
  law of, for a rigid body, 133
  of a system of particles, 71
  relation to moment of inertia, 63, 131
Anomaly, eccentric, 244
  mean, 260
  true, 39, 243
Anschütz-Kaempfe, H., gyrocompass, 155
Aphelion, 42
Archimedes, i, 54
Areal velocity, 36, 39, 72
Atwood's machine, 246

B

Baer law of river displacements, 164
Balmer series, 239
Beats, 108, 114
Bernoulli, Jacques, 53
  Jean, 53
Bicycle, 55
  gyroscopic effects, 157
Billiards, theory of game, 158
  high and low shots, 158, 250
  parabolic motion of ball, 160, 251
Block and tackle, 56
Body, rigid, 118, 133
Boltzmann, L., doubly cyclic systems, 225
  momentoids, 228
Brachistochrone, 95
Bridge, forces of support, 55
Buys-Ballot, law of, 164

C

Canonical, 220
  equations, 222
  variables, 222
Canonically conjugate, 222
Carathéodory, C., 174, 251
Cardan's suspension, 150
Carnot energy loss, 28, 29, 241
Centrifugal force, 59, 82, 163
  for increased rotation of the earth, 246
Chain, falling, 241
Chandler's period, 144
Circular frequency, 23, 87, 115
Cogredient, 14, 202
Collision, see *Impact*
Conservation of angular momentum, 73,
    79, 114
  of energy, 18, 31, 168, 189
  of momentum, 4, 79
Constraint, 48, 96
  principle of least, 210
Contact transformations, 220
Contragredient, 202
Contravariant, 202
Coriolis force, 59, 162
Corpuscular theory of light, 229
Coulomb, Ch. A., laws of friction, 81
Couple, 128
Coupled pendulums, 106, 248
Coupling coefficient, 107
Covariant, 14, 202
Curvature, 213
  of a trajectory, 33, 213
  principle of least, 212
Cyclic coordinates (variables), 197, 236
Cycloid, parametric equations, 94, 192
Cycloidal pendulum, 94, 192

D

Damping, aperiodic (critical), 104
  factor, 104
Decrement, logarithmic, 104
Differential of automobile, 254
Differential, perfect (exact), 46
Displacement, virtual, 50
Dissipation of energy, 168
Dissipative systems, 47, 168
Double pendulum, 111, 195
Drive mechanism of a piston engine, 49,
    51, 57

Index

Dynamics of a free particle, 38
  of a rigid body, 133
Dyne, 8

E

Ecliptic, 39
Einstein, A., general theory of relativity, 15, 209
Electron trajectory in the field of an ion, 242
Electron, variable mass, 30
Electron-volt, 243
Elliptical trajectory, for a central force proportional to the distance, 242
  in a Kepler type problem, 41, 179, 237, 242
Ellipticity, 143
Energy, 17, 18
  conservation of, 18, 31, 168, 189, 222
  equation (law) of, 17, 68, 222
  equation of, in relativity, 31
  free, 185
  inertia of, 31
  kinetic, 17, 122
  kinetic, for the rotation of a rigid body, 63, 122
  kinetic, in relativity, 31
  potential, 17, 45
  potential, of harmonic binding, 23
" English ", in billiards, 160
Equations of constraint, 50, 66
  rheonomous and scleronomous, 191
  time-dependent, 68
Equations of motion, various methods of integration, 16
Equipollence of forces in the statics of rigid bodies, 126
Erg, 8
Euler's circle, 143
  equations, 187
  equations of motion, 139, 226
  period, 143
  theorem, 190
  theory of polar fluctuations, 142
Eulerian angles, 196, 225
Evolute of a cycloid, 96

F

Fall, free, from a great distance, 20
  free, in air, 21
  free, near the earth, 19
  free, on rotating earth, 167, 253

Fermat's principle of least time, 207
Fermi, E., thermal neutrons, 260
Flattening of the earth, 143
Flywheel, calculation of, 245
Force, 4
  applied, 53
  couple, 128
  derivable from a potential, 46
  external, 70, 74
  fictitious, 59, 162, 192
  field, 17, 45
  generalized components of, 188, 192
  Hertz's ideas on, 5
  internal, 70, 74
  Kirchhoff's ideas on, 5
  lost, 61, 211
  moment of, see *torque*
  of reaction, 52
  parallelogram of, 6
  polygon, 125
  principle of superposition, 6
  units of, 7, 8
Foucault's gyrocompass, 154
  pendulum, 171, 253
Four-vector, 14
Frahm stabilization tank, 154
Freedom, degrees of, 48, 50
  of non-holonomic systems, 50
  of rigid bodies, 48, 118
Frequency of oscillation, 23, 87, 102
Friction, 54, 66, 81
  angle of, 82
  coefficient of, 81, 83
  cone of, 81
  kinetic or sliding, 54, 83, 158
  Coulomb's laws of, 81
  on an inclined plane, 82
  static, 54, 81, 84

G

Galilean transformation, 11
Galileo, law of inertia, 3, 10
  principle of virtual work, 53
Galvanometer, 247
Gauss, K. F., 8, 210, 215
Gauss' principle of least constraint, 210
Geodesics, 208, 214
Gibbs, J. W., notation for vector products, 38
Grassmann, H., notation for vector products, 38

Index

Gravitation, Newton's law of, 20, 39
  Einstein's theory of, 16
Gravitational constant, 20, 39
Gravity, acceleration due to, measured
    with a reversible pendulum, 92
  center of, 91
Gyrocompass, 154
Gyroscope, 150
Gyroscopic terms, 169
Gyrostabilizer 153

H

Hamel, G., 228, 261
Hamilton, W. R., algebra of quarternions,
    120
  hodograph, 34
Hamiltonian, 190, 217
Hamilton-Jacobi equation, 229
Hamilton's equations, 217–219
Hamilton's principle, 181, 232, 252
  illustration, 252
Hamilton's theory, 217
Heaviside, O., notation for vector pro-
    ducts, 38
Helical spring, oscillating, 110
Hertz, H., centrifugal force, 60
  concept of force, 5
  holonomic and non-holonomic con-
    ditions, 50, 185
  principle of least curvature, 212
Hesse's case of the unsymmetrical heavy
    top, 138
Hodograph, 34
  of planetary motion, 40, 242
Holonomic conditions, 50, 66
Huygens, C., center of oscillation of a
    pendulum, 91
  cycloidal pendulum, 94
Hydrogen atom, 238
Hypersurface 221

I

Impact, elastic, 24, 25, 240
  inelastic, 27
  inelastic, between an electron and an
    atom, 240
  in game of billiards, 159
Impulse, 159
Inertia, 3
  of energy, 31
  Galileo's law of, 3, 10

Inertia (contd.)
  moment of, see *moment of inertia*
  products of, 123, 246
Inertial forces, 59, 60, 76
  systems, 10
Integral variational principles of mecha-
    nics, 181
Intrinsic coordinates, 32, 36
Invariant, 14, 15, 16, 216, 219
Inversion (reflection through origin) of
    coordinate systems, 121
Isochronous pendulum, 88, 94

J

Jacobi's rule, 233
Joule, unit, 8

K

Kelvin, Lord, 169
Kepler's equation, 238, 243
  laws, 39–43, 235
  problem, 38–45, 71, 235
Kinematics in a plane, 32
  in space, 36
  of a rigid body, 118
Kinetic energy, see *energy*
Kirchhoff, G., concept of force, 5
Kirchner, F., nuclear disintegration of
    lithium, 242
Kowalewski's case of the unsymmetrical
    heavy top, 138

L

Lagrange, J. L., Mécanique analytique,
    i, 53
Lagrange's case of the three-body problem,
    174
  equations of the first kind, 66
  equations of the second kind, 185
  fictitious forces, 192
  indeterminate multipliers, 67
Lagrangian (function), 184, 209
  difficulty in defining it for general,
    especially non-mechanical systems,
    209
Legendre's standard form for elliptic
    integrals, 89
  transformation 226
Lever, 54
  inverse of, 55

Index

Light, corpuscular theory, 229
  independence of velocity of the reference frame, 11
  wave theory, 229
Linear momentum, see *momentum*
Line element, 213
Line of nodes, 136, 196
  advance of, 145
Liouville's theorem, 26
Lorentz, H. A., deformable electron, 15
  transformation, 12

M

Mass, 4
  center, 25, 70
  center, velocity of, 25, 70
  gravitational, 19
  inertial, 19
  longitudinal, 30
  reduced, 28, 64, 85
  relativisitc variation, 15, 30
  rest, 15
  transverse, 30
  units, 7, 8
  variable, 28
Maupertuis, P. L. N. de, principle of least action, 204
Mechanical system, 53
Michelson-Morley experiment, 15
Minkowski, H., proper time, 14, 209
Modulus of periodicity of an action integral, 238
Moment of acceleration, 35
  of a vector, 34
  of force, see *torque*
  of momentum, see *angular momentum*
  of velocity, 35
Momental ellipsoid, 124, 132
Moment of inertia, 62, 123
  law of parallel axes, 93
  of a compound pendulum, 91
  of a plane mass distribution, 250
  of a rigid body, 123
  of a sphere, 65
  principal, 124
Momentum, 4
  conservation of, 4, 79
  equation of, 4, 70, 133
  equation of, for a rigid body, 133
  in a collision of two masses, 25
  in relativity theory, 14

Momentum (contd.)
  moment of, or angular, see *angular momentum*
  of a rigid body, 130
Moon, acceleration due to earth's attraction, 241
  nodes, 146
  rocket to, 241
Multiplication of vectors, scalar, 7
  vectorial, 34

N

Newton, Sir, I., Philosophiae Naturalis Principia Mathematica, i, 3
  unit of force, 9
Newton's absolute time, 9, 11
  axioms, 3
  first law, 3
  fourth law, 6
  law of acceleration, 4
  law of gravitation, 16, 20, 39
  pail experiment, 9, 20
  second law, 4
  third law, 6
Non-holonomic conditions, 50, 244
  velocities, 141, 197, 226
Normal modes of oscillation, 107
Normal to a surface, 215
North Pole, celestial, 143
  geometric, 143
Nuclear disintegration of lithium, 242
Nutations, 146, 200

O

Oscillations, 87
  aperiodic, 104
  center of, 91
  forced, damped, 104
  forced, damped, resonance peak, 105, 247
  forced, undamped, 100
  free, damped, 103
  free, undamped, 22
  frequency of, 23, 87
  harmonic, 22, 87
  isochronous, 88, 94
  modulated, 106, 114
  of a balance wheel, 115
  period of, 23, 87, 90, 95
  quencher of, 250
  sympathetic, 106, 248
Osculating plane, 36, 214

286

Index

P

Parallelogram of forces, 6
Path, curvature of, 33
  prescribed, 65
  principle of the shortest, 208
  variation of, 182, 205
Pendulum, compound, 49, 91
  coupled, 106, 248
  cycloidal, 94, 192
  double, 111, 195
  forced motion of the point of suspension, 248
  Foucault's, 171
  isochronous, 88, 94
  reversible, 92
  seconds, 88
  simple, 49, 87
  spherical, 49, 96, 193
  spherical, for infinitesimal deflections, 247
Perihelion, 42
Periodicity, modulus of, 238
Phase difference of oscillations, 101, 105
Phase space, 238
Piston engine, drive mechanism of, 49, 51, 57
  double-acting, 245
Pitot tube, 20
Planck's elementary quantum of action, 181, 229, 238
Plane, inclined, 64
  inclined, vertically accelerated, 246
  inclined, with friction, 81
  invariable, 73, 136
  osculating, 36, 214
  rotating, 174, 251
Planetary motion, 38, 235
Poggendorff's experiment, 246
Poinsot method, 132
Point transformation, 201, 219
Polar coordinates, 39
Polar fluctuations, 142, 167
Polhode, (body cone), 143
Position coordinates, generalized, 185, 201
Potential, 45
  energy, see *energy*
  kinetic, 185
Power, 7
Precession of a spherical pendulum, 99
  of the equinoxes, 145
  pseudoregular, 137, 146, 200

Precession of a spherical pendulum (contd.)
  regular, 134, 142, 200
  under no forces, 145
Principal axes, transformation to, 124
Principal moments of inertia, 124
Principal normal, 36 214
Principle, d'Alembert's, 59, 61
  Hamilton's, 181, 232
  of least action, 204, 230
  of least constraint, 210
  of least curvature, 212
  of least time, 207
  of Maupertuis, 204, 230
  of the shortest path, 208
  of virtual work, 51
Products of inertia, 123, 246
Proper time, 14, 209

Q

Quantum numbers, 238
  theory, (old), 238
Quaternion algebra, 120

R

Radius of gyration, 91
Reaction, principle of action and, 6
  application to collisions of particles, 24
Reduction of a system of forces, 126
Reference frame or system, 9, 10
Reference point, change of, in the theory of a rigid body, 127
Reflection of coordinate system, 121
Relative motion, differential equations, 165
  in a plane, 251, 253
Relativity pinciple, in classical mechanics, 11
  of electrodynamics, 14
Relativity theory, general, 15, 209
  principle of energy, 31
  special, 5, 14, 79, 209
Resonance, 76, 102, 107, 116
  denominator, 102
  peaks in forced damped oscillations, 105, 247
Rest mass, 15
Reversible pendulum, 92,
Rheonomous conditions of constraint, 191
Rigid body, 62, 118, 133
Rolling wheel, 244
Rope, falling, 241

Index

Rotating straight line, 251, 253
Rotation, about a fixed axis, 62
  addition of, 120
  basic equation of, 63
  infinitesimal, 119
  of a rigid body, 118
  of a rigid body about a fixed axis, 62, 118
  permanent, of an unsymmetrical top, 146, 250
  virtual, 58, 71
Rotational couple, 128
  velocity, 119
Routh, E. J., Treatise, 150, 223
Routh's equations, 222
  function, 223

S

Scalar product, 7
Schlick, O., mass balancing, 76
  gyrostabilizer, 153
Schrödinger, E., 229
Schuler's law, 156
Scleronomous conditions of constraint, 191
Screw displacement, 119, 129
Secular equation, 109
Separation of variables, 231, 235
Sleigh, as an example of a non-holonomic system, 251
Sliding friction, 54
Spin, stable and unstable, 151
Stability of the rotation of a top, 250
  of a ship, 153
Static friction, 54
Statics in space, 37
  of a rigid body, 125
  plane, 34
Staude's case of the unsymmetrical heavy top, 138
System, closed, 79
  conservative, 47, 230
  cyclic, 222
  non-conservative (dissipative), 47, 231

T

Tautochrone, 95
Tensor, symmetrical, 123
  strain, 123
  stress, 123
  surface, 123
Three-body problem, 80
  Lagrange's case, 174

Time, absolute (Newton), 9, 11
  proper, 14, 209
Top, heavy symmetrical, 136, 196, 225
  heavy unsymmetrical, 138
  spherical, 125, 134
  symmetrical, 134, 225
  under no forces, 134, 146, 227
  unsymmetrical, 135, 146, 227
Torque, 35
  about an axis, 37, 58
  about a point, 37
  as a vector quantity, 241
  connection with virtual work, 58
  impulsive, 159
  polygon, 126
Trajectory, curvature of, 33
Transformation, angle-preserving, 27
  area-preserving, 26
  Galilean, 11
  Legendre, 226
  Lorentz, 12, 13
  orthogonal, 10
  unimodular, 258
Translation of a rigid body, 118
Turning stool, 74
Two-body problem of astronomy, 44, 80, 235

U

Units, absolute, 7, 8
  Giorgi's system, 8
  gravitational or practical, 7, 8

V

Variation of trajectory, 182, 205
Vector, 4
  algebra, 46
  analysis, 46
  axial, 120
  moment, 34
  notation for products, 38
  polar, 120
  scalar product, 7
  vector product, 34
Velocity, areal, 36, 39, 72
  coordinates, generalized, 185, 201
  decomposition, 32, 36
  of a rigid body in arbitrary motion, 119
Virtual work, 51
  displacement, 50
  rotation, 58

Index

W

Water drop in saturated atmosphere, 241
Watt, unit, 8
Wave theory of light, 229
Weight, units of, 7
" Wobbling " of a cylinder rolling on a plane support, 254
Work, 7
   in a virtual rotation, 58

Work,
   of the reactions, 52
   principle of virtual, 51
   units of, 8
   virtual, connection with torque, 58
World line element, 209
Wrench, 129

Y

Yo-yo, 246

图书在版编目（CIP）数据

力学 = Lectures on Theoretical Physics: Mechanics：英文 /（德）阿诺德·索末菲 (Arnold Sommerfeld) 著 . — 北京 : 世界图书出版有限公司北京分公司 , 2023.1
索末菲理论物理教程
ISBN 978–7–5192–9678–0

Ⅰ . ①力… Ⅱ . ①阿… Ⅲ . ①力学—教材—英文 Ⅳ . ① O3

中国版本图书馆 CIP 数据核字（2022）第 131080 号

中文书名	索末菲理论物理教程：力学
英文书名	Lectures on Theoretical Physics: Mechanics
著　　者	［德］阿诺德·索末菲（Arnold Sommerfeld）
策划编辑	陈　亮
责任编辑	陈　亮

出版发行	世界图书出版有限公司北京分公司
地　　址	北京市东城区朝内大街 137 号
邮　　编	100010
电　　话	010-64038355（发行）　 64033507（总编室）
网　　址	http://www.wpcbj.com.cn
邮　　箱	wpcbjst@vip.163.com
销　　售	新华书店
印　　刷	北京建宏印刷有限公司
开　　本	711mm×1245mm　1/24
印　　张	13.25
字　　数	289 千字
版　　次	2023 年 1 月第 1 版
印　　次	2023 年 1 月第 1 次印刷
版权登记	01-2022-0410
国际书号	ISBN 978-7-5192-9678-0
定　　价	109.00 元